Nonlinear Waves
in Bounded Media
The Mathematics of Resonance

Nonlinear Waves in Bounded Media

The Mathematics of Resonance

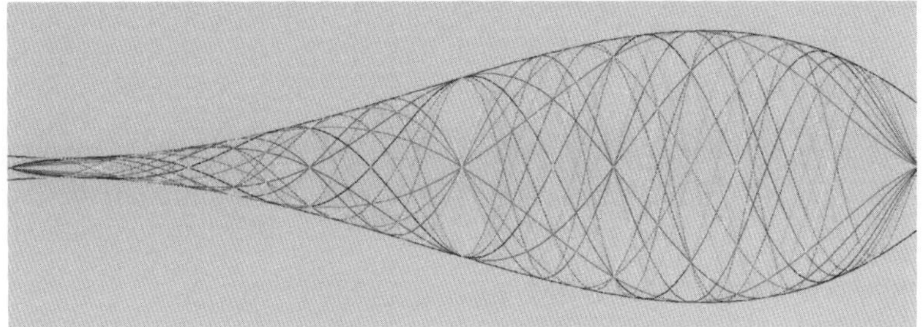

Michael P Mortell

University College Cork, Ireland

Brian R Seymour

University of British Columbia, Canada

World Scientific

NEW JERSEY · LONDON · SINGAPORE · BEIJING · SHANGHAI · HONG KONG · TAIPEI · CHENNAI · TOKYO

Published by

World Scientific Publishing Co. Pte. Ltd.

5 Toh Tuck Link, Singapore 596224

USA office: 27 Warren Street, Suite 401-402, Hackensack, NJ 07601

UK office: 57 Shelton Street, Covent Garden, London WC2H 9HE

British Library Cataloguing-in-Publication Data
A catalogue record for this book is available from the British Library.

NONLINEAR WAVES IN BOUNDED MEDIA
The Mathematics of Resonance

ISBN 978-981-3100-33-6

Typeset by Stallion Press
Email: enquiries@stallionpress.com

Printed in Singapore

To my wife
Pat

to my daughters
Deirdre and Siobhán

and grandchildren
Justin and Susana.

—Michael P. Mortell

To my wife
Rosemary

to my children
Mark, Jane and Richard

and grandchildren
Isabella, Jackson, Elsie-Jane, Jordyn and Linden.

—Brian R. Seymour

Preface

This book is intended to be self-contained and is aimed at graduate students in applied mathematics, physics and engineering, while the later chapters should also be of interest to researchers in these areas. It is expected that the reader has a basic knowledge of ordinary and partial differential equations (for example the material in Boyce and DiPrima, 2012 and Kevorkian, 2000), and should have no difficulty following Chapters 2–4.

Parts of Chapter 6 are covered in other books on waves that include: Billingam and King [2000], Whitham [1974], Courant and Friedrichs [1999] and Stoker [1957].

A knowledge of perturbation techniques, such as in Nayfeh [1973], Bender and Orszag [1999] and Johnson [2005] is also desirable as background to Chapters 7–13. Chapters 5 and 7–13 encompass much of our own research over many years.

Acknowledgments: D.E. Amundsen and E.A. Cox, who were our partners in much of the original research and who read, and commented on, parts of the manuscript. Also Bernhard Konrad assisted in some of the Latex typing and produced several figures.

Michael Mortell, Cork, Ireland
Brian Seymour, Vancouver, Canada
2016

Contents

Chapter 1

Introduction

In broad terms, a wave is a signal that propagates through a medium with a finite speed. Waves are ubiquitous in nature allowing us to see, hear and observe. They occur in a wide variety of media, producing such diverse phenomena as electrochemical and magneto-hydrodynamic waves. Here, though, we concentrate on three more common types that most people have some familiarity with: sound waves, water waves and elastic waves. Our focus is on such waves in bounded media so that dealing with reflections is an integral part of the problem. When a boundary is forced periodically this can lead to resonance phenomena when undamped linear theory predicts infinite amplitudes for certain applied frequencies. In such cases a nonlinear theory can often predict output amplitudes that are much larger than the input amplitudes, but are bounded. A practical example is resonant macrosonic synthesis (RMS), dealt with in Chapter 12. This has been applied in acoustic compressors, thermo-acoustic engines and thermo-acoustic refrigerators, see for example Lucas [1993].

Waves can be loosely divided into two main classes: hyperbolic, defined by the governing equation(s), and dispersive, defined by the assumed form of solution. The canonical hyperbolic equation is the *linear one-dimensional wave equation*

$$\frac{\partial^2 w}{\partial t^2} - c_0^2 \frac{\partial^2 w}{\partial x^2} = 0, \tag{1.1}$$

where c_0 is a constant, with the general solution

$$w(x,t) = f(x - c_0 t) + g(x + c_0 t) \tag{1.2}$$

in terms of arbitrary functions f and g. These functions describe waves traveling to the right (for f) and left (for g) with constant speed c_0.

Dispersive waves are defined by the form of solution:

$$\phi(x, t) = A \cos(kx - \omega t), \tag{1.3}$$

where $\omega = W(k)$. If a solution of the form (1.3) is substituted into a linear partial differential equation (p.d.e.), the condition for a non-trivial solution produces a relation of the form $G(\omega, k) = 0$. Solving for ω defines the function $W(k)$ and from (1.3) the wave speed is therefore $c(k) = \frac{W(k)}{k}$. Hence, different modes (with different k) travel with different speeds. An initial signal made up of many modes will spread or disperse. A prototype for linear dispersive waves is given by the linear Korteweg–de Vries (KdV) equation

$$\frac{\partial \phi}{\partial t} + c_0 \frac{\partial \phi}{\partial x} + \gamma \frac{\partial^3 \phi}{\partial x^3} = 0,$$

when we find

$$W(k) = c_0 k - \gamma k^3,$$

so that $c(k) = c_0 - \gamma k^2$.

The interest here is on nonlinear standing waves and resonant oscillations in bounded domains. These can be in either hyperbolic systems, such as an inviscid gas in a tube, or dispersive systems, such as water in a tank. In many cases, classical linear theory fails, predicting an infinite amplitude whereas a nonlinear theory gives a solution with a finite amplitude. In other cases linear theory gives a solution with finite amplitude, but differs from the solution given by nonlinear theory.

We ultimately deal with nonlinear waves in a domain of finite extent so that the waves reflect from boundaries and hence must pass through each other. In linear theory solutions can be superposed and hence the waves travel without interaction and for hyperbolic linear waves they also do not distort. However, for a fully nonlinear hyperbolic theory, the waves interact and distort and this leads to the fundamental difficulty, observed by Riemann [1858], that the wave traveling in one direction is affected by that traveling in the opposite direction and the characteristic equations cannot, in general, be integrated. A significant exception is a simple wave representing a nonlinear hyperbolic wave traveling into a uniform region. In this case one Riemann invariant is always constant and the characteristic equation can be integrated giving an exact solution describing a distorting wave that can develop shocks from continuous initial data. It emerges that there is a class of nonlinear hyperbolic problems where, to first order in the amplitude, the waves do not interact while the signal carried distorts. The motion is described as the superposition of noninteracting simple waves. One example is the small amplitude periodic motion of a gas in a closed tube driven by a piston at one end. When the applied frequency is in a resonant band, the motion contains shock waves that pass through each other at constant speed without interaction. The shape of the signal is determined by a nonlinear ordinary differential equation (o.d.e.) plus a shock condition. In physical terms there is a balance between dissipation due to shocks and the input of energy from the piston.

The later chapters of the book are centered on three sets of experiments. Firstly, Saenger and Hudson [1960] observed a gas in a closed tube excited by a reciprocating piston operating near a resonant frequency. Even for small piston amplitudes, they found shocks traveling back and forth that passed through each other without interacting. The second experiments are those of Chester and Bones [1968] who examined resonant oscillations of shallow water in a tank. The motion was continuous with high peaks separated by shallow troughs. Then Lawrenson, Lipkens, Lucas, Perkins and Van Doren [1998] considered resonant acoustic oscillations in closed tubes of variable cross section in shapes like cones, horns and bulbs. They found large amplitude, continuous pressure waves in the resonator. Each of these experiments requires a different nonlinear mathematical technique to explain the observations, and here we develop the appropriate mathematics.

A theme running through many examples is that the motion is fundamentally a linear standing wave $\phi(x,t) = f(x-c_0t) - f(x+c_0t)$, where f is periodic, while the signal f, carried by the wave with speed c_0, is determined by a nonlinear o.d.e.; i.e., the motion is determined by an iteration about a linear standing wave. This agrees with the observation of Saenger and Hudson [1960] that shocks travel with the linear sound speed and do not interact. The standing wave arises because the output at resonance is significantly larger than the input, and thus is a linear standing wave at the order of the output since it satisfies zero boundary conditions. For higher frequencies or amplitudes, when the distortion of the signal in one travel time is significant, a finite rate theory is needed. A shock can form in one travel time and a multiple scale approach does not work. This leads to the derivation of the "standard mapping" to determine the signal, and there is a direct analogy between acoustic resonance and chaotic dynamics. It emerges that the motion is a periodically forced simple wave, which is essentially the standard mapping. Similarly, the motion of resonantly forced shallow water in a tank is governed by a periodically forced KdV equation. Thus, the standard nonlinear p.d.e.s for unidirectional waves, the simple wave and the KdV equations, reappear with a forcing term. For resonant motions in an axisymmetric container with variable cross section, the essential observation is that the resonant frequencies are incommensurate, i.e., they are not multiples of the fundamental. Thus, in a perturbation scheme, the higher modes generated by quadratic and cubic terms are not resonant. This allows us to close the scheme at the third order, giving a hysteresis curve for the amplitude–frequency relationship. In contrast, for a tube with a constant bore all modes are multiples of the fundamental and thus resonate and form a shock.

For resonance in an axisymmetric tube of varying cross section, rather than looking for a wave-like solution we revert to separation of variables with a modal approximation. At first and second order in the perturbation expansion in powers of the amplitude the motion is a standing wave. The signal is then determined at the third order by a cubic equation representing the amplitude–frequency relation that exhibits hysteresis and contains the fundamental eigenvalue and eigenfunction. To calculate these quantities, the governing linear equations are transformed via a change of variables into the Webster horn equation. There are several useful families of functions that can approximate experimental shapes and are exact solutions of the Webster horn equation that yield exact eigenvalues and eigenfunctions. However, it is not always necessary to use these exact functions as the required eigenvalues and eigenfunctions can be calculated numerically for any given shape and then inserted into the amplitude–frequency equation.

The resonant sloshing of shallow water in a tank is governed by a periodically forced KdV equation. In this case the frequency dispersion spreads the wave, while nonlinearity steepens it. The balance results in a series of solitary waves whose number and height depend on the driver frequency within the resonant band.

Chapters 2–4 introduce some basic concepts of linear waves, giving examples in one, two and three dimensions. An important tool that helps assess the importance of different terms and is essential when using perturbation schemes is the concept of dimensionless variables. Dimensionless variables are the physical variables rescaled with relevant physical constants. For example, length might be measured in multiples of a wave length. This idea is introduced in Chapter 4. Chapter 5 introduces the use of approximate methods to find linear solutions, many of which are used later for nonlinear waves. Exact solutions of the Webster horn equation are derived and the connection with the N-soliton solutions of the KdV equation is shown.

Chapter 6 introduces nonlinear waves for the first time through the examples of exact solutions describing unidirectional kinematic and simple waves. In Chapter 7, we consider approximation methods for unidirectional nonlinear waves in systems with stratification or geometric nonuniformities. In these cases, exact solutions are not available, so expansion techniques, like nonlinear geometrical acoustics and multiscales, are used. The notion of the modulated simple wave, which can describe waves of arbitrary amplitude passing through stratified media, is introduced.

In Chapters 8–14, we deal with nonlinear waves in bounded media. Chapter 8 gives a series of examples of nonlinear standing waves for hyperbolic and dispersive systems using the ideas of noninteracting simple waves and multiscale expansions. All of the examples are worked out in detail. Chapter 9 deals with resonance in a closed straight tube when the time for shock formation is much longer than the traversal time of a wave in the tube. We refer to this as a "small rate" oscillation. Chapter 10 then deals with "finite rate" oscillations that result from forcing with higher frequencies or larger amplitudes. The solution is constructed in terms of solutions of a nonlinear difference equation, known to the chaos community as the "standard mapping". In order to construct periodic solutions of the standard mapping, the concept of critical points and invariant curves are introduced, and the analogy between shocks and chaos emerges from the analysis of the equation. The standard mapping is area-preserving and hence satisfies the equal area rule of shock dynamics. We also construct an exact solution of an area-preserving mapping, similar to the standard mapping, that contains an infinite number of shocks. The position and strength of each shock is explicitly calculated. The evolution of both small and finite rate oscillations is then examined in Chapter 11.

Chapters 9–11 involve resonant oscillations containing shocks; all examples are in uniform media. In contrast, Chapters 12 and 13 involve resonant oscillations with continuous outputs. In Chapter 12, we show that the shape of the resonator leads to incommensurate eigenvalues, i.e., the higher eigenvalues are not integer multiples of the fundamental. Consequently higher modes do not resonate and this results in a continuous output for all input frequencies. A practical example is RMS that is used to produce high pressure, but shockless, acoustic outputs. We also examine the effect of an inhomogeneity on a resonant motion. The penultimate chapter deals with resonant sloshing in a shallow tank. In this case the frequency dispersion balances the nonlinearity to give a continuous response. A forced KdV equation is derived and the results of numerical solutions are compared with the experimental output of Chester and Bones [1968]. Asymptotic solutions of the steady state, forced KdV equation are also derived and compared with experimental results. Chapter 14 deals briefly with resonance in an open tube.

Chapter 2

Physical Examples:
Basic Equations

In this chapter, we show how the one-dimensional wave equation arises in three different physical systems: longitudinal waves in an elastic panel, acoustic waves in a gas filled tube, and surface gravity waves in shallow water. In each case we first derive the appropriate *linear one-dimensional wave equation* (1.1). These three examples are used throughout the book to examine the effects of stratification, nonlinearity and fixed boundaries on these systems. Here, we consider the simplest forms of the equations that we can expand upon in later chapters to examine nonlinear wave propagation problems in stratified systems. In later applications $w(x,t)$ is identified with displacement, pressure, stress, density or velocity, depending on the context.

All of these wave problems can be written in the form of a simple hyperbolic system of equations. Then the condition that data can propagate off an initial curve is called the characteristic condition. This determines the fundamental wave speeds and the number of boundary and initial conditions required for a well posed problem.

2.1. Longitudinal Vibrations of an Elastic Panel

We consider a thin elastic panel or rod lying along the x-axis. With only a longitudinal force applied, longitudinal waves can propagate along the axis of the panel with no rotational displacement and there is just an axial displacement, $w(x,t)$, where x is the particle position

7

and t is time. Then the strain, or the relative length change from equilibrium, is $\lambda = \frac{\partial w}{\partial x}$.

According to Hooke's law of linear elasticity, the stress σ is proportional to the strain, so for an homogeneous panel,

$$\sigma = E_0\lambda, \tag{2.1}$$

where E_0 is the constant Young's modulus with the same dimensions as stress. For larger amplitude displacements the stress–strain law (2.1) is modified to include a nonlinear correction:

$$\sigma(\lambda) = E_0\lambda(1 + N\lambda + O(\lambda^2)), \tag{2.2}$$

where N is an $O(1)$ dimensionless material constant, and $N > 0$ corresponds to a "hardening" material while $N < 0$ is a "softening" material.

The equations of motion relating λ and the particle velocity $u = \frac{\partial w}{\partial t}$ are the horizontal force balance in the x direction and conservation of mass:

$$\rho_0\frac{\partial u}{\partial t} = \frac{\partial \sigma}{\partial x} \quad \text{and} \quad \frac{\partial u}{\partial x} = \frac{\partial \lambda}{\partial t}, \tag{2.3}$$

where ρ_0 is the constant density of the panel.

If $\lambda N \ll 1$ in (2.2), for small amplitude motions (2.1) and (2.3) imply that $w(x,t)$ satisfies (1.1) with the wave speed $c_0 = \sqrt{E_0/\rho_0}$. For an homogeneous rod or panel u, λ and σ all satisfy (1.1).

2.1.1. *Inhomogeneous Panel*

If the material of the panel is inhomogeneous with both density (ρ) and Young's modulus (E) dependent on position, (2.2) and (2.3) are modified, letting $\rho = \rho(x)$, with $\rho(0) = \rho_0$, and $E = E(x)$ with $E(0) = E_0$. The elastic material of the panel is then described by the stress–strain law

$$\sigma(\lambda, x) = E(x)\lambda(1 + N\lambda + O(\lambda^2)). \tag{2.4}$$

For small amplitude motions λ and u satisfy the linear, variable coefficient system

$$\frac{\partial u}{\partial t} = \frac{1}{\rho(x)}\frac{\partial \sigma}{\partial x} \quad \text{and} \quad \frac{\partial u}{\partial x} = \frac{\partial \lambda}{\partial t} = \frac{1}{E(x)}\frac{\partial \sigma}{\partial t}. \tag{2.5}$$

Eliminating σ, λ or u, the others satisfy

$$\frac{\partial^2 \sigma}{\partial t^2} - E\frac{\partial}{\partial x}\left(\frac{1}{\rho}\frac{\partial \sigma}{\partial x}\right) = 0, \quad \frac{\partial^2 \lambda}{\partial t^2} - \frac{\partial}{\partial x}\left(\frac{1}{\rho}\frac{\partial (E\lambda)}{\partial x}\right) = 0,$$

$$\frac{\partial^2 u}{\partial t^2} - \frac{1}{\rho}\frac{\partial}{\partial x}\left(E\frac{\partial u}{\partial x}\right) = 0. \tag{2.6}$$

To keep the calculations as simple as possible in some of the later examples, we pick the inhomogeneity to scale both density and Young's modulus in the same way, so that $\rho(x) = E(x)$ and (2.5) become, using (2.4)

$$\frac{\partial u}{\partial t} = \frac{1}{E(x)}\frac{\partial \sigma}{\partial x} = \frac{\partial \lambda}{\partial x} + \frac{E'}{E}\lambda \quad \text{and} \quad \frac{\partial u}{\partial x} = \frac{\partial \lambda}{\partial t}, \tag{2.7}$$

while (2.6) become

$$\frac{\partial^2 \sigma}{\partial t^2} - E\frac{\partial}{\partial x}\left(\frac{1}{E}\frac{\partial \sigma}{\partial x}\right) = 0, \quad \frac{\partial^2 \lambda}{\partial t^2} - \frac{\partial}{\partial x}\left(\frac{1}{E}\frac{\partial (E\lambda)}{\partial x}\right) = 0,$$

$$\frac{\partial^2 u}{\partial t^2} - \frac{1}{E}\frac{\partial}{\partial x}\left(E\frac{\partial u}{\partial x}\right) = 0. \tag{2.8}$$

Notice that in (2.8) σ and u satisfy the same equation with E replaced by $1/E$.

2.1.2. *Viscoelastic Panel*

If the material of the rod or panel is homogeneous but rate dependent, and so described as viscoelastic rather than elastic, then the equations of motion (2.3) are augmented by an equation of state of the form

$$\frac{\partial \sigma}{\partial t} = A(\sigma, \lambda)\frac{\partial \lambda}{\partial t} - B(\sigma, \lambda) \qquad (2.9)$$

for some prescribed functions A and B. This replaces (2.2) and in the small amplitude limit (2.9) can be written as

$$\frac{\partial \sigma}{\partial t} = E_0 \frac{\partial \lambda}{\partial t} - \frac{\sigma}{\tau_r} \qquad (2.10)$$

with $A = E_0$ and $B = \sigma/\tau_r$, where τ_r is a positive material constant, the relaxation time. Such a material is called an homogeneous Maxwell viscoelastic solid. Then σ and u both satisfy

$$\frac{\partial^2 \sigma}{\partial t^2} = c_0 \frac{\partial^2 \sigma}{\partial x^2} - \frac{1}{\tau_r}\frac{\partial \sigma}{\partial x}, \qquad (2.11)$$

where $c_0 = \sqrt{E_0/\rho_0}$. If $\tau_r \to \infty$, i.e., a long relaxation time, then (2.10) becomes (2.1), describing an homogeneous elastic material.

2.2. One-dimensional Motion of an Inviscid Gas in a Tube

The second example is one we will consider later in more depth, including nonlinear terms and different geometries. The one-dimensional motion of an inviscid gas contained in a straight tube with no external forcing is described in Eulerian coordinates by the nonlinear equations of conservation of mass and momentum:

$$\frac{\partial \rho}{\partial t} + \frac{\partial (u\rho)}{\partial x} = 0, \quad \rho\left(\frac{\partial u}{\partial t} + u\frac{\partial u}{\partial x}\right) + \frac{\partial p}{\partial x} = 0, \qquad (2.12)$$

where $u(x,t)$, $p(x,t)$, $\rho(x,t)$ are the axial velocity, pressure and density.

We will usually consider "small" variations in u, p and ρ from their values in an equilibrium reference state $(0, p_0, \rho_0)$. To do this it is convenient to introduce the *condensation*, $e(x,t) = \rho/\rho_0 - 1$, so that for small variations $|e| \ll 1$. Then, if the gas is polytropic, the equation of state can be expanded for small e as

$$\frac{p}{p_0} = \left(\frac{\rho}{\rho_0}\right)^{\gamma} = (1+e)^{\gamma} = 1 + \gamma e + \frac{\gamma(\gamma-1)}{2}e^2 + \cdots , \qquad (2.13)$$

where γ $(= 1.4$ for air$)$ is the gas constant. Then (2.12) can be approximated by

$$\frac{\partial e}{\partial t} + \frac{\partial(u[1+e])}{\partial x} = 0, \quad \frac{\partial u}{\partial t} + u\frac{\partial u}{\partial x} + c_0^2[1 + (\gamma-2)e]\frac{\partial e}{\partial x} = 0, \quad (2.14)$$

where $c_0 = \sqrt{\frac{\gamma p_0}{\rho_0}}$ is the linear sound speed in the reference state. For small amplitude waves, when $|u| \ll c_0$, (2.14) are approximated by

$$\frac{\partial e}{\partial t} + \frac{\partial u}{\partial x} = 0, \quad \frac{\partial u}{\partial t} + c_0^2\frac{\partial e}{\partial x} = 0. \qquad (2.15)$$

Eliminating either e or u from (2.15) shows that both e and u satisfy (1.1) with $c_0 = \sqrt{\frac{\gamma p_0}{\rho_0}}$.

2.2.1. *Tube with Variable Cross Section*

If the tube has a variable cross section, $s(x)$, the first of (2.12) becomes

$$\frac{\partial(s\rho)}{\partial t} + \frac{\partial}{\partial x}(su\rho) = 0, \qquad (2.16)$$

or, on using (2.13), in terms of e and u the system is approximated by

$$s\frac{\partial e}{\partial t} + \frac{\partial}{\partial x}(su[1+e]) = 0, \quad \frac{\partial u}{\partial t} + u\frac{\partial u}{\partial x} + c_0^2[1+(\gamma-2)e]\frac{\partial e}{\partial x} = 0. \quad (2.17)$$

The equivalent linear equations are

$$s\frac{\partial e}{\partial t} + \frac{\partial(su)}{\partial x} = 0, \quad \frac{\partial u}{\partial t} + c_0^2\frac{\partial e}{\partial x} = 0. \qquad (2.18)$$

Eliminating either e or u from (2.18), the other satisfies

$$\frac{\partial^2 e}{\partial t^2} - c_0^2\frac{1}{s}\frac{\partial(se_x)}{\partial x} = 0 \text{ or } \frac{\partial^2 v}{\partial t^2} - c_0^2 s\frac{\partial}{\partial x}\left(\frac{1}{s}v_x\right) = 0, \qquad (2.19)$$

where $v = su$ is the flux.

2.2.2. *Elastic Gas*

There is an equivalence between the equations of motion for an inviscid gas and those for an elastic solid (2.1) and (2.3). Defining the *excess pressure* as $\Sigma(x,t) = p - p_0$, the linear form of Eq. (2.13) can be written as

$$\Sigma = \gamma p_0 e, \tag{2.20}$$

while (2.15) are identical to (2.3) if we write

$$\Sigma = -\sigma, \ e = -\lambda \text{ and } \gamma p_0 = E_0. \tag{2.21}$$

We recall that λ is the strain, defined in terms of density by $\rho(1+\lambda) = \rho_0$, while e is the condensation, defined by $\rho = \rho_0(1+e)$. Then comparing (2.2) and (2.13) the second-order constants are related by $N = -(\gamma+1)/2$.

2.3. Shallow Water Waves — Hydraulic Flow

The third example is that of a surface gravity wave propagating through shallow water of depth $h(x,t)$, firstly over a flat bottom, $z = -H_0$, when continuity implies

$$\frac{\partial u}{\partial x} + \frac{\partial w}{\partial z} = 0, \tag{2.22}$$

where u and w are the particle velocities in the horizontal and vertical directions respectively. If the constant fluid density is ρ then the horizontal force balance in the x direction is

$$\rho\left(\frac{\partial u}{\partial t} + u\frac{\partial u}{\partial x}\right) = -\frac{\partial p}{\partial x}, \tag{2.23}$$

where p is pressure above atmospheric. The bottom and top boundary conditions are $w = 0$ on $z = -H_0$ and on the free surface $z = h(x,t)$

$$p = 0 \quad \text{and} \quad \frac{\partial h}{\partial t} + u\frac{\partial h}{\partial x} = w. \tag{2.24}$$

Assuming the fluid layer is shallow, the vertical pressure variation is hydrostatic, so $p(x,z,t) = \rho g(h - z)$, where g is the acceleration due gravity, then (2.23) implies that u is independent of z. Integrating (2.22) from $z = -H_0$ to $z = h(x,t)$ and using (2.24) implies that

$$\frac{\partial h}{\partial t} + \frac{\partial(hu)}{\partial x} = 0, \tag{2.25}$$

while, using the hydrostatic pressure assumption, (2.23) becomes

$$\frac{\partial u}{\partial t} + u\frac{\partial u}{\partial x} + g\frac{\partial h}{\partial x} = 0. \tag{2.26}$$

It is convenient for small amplitude waves to write the total water depth as $h(x,t) = H_0 + \eta(x,t)$, where $\eta(x,t)$ is the water surface height above $z = 0$. Then the governing Eqs. (2.25) and (2.26) are

$$\frac{\partial \eta}{\partial t} + \frac{\partial(u[H_0 + \eta])}{\partial x} = 0, \quad \frac{\partial u}{\partial t} + u\frac{\partial u}{\partial x} + g\frac{\partial \eta}{\partial x} = 0. \tag{2.27}$$

For small amplitude waves, (2.27) can then be linearized to give

$$\frac{\partial \eta}{\partial t} + H_0\frac{\partial u}{\partial x} = 0, \quad \frac{\partial u}{\partial t} + g\frac{\partial \eta}{\partial x} = 0. \tag{2.28}$$

Eliminating either u or η from (2.28), both variables satisfy (1.1) with the linear wave speed $c_0 = \sqrt{gH_0}$.

To define more precisely what shallow means in this context, if τ_p is the period of the surface wave, then the water is "shallow" if the surface wavelength $(c_0\tau_p)$ is large compared with the depth H_0, so $H_0 \ll c_0\tau_p$. Hence the term "long wave" is used to describe this limit.

We note here that there is an equivalence between Eqs. (2.25)–(2.26) and (2.12)–(2.13) for a polytropic gas with $\gamma = 2$. To see this define $\rho^* = \rho h$ and $p^* = \frac{1}{2}\rho gh^2$, then (2.25)–(2.26) become (2.12) with (ρ^*, p^*) associated with (ρ, p). But (ρ^*, p^*) are related through $p^* = \frac{g}{2\rho}(\rho^*)^2$, corresponding to $\gamma = 2$. This is often called the gas dynamic analogy for hydraulic flows.

2.3.1. *Variable Bottom Topography*

For a variable bottom topography, let $z = -H(x)$ with $H(0) = H_0$, then $h(x,t) = H(x) + \eta(x,t)$ is the total water depth and the governing equations are (2.26) with H_0 replaced by $H(x)$. The corresponding linear equations are

$$\frac{\partial \eta}{\partial t} + \frac{\partial (H(x)u)}{\partial x} = 0, \quad \frac{\partial u}{\partial t} + g\frac{\partial \eta}{\partial x} = 0. \tag{2.29}$$

Eliminating either η or u, the other satisfies

$$\frac{\partial^2 u}{\partial t^2} - \frac{\partial^2 (gHu)}{\partial x^2} = 0 \quad \text{or} \quad \frac{\partial^2 \eta}{\partial t^2} - \frac{\partial^2 (gH\eta)}{\partial x^2} = 0. \tag{2.30}$$

2.4. Characteristics

Most of the physical equations we consider fit into the general system

$$\frac{\partial \mathbf{u}}{\partial t} + \mathbf{A}(\mathbf{u}, x)\frac{\partial \mathbf{u}}{\partial x} = \mathbf{B}(\mathbf{u}, x), \tag{2.31}$$

where $\mathbf{u}$ and $\mathbf{B}$ are vectors and $\mathbf{A}$ is a square matrix with n rows. This system is *hyperbolic* when $\mathbf{A}(\mathbf{u}, x)$ has n real distinct eigenvalues each corresponding to a real wave speed $c(\mathbf{u}, x)$. Using the Cauchy–Kowalewski Theorem, see Courant and Hilbert [1962], the condition for initial data to propagate off an initial curve is that $\mathbf{A}$ has real distinct eigenvalues, i.e., the system is hyperbolic. Then the slopes of the characteristic curves along which information propagates, $\frac{dx}{dt} = c$ that define the wave speeds c, are found as the eigenvalues of the *characteristic condition*

$$\det \|\mathbf{A} - c\mathbf{I}\| = 0. \tag{2.32}$$

For example, Eq. (1.1) is of the form (2.31) by defining $v = w_x$ and $u = w_t$, then

$$\mathbf{u} = \begin{pmatrix} u \\ v \end{pmatrix}, \quad \mathbf{A} = \begin{pmatrix} 0 & -c_0^2 \\ -1 & 0 \end{pmatrix}, \quad \text{and} \quad \mathbf{B} = \begin{pmatrix} 0 \\ 0 \end{pmatrix}. \tag{2.33}$$

For a Maxwell viscoelastic material (see (2.9) and (2.10)) the stress (σ) and particle velocity (u) satisfy

$$\rho_0 \frac{\partial u}{\partial t} = \frac{\partial \sigma}{\partial x} \quad \text{and} \quad \frac{\partial \sigma}{\partial t} = E_0 \frac{\partial u}{\partial x} - \frac{\sigma}{\tau_r} \tag{2.34}$$

which are in the form (2.31) with

$$\mathbf{u} = \begin{pmatrix} u \\ \sigma \end{pmatrix}, \quad \mathbf{A} = \begin{pmatrix} 0 & -\frac{1}{\rho_0} \\ -E_0 & 0 \end{pmatrix}, \quad \text{and} \quad \mathbf{B} = \begin{pmatrix} 0 \\ -\frac{\sigma}{\tau_r} \end{pmatrix}. \qquad (2.35)$$

For (2.33) $\frac{dx}{dt} = \pm c_0$, while for (2.35) $\frac{dx}{dt} = \pm\sqrt{\frac{E_0}{\rho_0}}$. When there is stratification, as in systems (2.5), (2.18) and (2.29), the corresponding linear characteristic speeds are $\frac{dx}{dt} = \pm c_0 \sqrt{\frac{s(x)}{q(x)}}$, $\frac{dx}{dt} = \pm c_0$ and $\frac{dx}{dt} = \pm\sqrt{gH(x)}$ respectively. These represent waves traveling in positive and negative x directions. Note that since the inhomogeneity enters the systems in different ways, the dependence of the characteristic speeds on x is different in each case.

The number of characteristics leaving an initial curve in a positive time-like direction determines the number of conditions required for a well posed problem. Hence, for an initial-boundary value problem on $x \geq 0$, $t \geq 0$, for either system (2.33) or (2.35), two conditions are required along $t = 0$, while one is required at a boundary like $x = 0$ or $x = l$.

Chapter 3

Classical Linear Solutions

Here the general one-dimensional solution on an infinite region, the classical D'Alembert solution, together with extensions to a semi-infinite region and a finite domain are derived. The wave-like nature of these solutions, describing disturbances that propagate at finite speeds, is emphasized.

For a finite domain we also show the connection to the solution derived by separation of variables. This introduces important concepts such as resonant frequencies, the fundamental harmonic and normal modes. Different boundary conditions produce different modes and resonant frequencies, so we review boundary conditions at interfaces between different materials.

The concepts of reflected and transmitted waves are introduced and used to construct a boundary condition representing the radiation of energy from the end of a partially open pipe or bonded elastic rods. An important example considered in later chapters is the forced periodic motion of a gas in a tube. Resonant and non-resonant solutions are constructed using a difference equation and it is shown how these solutions are related to the normal mode form found by separation of variables.

3.1. General Solution

The general solution to the linear wave equation (1.1) can be obtained by introducing the new variables

$$\alpha = x - ct \quad \text{and} \quad \beta = x + ct,$$

and transforming the equation for $w(\alpha, \beta)$ by a simple application of the chain rule into

$$\frac{\partial^2 w}{\partial \alpha \partial \beta} = 0. \tag{3.1}$$

The general solution to (3.1), which can be checked by differentiation, is

$$w(\alpha, \beta) = A(\alpha) + B(\beta) = A(x - ct) + B(x + ct), \tag{3.2}$$

where A and B are arbitrary functions. The derivation is simple. Integration of (3.1) w.r.t α implies that $\frac{\partial w}{\partial \beta} = B'(\beta)$ and (3.2) follows on integrating w.r.t. β. The functions A and B are determined by initial and boundary conditions.

The general solution (3.2) represents two waves traveling in opposite directions, where A has velocity c and B has velocity $-c$. Information is propagated along two families of lines $x \pm ct = $ constant, called the *characteristics*.

Another useful interpretation of the solution to (1.1) can be found by rewriting it in the form

$$\left(\frac{\partial}{\partial t} - c\frac{\partial}{\partial x} \right) \left(\frac{\partial}{\partial t} + c\frac{\partial}{\partial x} \right) w = 0. \tag{3.3}$$

It is clear that a solution of $\frac{\partial w}{\partial t} + c\frac{\partial w}{\partial x} = 0$ is a solution of (3.3), as is a solution of $\frac{\partial w}{\partial t} - c\frac{\partial w}{\partial x} = 0$. If we interpret $\frac{\partial w}{\partial t} + c\frac{\partial w}{\partial x} = 0$ as a directional derivative along the curve $\alpha = $ constant, it can be split into two ordinary differential equations along the curve $\alpha = \overline{\alpha} = $ constant, as

$$\frac{dw}{dt}\Big|_{\overline{\alpha}} = 0 \text{ on } \frac{dx}{dt}\Big|_{\overline{\alpha}} = c.$$

Then $w = g(\overline{\alpha})$ on $x - ct = \overline{\alpha}$, for arbitrary g. For the initial condition $w(x, 0) = G(x)$, since $\overline{\alpha} = x$ on $t = 0$, then $g = G$ and the solution is $w = G(x - ct)$. This represents a wave traveling to the right with speed c and undistorted shape. This solution can be called a *linear simple wave* and it has a nonlinear generalization, as we will see later.

3.2. D'Alembert's Solution

For a particular problem, the functions $A(\alpha)$ and $B(\beta)$ in (3.2) are found from prescribed initial and/or boundary conditions. The simplest example is the pure initial value problem on the infinite line with prescribed values of $w(x,0)$ and $w_t(x,0)$. As a general rule we apply the same number of conditions at the boundary of a region in $x-t$ space as there are characteristics entering the region in the positive t direction on that boundary. The initial "boundary" $t=0$ has two characteristics entering the region $t>0$ and thus two conditions are specified on $t=0$. Hence, using the shorthand notation for a partial derivative $w_t \equiv \frac{\partial w}{\partial t}$, the full problem is stated as:

$$
\begin{aligned}
w_{tt} &= c^2 w_{xx}, & -\infty < x < \infty, \quad t > 0, \\
w(x,0) &= \phi(x), & -\infty < x < \infty, \quad t = 0, \\
w_t(x,0) &= \psi(x), & -\infty < x < \infty, \quad t = 0.
\end{aligned}
\tag{3.4}
$$

In this context, an infinite interval simply means an interval that is long compared with the region over which non-zero initial data is prescribed, so the waves do not reach the "ends" of the medium during the time of interest.

By substituting the initial conditions from (3.4) into (3.2), we find

$$
A(x) + B(x) = \phi(x) \text{ and } -cA'(x) + cB'(x) = \psi(x)
$$

and then it is easy to show that

$$
A(z) = \frac{1}{2}\phi(z) - \frac{1}{2c}\int_a^z \psi(s)ds, \; B(z) = \frac{1}{2}\phi(z) + \frac{1}{2c}\int_a^z \psi(s)ds,
$$

for an arbitrary point a. Then

$$
\begin{aligned}
w(x,t) &= A(x-ct) + B(x+ct) \\
&= \frac{1}{2}\left[\phi(x-ct) + \phi(x+ct)\right] + \frac{1}{2c}\int_{x-ct}^{x+ct} \psi(s)ds.
\end{aligned}
\tag{3.5}
$$

This is the classical *D'Alembert's solution* from which we can make several observations.

The solution at the point $\mathbf{P}:(x,t)$, $t>0$, is completely determined by the data inside the triangular region enclosed by the

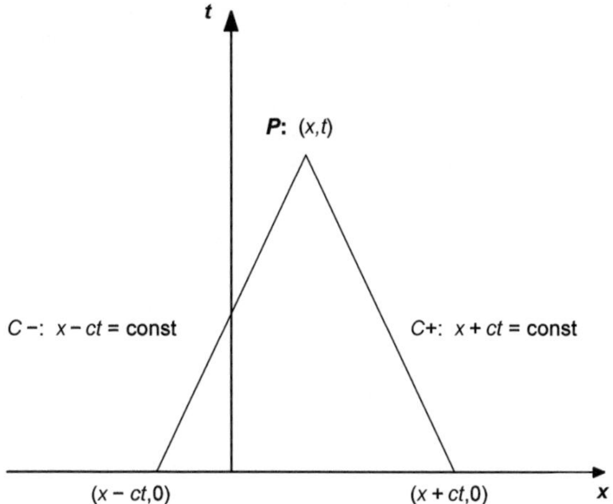

Fig. 3.1.   Domain of dependence.

two characteristic lines $C^{\pm}$ connecting (x,t), with $(x-ct,0)$, and $(x+ct,0)$. This region is called the *domain of dependence*, see Fig. 3.1.

For any point $(x,0)$ on the x-axis, there are two characteristic lines $C^{\pm}$ emanating from this point into the region $t > 0$. Any point inside this region, called the *range of influence*, is affected by the initial data at $(x,0)$, see Fig. 3.2.

In order to better understand the behavior of solutions to the wave equation, it is instructive to examine solutions of several initial-boundary value problems.

3.2.1. *Solution on the Semi-infinite Line*

For certain boundary conditions at $x = 0$, D'Alembert's solution can be adapted for an initial-boundary value problem on a semi-infinite domain $x > 0$ using the method of images. By this we mean that we extend the data on $x > 0$ to $x < 0$ and then simply use (3.5).

Consider the region $0 < x < \infty$ with the same initial conditions as (3.4), so that

$$w(x,0) = \phi(x), \quad 0 < x < \infty,$$
$$w_t(x,0) = \psi(x), \quad 0 < x < \infty,$$

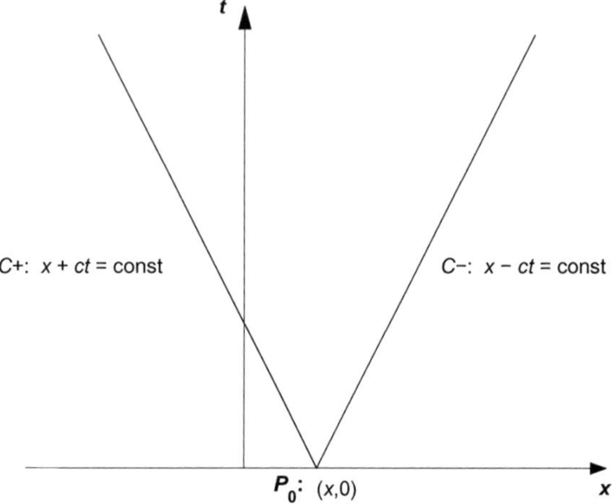

Fig. 3.2.   Range of influence.

with $\phi(0) = 0$, but now specify a reflecting boundary condition on $x = 0$: $u(0, t) = 0$, $t > 0$. Recall the previous comment about the number of conditions needed on a boundary.

To utilize D'Alembert's solution we extend both ϕ and ψ to $\widehat{\phi}$ and $\widehat{\psi}$, odd functions about $x = 0$, and hence defined for all x as

$$\widehat{\phi}(x) = \begin{cases} \phi(x), & x > 0 \\ -\phi(-x), & x < 0, \end{cases} \qquad \widehat{\psi}(x) = \begin{cases} \psi(x), & x > 0 \\ -\psi(-x), & x < 0. \end{cases}$$

Then, by (3.5), on $-\infty < x < \infty$:

$$w(x, t) = \frac{1}{2}\left[\widehat{\phi}(x - ct) + \widehat{\phi}(x + ct)\right] + \frac{1}{2c}\int_{x-ct}^{x+ct} \widehat{\psi}(s)ds, \qquad (3.6)$$

or

$$w(x, t) = \begin{cases} \frac{1}{2}[\phi(x - ct) + \phi(x + ct)] + \frac{1}{2c}\int_{x-ct}^{x+ct} \psi(s)ds, & x > ct, \\ \frac{1}{2}[\phi(x - ct) - \phi(ct + x)] + \frac{1}{2c}\int_{ct-x}^{x+ct} \psi(s)ds, & 0 < x < ct. \end{cases}$$
$$(3.7)$$

If $x > ct$, the arguments $x + ct$ and $x - ct$ of ϕ in (3.6) are both positive and then $\widehat{\phi} \equiv \phi$. If $0 < x < ct$ then $x - ct < 0$, in which case $\phi(x) = -\phi(-x)$ or $\phi(x - ct) = -\phi(ct - x)$. Similar observations apply to $\int_{x-ct}^{x+ct} \widehat{\psi}(s)ds$ in (3.6) to give $\int_{ct-x}^{x+ct} \psi(s)ds$ in (3.7).

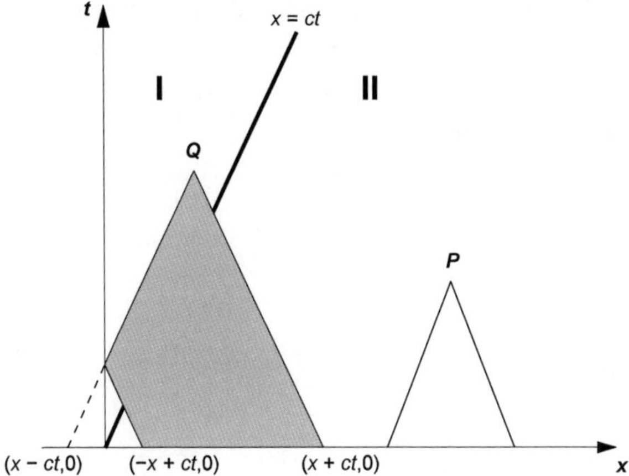

Fig. 3.3.   Domain of dependence with reflection.

The solution (3.7) can be better understood by examining the characteristics. The characteristic line from the origin, $x = ct$, separates the first quadrant into two parts, I $(0 < x < ct)$ and II $(x > ct)$. In Region II, the presence of the boundary $x = 0$ is not felt since any information from this boundary, traveling with speed c, never reaches this region as it always lies to the left of $x = ct$. In Region I, the solution depends on the direct left-travelling characteristic from $t = 0$ and the right-traveling characteristic for the wave reflected from the boundary at $x = 0$ that started at $(ct - x, 0)$. The domain of dependence at $\mathbf{Q} :(x, t)$ becomes the shaded area shown in Fig. 3.3, while to the right, the domain of dependence at $\mathbf{P} :(x, t)$ is unaffected by the boundary.

Note that the method of images, extending the data on $x > 0$ to $x < 0$, can only be used for the simple boundary conditions $u(0, t) = 0$, when the extensions are odd about $x = 0$, and $u_x(0, t) = 0$, when the extensions are even.

3.2.2. *Solution in a Finite Domain*

The method of images can also be used to investigate the solution on a finite domain $[0, l]$ for simple reflective boundary conditions like $w(l, t) = 0$, or $w_x(0, t) = 0$. However, it is cumbersome keeping track of all reflections for large times, so the method of separation of variables (see Sec. 3.3) is often the preferred method. We give an outline of how to proceed using characteristics before considering the problem using separation of variables.

The initial and boundary conditions for the simplest fixed end problem are:

$$
\begin{aligned}
w(x,0) &= \phi(x), & 0 &< x < \ell, \\
w_t(x,0) &= \psi(x), & 0 &< x < \ell, \\
w(0,t) &= u(\ell,t) = 0, & t &> 0.
\end{aligned}
\tag{3.8}
$$

Note that there are two conditions on $t = 0$, and one each on $x = 0$ and $x = l$, where only one characteristic enters $0 \leq x \leq l$. Again we extend both ϕ and ψ to $\widehat{\phi}$ and $\widehat{\psi}$ to be odd about $x = 0, l$, and are then odd on $-l < x < \ell$, and then extended periodically to $(-\infty, \infty)$ to give periodic functions with period $2l$. The solution can then be obtained using D'Alembert's solution (3.5):

$$
w(x,t) = \frac{1}{2}\left[\widehat{\phi}(x - ct) + \widehat{\phi}(x + ct)\right] + \frac{1}{2c}\int_{x-ct}^{x+ct} \widehat{\psi}(s)ds.
$$

The solution at any point P_0 is found by following the characteristics back through multiple reflections to the initial line $t = 0$, as shown in Fig. 3.4.

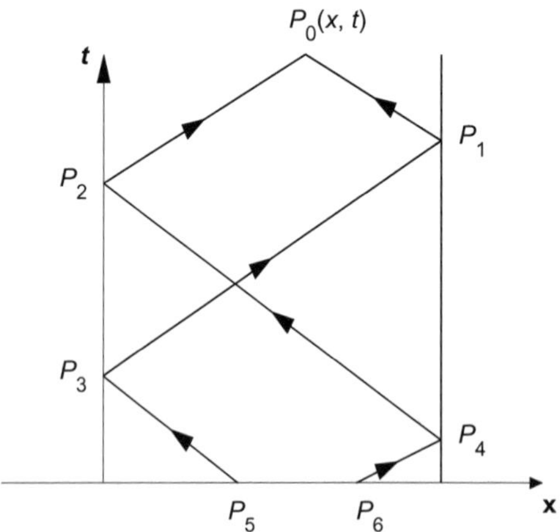

Fig. 3.4.   Characteristics showing reflections.

3.3. Free Vibrations of An Elastic Panel

We use a completely different approach, the method of separation of variables, to reconsider the problem (3.8), now in the context of free vibrations of an elastic panel. This method is not as intuitive for wave problems as methods that use characteristics curves, as the wave-like nature of disturbances is usually disguised in the final form of solution. However, it is often the most appropriate for this type of problem, and it introduces the important concepts of harmonics, eigenvalues and eigenfunction expansions.

Consider the vibrations of a panel of length ℓ with both ends fixed and without an external force. As in (3.8), the initial displacement is given as $\phi(x)$, and the initial velocity is $\psi(x)$. The motion is described by (1.1) for the displacement $w(x,t)$ subject to the fixed end conditions $w(0,t) = w(l,t) = 0$.

We look for separable solutions of the form

$$w(x,t) = X(x)T(t),$$

and substitute this into (1.1) yielding

$$\frac{X''}{X} = \frac{1}{c^2}\frac{T''}{T} = -\kappa^2, \tag{3.9}$$

where $\kappa^2 > 0$ is an unknown separation constant. Acceptable values of κ will depend on the boundary conditions.

Since X only depends on x and T only on t, we can separate (3.9) as two second order o.d.e.s:

$$X'' + \kappa^2 X = 0, \ T'' + c^2\kappa^2 T = 0. \tag{3.10}$$

In order to satisfy the boundary conditions we require

$$X(0) = X(l) = 0. \tag{3.11}$$

A nontrivial solution (i.e., $u(x,t) \neq 0$) of (3.10) for $X(x)$ exists only for $\kappa > 0$ when the general solution is

$$X(x) = C_1\cos(\kappa x) + C_2\sin(\kappa x),$$

where C_1 and C_2 are arbitrary constants. The boundary condition $X(0) = 0$ then yields $C_1 = 0$, while $X(\ell) = 0$ yields

$$C_2 \sin(\kappa\ell) = 0,$$

which implies that $\kappa\ell = n\pi$, integer n. This yields the set of nontrivial solutions

$$X_n(x) = \sin(k_n x), \ \ k_n = \kappa_n = \frac{n\pi}{\ell}, \ \ \ n = 1, 2, \ldots$$

The general solution of the second of (3.10) is then

$$T_n(t) = A_n \cos(\omega_n t) + B_n \sin(\omega_n t),$$

where $\omega_n = ck_n$; A_n and B_n are arbitrary constants. Combining these results, since the equations are linear, produces the solution to (1.1), as

$$w(x, t) = \sum_{n=1}^{\infty} \left(A_n \cos \frac{n\pi ct}{\ell} + B_n \sin \frac{n\pi ct}{\ell} \right) \sin \frac{n\pi x}{\ell}, \qquad (3.12)$$

which is the sum of separable solutions. The solution (3.12) satisfies the equation and boundary conditions, while the constants A_n and B_n are determined from the initial conditions as follows:

$$\phi(x) = w(x, 0) = \sum_{n=1}^{\infty} A_n X_n(x), \ \ \psi(x) = w_t(x, 0) = \sum_{n=1}^{\infty} \omega_n B_n X_n(x).$$

Since the X_n form an orthogonal set, i.e., $\int_0^\ell X_m X_n dx = \frac{2}{l}\delta_{mn}$, where δ_{mn} is the Kronecker delta, we obtain

$$A_n = \frac{2}{\ell} \int_0^\ell \phi(s) X_n(s) ds, \ \ B_n = \frac{2}{\omega_n \ell} \int_0^\ell \psi(s) X_n(s) ds. \qquad (3.13)$$

A_n and B_n are the coefficients in the Fourier sine series for $\phi(x)$ and $\psi(x)$ respectively; they are odd functions about $x = 0$. The solution (3.12) and (3.13) describes a standing wave on the finite panel consisting of a *fundamental harmonic* with frequency $\omega_1 = \pi c/\ell$ and higher modes, or overtones, at frequencies $\omega_n = n\pi c/\ell$ ($n \geq 2$).

In the solution set for $X(x)$, X_n are the *eigenfunctions*, $k_n = \frac{n\pi}{\ell}$ are the *eigenvalues* and the full series (3.12) is the *eigenfunction expansion*. The period of the fundamental harmonic is $\frac{2\pi}{\omega_1} = \frac{2l}{c}$. This is the round-trip travel time in the panel.

Of course the solution (3.12) and (3.13) must fit into the form of the general solution (3.2), and it is easily rearranged into the D'Alembert form

$$w(x,t) = \sum_{n=1}^{\infty} \frac{1}{2}\sqrt{A_n^2 + B_n^2}$$

$$\times \left[\sin\left(\frac{n\pi}{\ell}(x - ct + \alpha_n)\right) + \sin\left(\frac{n\pi}{\ell}(x + ct - \alpha_n)\right)\right],$$

where

$$\sin\alpha_n = \frac{B_n}{\sqrt{A_n^2 + B_n^2}}, \quad \cos\alpha_n = \frac{A_n}{\sqrt{A_n^2 + B_n^2}}.$$

3.4. Reflection of a Plane Stress Wave from a Boundary

Consider a stress wave traveling from left to right incident on the boundary $x = 0$ that divides elastic materials with different, but constant, properties, see Fig. 3.5. The material occupying $x < 0$ (called Region I) has density and Young's modulus (ρ_1, E_1), while the density and Young's modulus for $x > 0$ (called Region II) are (ρ_2, E_2). The corresponding wave speeds are $c_n = \sqrt{\frac{E_n}{\rho_n}}$, $n = 1, 2$.

The particle velocity at x is $u(x,t)$, satisfying $u_{tt} = c_n^2 u_{xx}$ in each region. Hence, we can write

$$u(x,t) = \phi_n(t + x/c_n) - \psi_n(t - x/c_n), \tag{3.14}$$

and from (2.3) the corresponding solution for the stress is

$$\sigma(x,t) = i_n[\phi_n(t + x/c_n) + \psi_n(t - x/c_n)], \tag{3.15}$$

where the local impedance is $i_n = \sqrt{E_n \rho_n}$. These solutions can be verified by direct substitution.

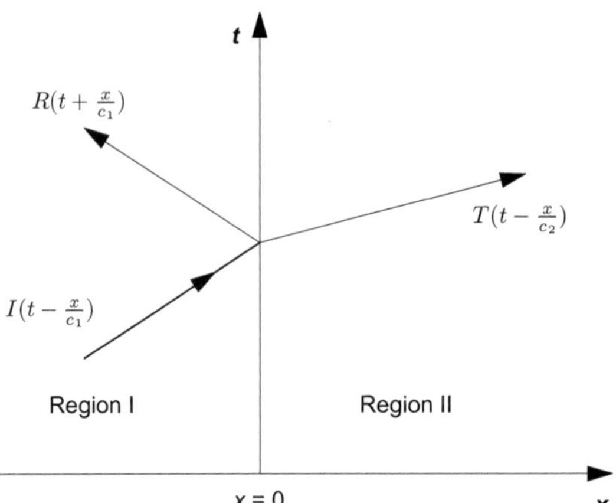

Fig. 3.5.   Reflected and transmitted waves.

Since there is no wave traveling to the left from $x = \infty$ in Region II, $\phi_2(t + x/c_2) \equiv 0$. We denote the wave incident on $x = 0$ from $x < 0$ as $\psi_1(t - x/c_1) = I(t - x/c_1)$, the reflected wave $\phi_1(t + x/c_1) = R(t + x/c_1)$ and the transmitted wave as $\psi_2(t - x/c_2) = T(t - x/c_2)$, as shown in Fig. 3.5.

Hence, the velocity and stress in each region may be written as

$$u(x,t) = \begin{cases} R(t + x/c_1) - I(t - x/c_1), & -\infty < x < 0 \\ -T(t - x/c_2), & 0 < x < \infty, \end{cases}$$

$$\sigma(x,t) = \begin{cases} i_1 R(t + x/c_1) + i_1 I(t - x/c_1), & -\infty < x < 0 \\ i_2 T(t - x/c_2), & 0 < x < \infty, \end{cases}$$

where I is a known incident wave function. The interface conditions at $x = 0$ are continuity of velocity and stress, so that

$$u(0^-, t) = u(0^+, t), \quad \text{and} \quad \sigma(0^-, t) = \sigma(0^+, t),$$

which translate into

$$I(t) - R(t) = T(t), \; I(t) + R(t) = \frac{i_2}{i_1} T(t).$$

Solving for the reflected and transmitted waves:

$$R(t) = \frac{i_2 - i_1}{i_2 + i_1} I(t), \; T(t) = \frac{2 i_1 I(t)}{i_1 + i_2}, \tag{3.16}$$

and hence the full solution is

$$u(x,t) = \begin{cases} (\frac{i_2-i_1}{i_2+i_1})I(t+x/c_1) - I(t-x/c_1), & -\infty < x < 0 \\ -\frac{2i_1}{i_1+i_2}I(t-x/c_2), & 0 < x < \infty, \end{cases}$$

$$(3.17)$$

$$\sigma(x,t) = \begin{cases} i_1(\frac{i_2-i_1}{i_2+i_1})I(t+x/c_1) + i_1 I(t-x/c_1), & -\infty < x < 0 \\ \frac{2i_1 i_2}{i_1+i_2}I(t-x/c_2), & 0 < x < \infty. \end{cases}$$

$$(3.18)$$

We note the special cases:

(i) Matching impedances, $i_2 = i_1$; then there is no reflected wave, i.e., $R = 0$, and the transmitted wave has the same amplitude as the incident wave since (3.16) implies $T(t) = I(t)$.

(ii) When $\frac{i_1}{i_2} \to 0$, the density of Region I is much less than that in Region II, then (3.16) becomes

$$R(t) = I(t), \; T(t) = 0.$$

The transmitted wave has zero amplitude as there is not enough energy in the disturbance in Region I to affect Region II. This corresponds to a 'fixed' or 'closed end' boundary condition.

(iii) When $\frac{i_1}{i_2} \to \infty$, the density in Region I is much greater than that in Region II, then (3.16) becomes

$$R(t) = -I(t), \; T(t) = 2I(t).$$

This corresponds to the boundary at $x = 0$ being stress-free, as can be seen from the second of (3.18) since $\frac{i_1 i_2}{i_1+i_2} \to 0$, and then the amplitude of the transmitted wave is double that of the incident wave.

(iv) At $x = 0^-$, just in Region I, we can relate σ and u to produce what is referred to as an impedance boundary condition. Using

(3.16)–(3.18)

$$u(0^-,t) = \left(\frac{i_2 - i_1}{i_2 + i_1}\right) I(t) - I(t) = -\frac{2i_1}{i_2 + i_1} I(t),$$

$$\sigma(0^-,t) = i_1 \left(\frac{i_2 - i_1}{i_2 + i_1}\right) I(t) + i_1 I(t) = \frac{2i_1 i_2}{i_2 + i_1} I(t)$$

and hence

$$\sigma(0^-,t) = -i_2 u(0^-,t). \tag{3.19}$$

A stress-free end corresponds to $i_2 = 0$, while a fixed end corresponds to $i_2 \to \infty$.

(v) A similar argument to (iv) holds for the case when there is only an incident wave from the right, so $\psi_1 = 0$, with ϕ_2 known. Then at $x = 0^+$ the impedance boundary condition is

$$\sigma(0^+,t) = i_1 u(0^+,t). \tag{3.20}$$

Chapter 4

Linear Physical Examples

The basic examples in Chapters 2 and 3 are described in terms of standard dimensional variables. However, to define and utilize naturally occurring small parameters, such as the ratio of the period of an oscillation to a relaxation time, all variables must be redefined in terms of the appropriate scales for the phenomena to be investigated. Hence, we normally work with *dimensionless* variables rather than the original physical variables. This simplifies the equations and introduces physically relevant dimensionless parameters that can be compared with unity. This is essential when using perturbation schemes. The process and choices of length and time scales are described in Sec. 4.1.

We then present some examples that expand on the introductory material with an emphasis on the wave-like behavior of the solutions in the geometries that are investigated in the nonlinear situation in later chapters. We include some basic results for linear spherical and cylindrical waves, and for completeness add some background on the Dirac delta function.

4.1. Dimensionless Variables

It is often more convenient to work with *dimensionless* variables rather than the original physical variables. Dimensionless variables are simply the physical variables rescaled with some convenient physical constants. Length might be measured in multiples of a wave length or relaxation length rather than in centimeters. The choice

29

depends on what scales are important for a particular problem. For example, for waves in a Maxwell solid of length l, u and σ satisfy (2.35) for $0 \leq x \leq l$, with velocity boundary conditions $u(0,t) = ah(t/\tau_p)$, $u(l,t) = 0$, where $\|h\| = O(1)$ and τ_p is the period or pulse length, when $h \equiv 0$ for $t > \tau_p$. The boundary condition at $x = 0$ introduces two constants, a velocity amplitude a and τ_p, while the equations contain three constants, Young's modulus, E_0, density, ρ_0 and relaxation time τ_r; the wave speed is $c_0 = \sqrt{\frac{E_0}{\rho_0}}$.

The dependent and independent variables are scaled with typical dimensional constants. Then both the governing equations and boundary condition are rewritten in dimensionless variables and any constants remaining will be dimensionless parameters. Since E_0 has the same dimensions as stress and c_0 that of velocity, these are obvious constants to scale σ and u. Additionally, to maintain the wave-like nature of the system, if t is scaled with some constant T, then x must be scaled with $L = c_0 T$. The values of L and T depend on the problem, but only one is specified independently.

Hence we define the new dimensionless variables $\widehat{x} = x/c_0 T$, $\widehat{t} = t/T$, $\widehat{\sigma} = \sigma/E_0$ and $\widehat{u} = u/c_0$. Then, for example, (2.34) becomes

$$\frac{\partial \widehat{u}}{\partial \widehat{t}} = \frac{\partial \widehat{\sigma}}{\partial \widehat{x}} \quad \text{and} \quad \frac{\partial \widehat{\sigma}}{\partial \widehat{t}} = \frac{\partial \widehat{u}}{\partial \widehat{x}} - \epsilon \widehat{\sigma},$$

where $\epsilon = T/\tau_r$. The boundary condition at $\widehat{x} = 0$ becomes $\widehat{u}(0,\widehat{t}) = Mh(\omega\widehat{t})$, where $\omega = T/\tau_p$, $M = a/c_0$ and $0 \leq \widehat{x} \leq d = l/L$. The system now contains four dimensionless parameters: d, M, ϵ and ω. M is an amplitude parameter, like the Mach number, and to use the linear equations we usually require $M \ll 1$. The values and interpretations of d, ϵ and ω depend on the choices of either T or L.

For a finite material usually $L = l$, so $d = 1$ and $T = lc_0^{-1}$ is the travel time across $[0, 1]$. Then $\epsilon = l/c_0\tau_r$ is the ratio of the panel length to the viscoelastic relaxation length, $c_0\tau_r$, and $\omega = l/c_0\tau_p = l/\xi$ is the ratio of the material length to the applied wavelength, $\xi = c_0\tau_p$.

When $0 \leq x < \infty$ there is no natural travel time, but there are still two choices for T: τ_p and τ_r. If $T = \tau_p$, so the main interest is in times of the same order as the period defined by the boundary condition, then $\omega = 1$ and $\epsilon = \tau_p/\tau_r$ is the ratio of the period of the input to the relaxation time. Since $\widehat{u}(0,\widehat{t}) = Mh(\widehat{t})$ there is no time parameter in the boundary condition, but ϵ occurs in the equations. In this situation an oscillation with $\epsilon \ll 1$ is called a "high frequency" disturbance since the frequency defined by the input is much greater than the frequency defined by the relaxation. Alternatively, picking $T = \tau_r$, so the interest is in solutions valid over the relaxation time, then $\epsilon = 1$. In this case $\omega = \tau_r/\tau_p$ occurs in the boundary condition, $\widehat{u}(0,\widehat{t}) = Mh(\omega\widehat{t})$, but there is no time parameter in the equation.

Wave problems in continuously varying media contain a length scale defined by the material stratification. For example, stress waves in an elastic panel described by Eqs.(2.8) are modified by the variable Young's modulus $E(x/D)$. Then the stratification length $\|E/\frac{dE}{dx}\| = D$ is a measure of the variation in the material properties and has the dimension of length, while the input at $x = 0$ introduces the wavelength $\xi = c_0\tau_p$, where $c_0 = \sqrt{\frac{E(0)}{\rho(0)}}$. Again there are two options: picking $L = D$ and picking $L = \xi$.

We illustrate the difference with a simple example, the Webster horn equation, which we consider in detail later. From Eqs.(2.8), solutions of

$$\frac{\partial^2 u}{\partial t^2} - \frac{1}{E(x/D)}\frac{\partial(E(x/D)u_x)}{\partial x} = 0, \tag{4.1}$$

govern the particle velocity $u(x,t)$ in longitudinal stress waves in an elastic panel with variable Young's modulus $E(x/D)$. A typical boundary condition is $u(0,t) = ah(t/\tau_p)$. Picking $L = D$, so that $\widehat{x} = x/D$, $\widehat{t} = c_0 t/D$ and $\widehat{u} = u/c_0$ then (4.1) becomes

$$\frac{\partial^2 \widehat{u}}{\partial \widehat{t}^2} - \frac{1}{E(\widehat{x})}\frac{\partial}{\partial \widehat{x}}(E(\widehat{x})\widehat{u}_{\widehat{x}}) = 0, \tag{4.2}$$

while the boundary condition is now $\widehat{u}(0,\widehat{t}) = Mh(\omega\widehat{t})$ with $M = a/c_0$ and $\omega = D/c_0\tau_p = D/\xi$, which is dimensionless. When $\omega \gg 1$, the applied input is of "high frequency" relative to the frequency defined by the stratification of the medium. In this case, there is a parameter in the boundary condition, but not in the equation.

Alternatively, picking $L = \xi$ and hence $\hat{t} = c_0 t/\xi$, there is no parameter in the boundary condition, but the stratification function is scaled as $E(\varepsilon\hat{x})$, so when $\varepsilon = \xi/D \ll 1$ then E is considered a slowly varying function of $\hat{x}$. In this case (4.1) becomes

$$\frac{\partial^2 \hat{u}}{\partial \hat{t}^2} - \frac{1}{E(\varepsilon\hat{x})} \frac{\partial}{\partial \hat{x}}(E(\varepsilon\hat{x})\hat{u}_{\hat{x}}) = 0, \qquad (4.3)$$

while the boundary condition is $\hat{u}(0,\hat{t}) = Mh(\hat{t})$.

4.2. Periodic Forcing of a Gas in a Closed Tube

This is a classical problem that we will investigate in greater depth in Chapter 9. Here, we consider a simple straight tube of constant cross-sectional area and length l lying along the x-axis. At the end $x = l$ a flat piston oscillates periodically with period τ_p and generates a periodic motion of the gas, while the end $x = 0$ is closed. The notation and basic equations describing the motion are derived in Sec. 2.2. If the piston motion is $x = l + X(t)$ and $|X| \ll l$, so that the velocity of the piston $\frac{dX}{dt}$ is small compared with the sound speed $c_0 = \sqrt{\frac{\gamma p_0}{\rho_0}}$, small amplitude sound waves will be generated. The deviations of particle velocity, pressure and density from ambient levels, $u(x,t)$, $p(x,t)$, $\rho(x,t)$, all satisfy the wave equation (1.1).

The boundary conditions can be written in terms of the velocity as

$$u(0,t) = 0 \text{ and } u(l,t) = a\sin(2\pi t/\tau_p), \qquad (4.4)$$

where $\frac{dX}{dt}(l,t) = a\sin(2\pi t/\tau_p)$.

The problem is simplified if it is first converted into dimensionless variables. Taking $L = l$, and $T = lc_0^{-1}$ and c_0 as the velocity scale, then in dimensionless variables (4.4) becomes

$$u(0,t) = 0 \text{ and } u(1,t) = M\sin(2\pi\omega t), \qquad (4.5)$$

where the dimensionless frequency is $\omega = l/c_0\tau_p$ and $M = a/c_0 \ll 1$ is the Mach number. Then $u(x,t)$ satisfies (1.1) with $c_0 = 1$.

Solving the linear wave equation is often easier in characteristic variables, so we define

$$\alpha = t - x \text{ and } \beta = t + x - 1.$$

As α and β are constant along their respective characteristics, it is always possible to add an arbitrary constant to each variable. In this case the constants are chosen so that $\alpha = t$ on $x = 0$ and $\beta = t$ on $x = 1$, and then α and β measure the time at which the relevant characteristic leaves the boundary. We can now write the general solution in the form

$$u(x,t) = F(\alpha) + G(\beta). \tag{4.6}$$

The boundary conditions (4.5) imply that

$$F(t) + G(t - 1) = 0, \text{ and } F(t - 1) + G(t) = M\sin(2\pi\omega t). \tag{4.7}$$

By eliminating F, these combine to give a linear difference equation for G on $x = 1$:

$$G(t) = G(t - 2) + M\sin(2\pi\omega t). \tag{4.8}$$

It will be useful later, when considering the nonlinear problem, to interpret the terms in Eq. (4.8). The difference in the arguments of G is 2, which is the dimensionless travel time from $x = 1$ to $x = 0$ and back. The difference equation states that the signal $G(t)$ that leaves $x = 1$ at time t is the signal that left $x = 1$ at the previous time $t - 2$ augmented by piston input, $M\sin(2\pi\omega t)$.

If we look for a solution to (4.8) of the form

$$G(t) = A\sin(2\pi\omega t) + B\cos(2\pi\omega t),$$

substituting in (4.8), and equating coefficients of $\sin(2\pi\omega t)$ and $\cos(2\pi\omega t)$, yields

$$A = \frac{M}{2} \text{ and } B = -\frac{M}{2}\cot(2\pi\omega). \tag{4.9}$$

Then

$$G(t) = -\frac{M}{2\sin(2\pi\omega)}\cos(2\pi\omega t + 2\pi\omega), \tag{4.10}$$

provided $\omega \neq \omega_n = n/2$, for integer n. Note that the higher resonant frequencies are integer multiples of the fundamental. Since $|G| \to \infty$ as $\omega \to \omega_n$, the frequencies ω_n are called the *resonant frequencies* of the system. A typical response curve for $|G|$ around $\omega = 1/2$ is shown as the curve $k = 1$ in Fig. 4.1. Hence, when the system is in resonance, Eq. (4.8) can be interpreted as the case when the dimensionless travel time in the tube, 2, coincides with the period of the input, $\frac{1}{\omega}$, so that $\omega = \frac{1}{2}$. Then there is no periodic solution of period 2 of (4.8) since the arguments of G are the same. We will return to this when considering the equivalent nonlinear problem.

Combining (4.6) and (4.10) yields

$$u(x,t) = \frac{M}{2\sin(2\pi\omega)}[\cos(2\pi\omega(t - x)) - \cos(2\pi\omega(t + x))], \quad \omega \neq \frac{n}{2}, \tag{4.11}$$

which is the superposition of waves traveling in opposite directions. This solution (4.11) can also be written in the *normal mode* form

$$u(x,t) = \frac{M}{\sin(2\pi\omega)}\sin(2\pi\omega t)\sin(2\pi\omega x), \quad \omega \neq \frac{n}{2}. \tag{4.12}$$

At this point it is instructive to return to the original physical variables, when the particle velocity (4.11) becomes

$$u(x,t) = \frac{a}{2\sin\left(\frac{2\pi l}{c_0\tau_p}\right)}\left[\cos\left(\frac{2\pi}{c_0\tau_p}(c_0 t - x)\right) - \cos\left(\frac{2\pi}{c_0\tau_p}(c_0 t + x)\right)\right] \tag{4.13}$$

$$= \frac{a}{\sin\left(\frac{2\pi l}{c_0\tau_p}\right)}\sin\left(\frac{2\pi}{\tau_p}t\right)\sin\left(\frac{2\pi}{c_0\tau_p}x\right), \quad \frac{l}{c_0\tau_p} \neq \frac{n}{2}. \tag{4.14}$$

The solution in the form (4.14) could also be found using separation of variables, as in 3.3 for an elastic panel.

4.2.1. *Damping through Radiation*

When the end $x = 0$ is closed there is no loss of energy to the adjacent outside medium. However, when the end is 'partially open' or 'partially closed' in the sense that it is allowed to radiate energy out of the tube into the surrounding air, we anticipate that the oscillation is damped. Across the boundary $x = 0$ both pressure and velocity are continuous, so, as shown by (2.21) and (3.20) in Sec. 3.4 for an elastic panel, there is an impedance boundary condition of the form

$$e(0^+, t) = -iu(0^+, t), \qquad (4.15)$$

where $i \geq 0$ and we note that, compared with (3.20), $e = -\sigma$. Then since $e = F(\alpha) - G(\beta)$ and u is given by (4.6), the reflected wave at $x = 0$ (F) is related to the incident wave (G) by

$$F(t) + kG(t - 1) = 0, \qquad (4.16)$$

where $k = -(1-i)/(1+i)$. Condition (4.16) replaces the first of (4.7). Note that $-1 \leq k \leq 1$ and $i \to \infty$ $(k = 1)$ corresponds to a closed end, while $i = 0$ $(k = -1)$ could model an open end, see Seymour and Mortell [1973b].

Equations (4.16) and the second of (4.7) imply that the linear difference equation for G becomes

$$G(t) = kG(t - 2) + M \sin(2\pi\omega t). \qquad (4.17)$$

When there is no piston movement (i.e., no forcing, $M = 0$) and the end $x = 1$ is partially closed, defining $G_n = G(t + 2n)$, (4.17) implies that $G_n(t) = k^n G(t)$. When $|k| < 1$, $G_n(\eta) \to 0$ as $n \to \infty$ and any initial disturbance in the tube simply decays in time as it traverses back and forth since there is a loss of energy each time the wave hits $x = 0$.

When $M \neq 0$ an analysis similar to the case $k = 1$ gives

$$A = \frac{M(1 - k \cos 4\pi\omega)}{(k - 1)^2 + 4k \sin^2 2\pi\omega} \text{ and } B = \frac{-kM \sin 4\pi\omega}{(k - 1)^2 + 4k \sin^2 2\pi\omega}, \qquad (4.18)$$

so that

$$G(t) = M \frac{\sin 2\pi\omega t - k \sin(2\pi\omega t + 4\pi\omega)}{(k - 1)^2 + 4k \sin^2 2\pi\omega}. \qquad (4.19)$$

The damping now produces finite amplitude responses for all ω. Figure 4.1 shows the response $|G|$ around $\omega = 1/2$ for $k = 1$ (undamped), 0.7 and 0.5. As k decreases the amplitude decreases.

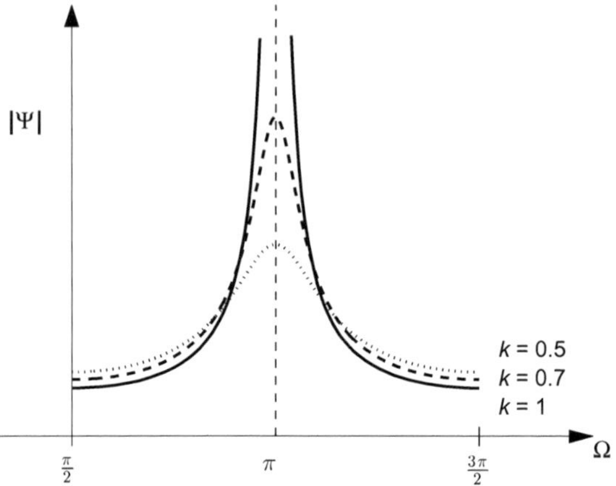

Fig. 4.1.   Response curve with and without damping.

It is readily checked from (4.18) that for a closed end, when $k = 1$, $A = \frac{M}{2}$ and $B = -\frac{M}{2}\cot 2\pi\omega$, as in (4.9).

For the case when the tube is open at $x = 0$, so that $k = -1$, then (4.18) implies that

$$A = \frac{M}{2} \quad \text{and} \quad B = \frac{M}{2}\tan 2\pi\omega,$$

while (4.19) gives

$$G(t) = \frac{M}{2\cos(2\pi\omega)}\sin(2\pi\omega t + 2\pi\omega), \tag{4.20}$$

provided $\omega \neq \omega_n = \frac{1}{4} + \frac{n}{2}$, for integer n. The frequencies ω_n are the resonant frequencies of this system. Combining (4.16), (4.6) and (4.20) yields

$$u(x,t) = \frac{M}{2\cos(2\pi\omega)}[\sin(2\pi\omega(t + x)) + \sin(2\pi\omega(t - x))], \tag{4.21}$$

$$= \frac{M}{\cos(2\pi\omega)}\cos(2\pi x\omega)\sin(2\pi t\omega), \quad \omega \neq \frac{1}{4} + \frac{n}{2}. \tag{4.22}$$

The solution in the form (4.22) for an elastic panel with a free end at $x = 0$ could also be found using separation of variables, as in Sec. 3.3.

4.3. Spherical and Cylindrical Waves in a Gas

In later chapters we consider nonlinear acoustic waves with either spherical or cylindrical symmetry. Here, as an introduction, we introduce the basic equations and some simple linear results for such geometries.

4.3.1. *Spherical Waves*

The equations describing the motion of a polytropic gas in Cartesian coordinates are given in (2.14). The same equations in spherical coordinates describing disturbances that are symmetric about the origin, $r = 0$, with dependent variables $u(r, t)$, the radial particle velocity, and $c(r, t) = c_0 + a(r, t)$, the sound speed in the gas, are

$$u_t + uu_r + k^{-1}cc_r = 0, \ c_t + uc_r + kc(u_r + 2u/r) = 0, \qquad (4.23)$$

where $k = (\gamma - 1)/2$ and $c_0 = \sqrt{\frac{\gamma p_0}{\rho_0}}$ is the ambient sound speed. Picking dimensionless variables with $L = R_0$ and $T = R_0/c_0$, for some typical radius R_0, (4.23) remain unchanged, but $c(r, t) = 1 + a(r, t)$.

For linear motions (4.23) becomes

$$u_t + k^{-1}a_r = 0, \ a_t + k(u_r + 2u/r) = 0. \qquad (4.24)$$

We write the radial velocity in terms of the velocity potential $\varphi(r, t)$ by $u(r, t) = \frac{\partial \varphi}{\partial r}$ and the (dimensionless) excess pressure as $p - 1 = -\frac{\partial \varphi}{\partial t}$. Eliminating a, the equation for $\varphi(r, t)$ is

$$\varphi_{tt} = \frac{1}{r^2}(r^2\varphi_r)_r \qquad (4.25)$$

which can be rewritten as

$$(r\varphi)_{tt} = (r\varphi)_{rr}, \qquad (4.26)$$

so that the general solution is

$$\varphi(r, t) = \frac{1}{r}\phi(r - t) + \frac{1}{r}\psi(r + t), \qquad (4.27)$$

for arbitrary φ and ψ.

As a simple example we take the case of an outgoing wave from $r = 1$ ($r = R_0$, in dimensional variables) i.e., prescribe $\varphi(1, t) = \Phi(t)$, $t > 0$. Then the solution is

$$\varphi(r, t) = \frac{1}{r}\Phi(t - r + 1). \tag{4.28}$$

The amplitude of the wave is attenuated by the factor $\frac{1}{r}$ as the wave propagates into the region $r > 1$.

Another example is the *bursting balloon*. The equation to be solved is (4.25) with the initial conditions

$$\varphi(r, 0) = 0 \quad \text{and} \quad -\varphi_t(r, 0) = F(r). \tag{4.29}$$

Here, $F(r)$ is a given function, and the second initial condition is equivalent to specifying the variation in the initial pressure over the ambient.

The solution of (4.25) is given by (4.27), and $\varphi(r, 0) = 0$ implies $\psi(t) = -\phi(t)$, so that $\varphi(r, t) = \frac{1}{r}[\phi(r - t) - \phi(r + t)]$, and $\varphi_t = -\frac{1}{r}[\phi'(r - t) + \phi'(r + t)]$. Then $F(r) = -\varphi_t(r, 0) = \frac{2}{r}\phi'(r)$, and this determines $\phi(r)$. As an example, we take

$$F(r) = \begin{cases} W, & r \leq 1 \\ 0, & r > 1, \end{cases} \tag{4.30}$$

where W is constant. This is equivalent to specifying that initially there is a constant pressure inside the balloon in excess of that outside. So, from (4.27),

$$\phi'(r) = \frac{r}{2}F(r) = \begin{cases} \dfrac{W}{2}r, & 0 \leq r \leq 1 \\ 0, & r > 1. \end{cases} \tag{4.31}$$

For $r > 1$, i.e., outside the original balloon, the pressure wave is outgoing and $p - 1 = -\frac{\partial \varphi}{\partial t} = \frac{1}{r}\phi'(r - t)$. From Fig. 4.2 it is seen that

$\varphi = 0$ for $t < r - 1$, i.e., ahead of the wave;

$\varphi \neq 0$ for $r - 1 < t < r + 1$, i.e., between the front and tail of the wave;

$\varphi = 0$ for $t > r + 1$, i.e., behind the wave tail.

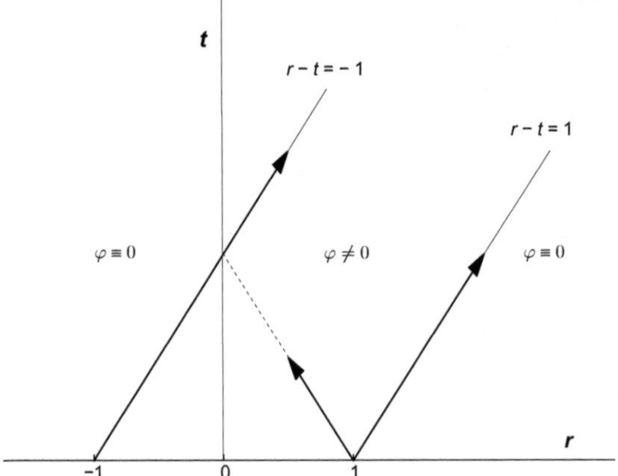

Fig. 4.2.   Bounding characteristics.

Here, $r - t = 1$ is the wave front and $r - t = -1$ is the tail of the
wave. Then, between the front and the tail

$$p - 1 = \frac{1}{r}\varphi'(r - t) = \frac{W}{2r}(r - t), \ \ r - 1 < t < r + 1$$

and $p - 1 = 0$, otherwise.

At the front of the wave when $t = r - 1$, $p - 1 = \frac{W}{2r}$, while at
the tail of the wave $p - 1 = -\frac{W}{2r}$ and the signal decays like $\frac{1}{r}$ as
$r \to \infty$. There is a sharp front and a sharp tail to the wave as shown
in Fig. 4.3. From (4.27) the rate of flow out of a sphere of radius r is

$$4\pi r^2 \varphi_r = -4\pi\phi(r - t) + 4\pi r\phi'(r - t) \tag{4.32}$$

and we require ϕ and ϕ' to be finite at $r = 0$. The source strength at
$r = 0$ is defined by

$$\lim_{r \to 0} 4\pi r^2 \varphi_r = Q(t).$$

Therefore from (4.32)

$$Q(t) = -4\pi\phi(-t) \ \ \text{or} \ \ \phi(\xi) = -\frac{1}{4\pi}Q(-\xi)$$

and

$$\varphi(r, t) = \frac{1}{r}\phi(r - t) = -\frac{1}{4\pi r}Q(t - r), \ \ t - r > 0.$$

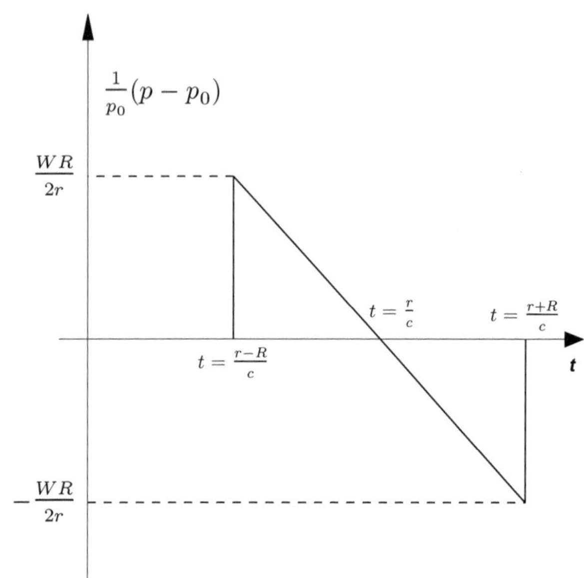

Fig. 4.3.   Structure of pressure wave.

We note that if $Q(t) = \delta(t)$, the Dirac delta function, the Green's function for the spherical wave equation, i.e., the solution to a delta function input, is

$$\varphi(r, t) = -\frac{1}{4\pi r}\delta(t - r).$$

The definition and some properties of the delta function follow in Sec. 4.3.2.

4.3.2. *The Dirac Delta Function*

At an elementary level we can consider the identity

$$f(t) = \int_{-\infty}^{\infty} \delta(t - \tau)f(\tau)d\tau$$

as being the defining property of the delta function $\delta(t)$. However, often a little more detail is required, and an outline of further useful properties is given here.

We define a test function $\varphi(t)$ to be infinitely differentiable on any finite interval in $-\infty < t < \infty$ with it and all derivatives identically zero outside some finite interval, i.e., it has compact support. The set of all test functions is denoted by D. We say $\varphi_n \to \varphi$ in D as $n \to \infty$ if each φ_n is a test function, all φ_n have compact support independent of n, and each derivative of φ_n converges uniformly to the corresponding derivative of φ.

For continuous functions f,

$$f\langle\varphi\rangle = \int_{-\infty}^{\infty} f(t)\varphi(t)dt \tag{4.33}$$

converges for each φ in D. For fixed f, (4.33) assigns a number to each element of D i.e., it defines a functional on D. The functional is linear if

$$f\langle a\varphi + b\theta\rangle = af\langle\varphi\rangle + bf\langle\theta\rangle$$

and is continuous if $f\langle\varphi_n\rangle \to f\langle\varphi\rangle$.

Definition. A distribution is a continuous linear functional on D.

Let $f(x)$ be defined by

$$f(\varphi) \equiv \int_{-\infty}^{\infty} f(x)\varphi(x)dx, \quad \varphi(x) \in D.$$

Suppose $f(x) \equiv \delta(x)$, the delta function, and $\delta(\varphi) = \int_{-\infty}^{\infty} \delta(x)\varphi(x)dx = \varphi(0)$, then $\delta(\varphi) = \varphi(0)$ identifies the delta function $\delta(x)$.

We define the *derivative* of a distribution by $T'(\varphi) = -T(\varphi')$. This is motivated by integration by parts

$$T'(\varphi) = \int_{-\infty}^{\infty} f'(x)\varphi(x)dx$$

$$= \varphi(x)f(x)]_{-\infty}^{\infty} - \int_{-\infty}^{\infty} f(x)\varphi'(x)dx$$

$$= -T(\varphi'), \text{using the compact support of } \varphi.$$

Thus, we have $(\delta', \varphi) = -(\delta, \varphi') = -\varphi'(0)$ and

$$(\delta'', \varphi) = -(\delta', \varphi') = (\delta, \varphi'') = \varphi''(0).$$

Definition. Heaviside step function $H(x)$ is defined by

$$H(x) = \begin{cases} 1, & x > 0 \\ 0, & x < 0. \end{cases}$$

Then $\frac{d}{dx}H(x) = \delta(x)$, since

$$\int_{-\infty}^{\infty} \frac{d}{dx}H(x)\varphi(x)dx \equiv T'(\varphi) = -T(\varphi')$$

$$= -\int_{-\infty}^{\infty} H(x)\varphi'(x)dx = -\int_{0}^{\infty} \varphi'(x)dx$$

$$= \varphi(0) = \delta(\varphi).$$

4.3.2.1. *The delta function and the Green's function*

Consider the ordinary differential equation

$$\frac{du}{dt} + au = \delta(t - t_0), \quad t > 0 \tag{4.34}$$

with $u(0) = 0$. We note immediately that, since $\frac{du}{dt}$ behaves like a delta function, $u(t)$ behaves like a Heaviside step function, since $\frac{d}{dx}H(x) = \delta(x)$. Integrating (4.34) over the interval $t_0 - \varepsilon < t < t_0 + \varepsilon$,

$$\int_{t_0-\varepsilon}^{t_0+\varepsilon} \frac{du}{dt}dt + a\int_{t_0-\varepsilon}^{t_0+\varepsilon} u(t)dt = \int_{t_0-\varepsilon}^{t_0+\varepsilon} \delta(t - t_0)dt = 1,$$

then $u(t_0^+) - u(t_0^-) = 1$, as $\varepsilon \to 0$, since $u(t)$ is finite. Now (4.34) becomes

$$\frac{du}{dt} + au = 0, \quad 0 < t < t_0$$

$$u(0) = 0$$

and

$$\frac{du}{dt} + au = 0, \quad t > t_0$$

$$u(t_0^+) = 1 + u(t_0^-)$$

and the latter is the new "initial" condition at $t = t_0$. The solutions are

$$u(t) = \begin{cases} 0 & t < t_0 \\ Ae^{at} & t > t_0, \end{cases}$$

but $u(t_0^+) = 1$, since $u(t_0^-) = 0$, hence $A = e^{at_0}$ and

$$u(t) = e^{-a(t-t_0)} H(t - t_0). \tag{4.35}$$

We call (4.35) the Green's function $G(t-t_0)$. The Green's function is the response to the delta function input $\delta(t-t_0)$, taking into account the initial or boundary conditions.

4.3.2.2. *The delta function in polar coordinates*

In three dimension, the delta function is $\frac{\delta(r)}{4\pi r^2}$, since

$$\int_{\text{sphere}} \frac{\delta(r)}{4\pi r^2} dV = \frac{1}{4\pi} \int_{\text{sphere}} \frac{\delta(r)}{r^2} r^2 \sin\theta d\theta d\lambda$$

$$= \frac{1}{4\pi} \int_0^{2\pi} d\lambda \int_0^{\pi} \sin\theta d\theta \int_{0^-}^{R} \delta(r) dr$$

$$= 1.$$

In two dimension, the delta function is $\frac{\delta(r)}{2\pi r}$, since

$$\int_{\text{Area}} \frac{\delta(r)}{2\pi r} dA = \int_{\theta=0}^{2\pi} d\theta \int_{0^-}^{R} \frac{\delta(r)}{2\pi r} r dr = 1.$$

4.3.3. *Cylindrical Wave — Symmetry About a Line*

The equations describing the motion of a polytropic gas that is symmetric about the line, $r = 0$, written in (dimensionless) cylindrical coordinates are

$$u_t + uu_r + (1+e)^{\gamma-2} e_r = 0, \quad e_t + ue_r + (1+e)u_r + u/r = 0, \tag{4.36}$$

where u is particle velocity and $e = \rho/\rho_0 - 1$ is the condensation. For linear motions (4.36) becomes

$$u_t + e_r = 0, \ \ e_t + u_r + u/r = 0 \tag{4.37}$$

which can be converted to an equation for the velocity potential $\varphi(r,t)$, where $u(r,t) = \frac{\partial \varphi}{\partial r}$, to give

$$\varphi_{tt} = \left(\varphi_{rr} + \frac{1}{r}\varphi_r \right). \tag{4.38}$$

If we have a line source of strength $q(t)$ per unit length, the element of length dz has strength $q(t)dz$. We think of this as a point source with spherical symmetry. Then, from the spherical result, $d\varphi = -\frac{1}{4\pi R}q(t - R)dz$, where $r^2 + z^2 = R^2$, see Fig. 4.4, and

$$\varphi(r,t) = -\frac{1}{4\pi}\int_{-\infty}^{\infty} \frac{1}{R}\, q(t-R)dz \tag{4.39}$$

$$= -\frac{1}{2\pi}\int_{0}^{\infty} \frac{1}{R}\, q(t-R)dz$$

by symmetry. The change of variables $z = r\sinh\xi, \ \ R = r\cosh\xi$ then yields the representation

$$\varphi(r,t) = -\frac{1}{2\pi}\int_{0}^{\infty} q(t - r\cosh\xi)d\xi. \tag{4.40}$$

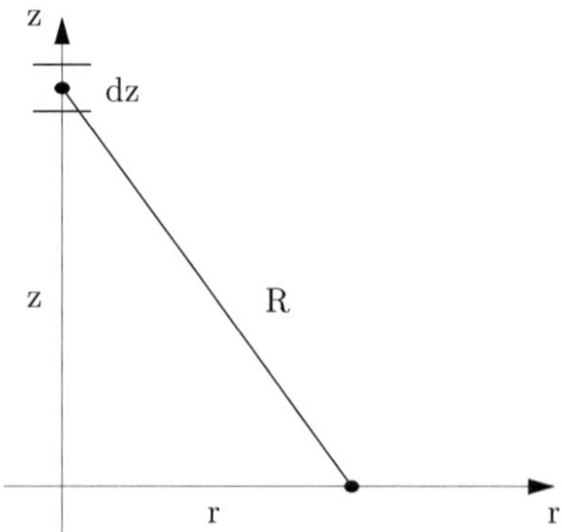

Fig. 4.4.   Point source geometry.

This is the general solution for an *outgoing* wave from a line source of strength $q(t)$ per unit length. If $q(t)$ is a periodic source, i.e., $q(t) = e^{-i\omega t}$, then

$$\varphi(r, t) = -\frac{1}{2\pi} e^{-i\omega t} \int_0^\infty e^{i\omega r \cosh \xi} d\xi$$

$$= -\frac{i}{4} H_0^{(1)}(\omega r) e^{-i\omega t},$$

since the Hankel function is defined by

$$H_0^{(1)}(z) = \frac{1}{i\pi} \int_{-\infty}^\infty e^{iz \cosh \xi} d\xi,$$

see Lebedev [1972] p. 117. Let $\tau = t - r \cosh \xi$, then

$$\varphi = -\frac{1}{2\pi} \int_{-\infty}^{t-r} q(\tau) \frac{d\tau}{r \sinh \xi}.$$

Since $\sinh^2 \xi = \cosh^2 \xi - 1$, $r \sinh \xi = \sqrt{(t - \tau)^2 - r^2}$ and

$$\varphi = -\frac{1}{2\pi} \int_{-\infty}^{t-r} \frac{q(\tau)}{(t - \tau)^2 - r^2} d\tau, \quad t - r > 0. \tag{4.41}$$

This is the solution of the two-dimensional cylindrical wave equation with a source strength of $q(t)$ per unit length.

If $q(t) = \delta(t)$, then

$$\varphi(r, t) = -\frac{1}{2\pi} \int_{-\infty}^{t-r} \frac{\delta(\tau)}{\sqrt{(t - \tau)^2 - r^2}} d\tau, \quad t - r > 0$$

$$= -\frac{1}{2\pi} \frac{1}{\sqrt{t^2 - r^2}} H(t - r),$$

where H is the Heaviside step-function. This is the Green's function for the cylindrical wave equation.

For a general forcing function $\phi(t) = \int_{-\infty}^{\infty} \phi(\tau)\delta(t-\tau)d\tau$, by the definition of the δ-function, and (4.41)

$$\varphi(r,t) = -\frac{1}{2\pi}\int_{-\infty}^{\infty}\frac{\phi(\tau)}{\sqrt{(t-\tau)^2-r^2}}H(t-\tau-r)d\tau, \quad (4.42)$$

$$= -\frac{1}{2\pi}\int_{-\infty}^{t-r}\frac{\phi(\tau)}{\sqrt{(t-\tau)^2-r^2}}d\tau. \quad (4.43)$$

Suppose $q(t) \equiv 0$ outside $0 \le t \le T$, i.e., there is a beginning and an end in time to the line source forcing. Then, from (4.41),

$$\varphi(r,t) = -\frac{1}{2\pi}\int_0^T\frac{q(\tau)}{\sqrt{(t-\tau)^2-r^2}}d\tau, \quad 0 < t-r < T.$$

Let $t \to \infty$ with r fixed, then

$$\varphi(r,t) \sim -\frac{1}{2\pi t}\int_0^T q(\tau)d\tau,$$

i.e., $\varphi(r,t) = O(\frac{1}{t})$ as $t \to \infty$, r fixed and there is no sharp tail to the wave. This is in contrast to the one-dimensional plane wave and the spherical wave.

From (4.40)

$$\varphi_r = \frac{1}{2\pi}\int_0^{\infty}\cosh\xi\, q'(t-r\cosh\xi)d\xi.$$

Let $\cosh\xi = \frac{1}{r}(t-\tau)$, then

$$\varphi_r = \frac{1}{2\pi r}\int_{-\infty}^{t-r}\frac{(t-\tau)q'(\tau)d\tau}{\sqrt{(t-\tau)^2-r^2}},$$

since when $\xi = 0$, $\tau = t - r$ and $\tau \to -\infty$ as $\xi \to \infty$. As $r \to 0$,

$$\varphi_r \sim \frac{1}{2\pi r}\int_{-\infty}^{t}q'(\tau)d\tau = \frac{q(t)}{2\pi r}, \quad \text{with } q(-\infty) = 0.$$

Integration then yields the singularity at $r = 0$:

$$\varphi(r,t) \sim \frac{q(t)}{2\pi}\ln r, \quad r \to 0;$$

i.e., the solution of (4.38) has a logarithmic singularity at the origin.

We now consider (4.38) in Cartesian coordinates with time periodic solutions, then $\varphi(x, y, t)$ satisfies

$$\varphi_{tt} = \varphi_{xx} + \varphi_{yy}, \quad -\infty < x, y < \infty, \quad t > 0.$$

If $\varphi(x, y, t) = \psi(x, y)e^{-i\omega t}$, then $\psi(x, y)$ satisfies the Helmholtz equation

$$\psi_{xx} + \psi_{yy} + \omega^2 \psi = 0.$$

For cylindrical waves, $\psi(x, y) = \psi(r)$, $r^2 = x^2 + y^2$ and

$$\psi'' + \frac{1}{r}\psi' + \omega^2 \psi = 0.$$

This is a Bessel equation and we take the solution $\psi = c_1 H_0^{(1)}(\omega r)$, where $H_0^{(1)}(z) = J_0(z) + iY_0(z)$ is the Hankel function of the first kind. We have chosen the Hankel function $H_0^{(1)}$ since it gives the correct outgoing wave behavior for large z, and has a logarithmic singularity at the origin. These conditions give the Green's function.

$$\varphi(r, t) = e^{-i\omega t}\psi(r) = c_1 H_0^{(1)}(\omega r)e^{-i\omega t} \sim c_1 \left(\frac{2}{\pi \omega r}\right)^{\frac{1}{2}} e^{i\omega\left(r - t - \frac{\pi}{4}\right)},$$

$$\omega r \gg 1,$$

see Lebedev [1972] p. 135. This represents an outgoing wave for large ωr, and is referred to as a radiation condition. We must now check the condition at the origin — does the formula give a logarithmic singularity?

Now $c_1 H_0^{(1)}(\omega r) \sim c_1 \frac{2i}{\pi} \ln(\omega r)$, $\omega r \to 0$, see Lebedev [1972] p. 135. Since we require the coefficient of $\ln(\omega r)$ to be $\frac{1}{2\pi}$, $c_1 = -\frac{i}{4}$ and

$$\varphi(r, t) = -\frac{i}{4} H_0^{(1)}(\omega r)e^{-i\omega t}.$$

This is the same result as deduced from (4.39) and represents the outgoing free-space Green's function for the wave equation in two dimensions.

4.3.4. *Spherical Means*

The wave equation in three dimensions with sound speed unity is

$$\varphi_{tt} = \varphi_{xx} + \varphi_{yy} + \varphi_{zz}. \tag{4.44}$$

We put a point source at $\underline{\xi} = (\xi, \eta, \zeta)$ and then the response at $\underline{x} = (x, y, z)$ is

$$\frac{1}{r}\phi(r - t), \quad \text{where} \quad r = |\underline{x} - \underline{\xi}|,$$

while for a source of *unit* strength, the response is $-\frac{1}{4\pi r}\delta(r - t)$. For a distribution of sources, $h(\underline{\xi})$, over the volume V,

$$\varphi(\underline{x}, t) = \frac{1}{4\pi} \iiint_{V_{\underline{\xi}}} h(\underline{\xi}) \frac{1}{r} \delta(r - t) d\underline{\xi} \tag{4.45}$$

is a solution of (4.44).

We introduce polar coordinates (r, θ, λ) centered at $\underline{x} : \xi = x + r\sin\theta\cos\lambda$, $\eta = y + r\sin\theta\sin\lambda$, $\zeta = z + r\cos\theta$, and then (4.45) becomes

$$\varphi(\underline{x}, t) = \frac{1}{4\pi} \iiint_{V_{\underline{\xi}}} h(\xi, \eta, \zeta) \frac{1}{r} \delta(r - t) r^2 \sin\theta \, dr \, d\theta \, d\lambda \tag{4.46}$$

$$= \frac{t}{4\pi} \iint_{S_t} h(\xi, \eta, \zeta) \sin\theta \, d\theta \, d\lambda, \tag{4.47}$$

on integrating w.r.t. r using the δ−function, where S_t is the surface of the sphere of radius t, i.e., $(x - \xi)^2 + (y - \eta)^2 + (z - \zeta)^2 = t^2$, or $r - t = 0$. Now the element of surface area $dS_t = t^2 \sin\theta \, d\theta \, d\lambda$, and thus

$$\varphi(\underline{x}, t) = t \cdot \frac{1}{4\pi t^2} \iint_{S_t} h(\xi, \eta, \zeta) dS_t = t \, M_t[h], \tag{4.48}$$

where $M_t[h]$ is the mean value of h over the surface of the sphere centered at $\underline{x}$ with radius t, since the surface area of the sphere is $4\pi t^2$.

From (4.48), the initial value $\varphi(\underline{x}, 0) = 0$, and

$$\varphi_t = M_t[h] + t\frac{d}{dt}\{M_t[h]\}.$$

Then

$$\varphi_t(\underline{x}, 0) = M_t[h]\Big|_{t=0} = h(\underline{x})$$

since $\underline{\xi} \to \underline{x}$ as $t \to 0$. Hence the solution of the three-dimensional wave Eq. (4.44), with the initial conditions $\varphi(\underline{x}, 0) = 0$ and $\varphi_t(\underline{x}, 0) = h(\underline{x})$ is given by (4.48).

We need to solve the three-dimensional wave Eq. (4.44) with the general initial conditions

$$\varphi(\underline{x}, 0) = \phi(\underline{x}), \quad \text{and} \quad \varphi_t(\underline{x}, 0) = \psi(\underline{x}). \tag{4.49}$$

Let $\varphi(\underline{x}, t) = \frac{\partial v}{\partial t}(\underline{x}, t) + w(\underline{x}, t)$, where $v(\underline{x}, t)$ satisfies (4.44) subject to $v(\underline{x}, 0) = 0$, $\frac{\partial v}{\partial t}(\underline{x}, 0) = \phi(\underline{x})$ and $w(x, t)$ satisfies (4.44) subject to $w(\underline{x}, 0) = 0$, $\frac{\partial w}{\partial t}(\underline{x}, 0) = \psi(\underline{x})$. Then $\varphi(\underline{x}, t)$ satisfies the wave Eq. (4.44), since both v and w satisfy it. Also $\varphi(\underline{x}, 0) = \frac{\partial v}{\partial t}(\underline{x}, 0) + w(\underline{x}, 0) = \phi(\underline{x})$ and

$$\begin{aligned} \varphi_t(\underline{x}, 0) &= v_{tt}(\underline{x}, 0) + w_t(\underline{x}, 0) & (4.50) \\ &= [v_{xx} + v_{yy} + v_{zz}]_{t=0} + \psi(\underline{x}) & (4.51) \\ &= \psi(\underline{x}), \end{aligned}$$

since $v(\underline{x}, 0) = 0$. We note that $v(\underline{x}, t) = tM_t[\phi]$ and $w(\underline{x}, t) = tM_t[\psi]$, so that

$$\varphi(\underline{x}, t) = \frac{\partial v}{\partial t} + w = \frac{\partial}{\partial t}\{tM_t[\phi]\} + tM_t[\psi] \tag{4.52}$$

is the solution of (4.44) subject to the initial condition (4.49).

Let, $(\phi, \psi) \neq 0$ over a volume V and zero otherwise, and let t_1 be the time to travel from the observation point $\underline{x}$ to the nearest point of $V(\underline{\xi})$. Then (4.52) implies $\varphi(\underline{x}, t) = 0$ for $t < t_1$, since the surface over which the mean is taken does not intersect $V(\underline{\xi})$. Similarly, if t_2 is the time to travel from $\underline{x}$ to the farthest point of V, then $\varphi(\underline{x}, t) \equiv 0$ for $t > t_2$. For $t_1 < t < t_2$, $\varphi(\underline{x}, t) \neq 0$. This shows again that there is a sharp front and a sharp tail to the wave.

4.3.5. *Method of Descent*

We wish to solve the two-dimensional wave equation

$$\varphi_{tt} = \varphi_{xx} + \varphi_{yy} \tag{4.53}$$

subject to the initial conditions $\varphi(\underline{x}, t) = \phi(\underline{x})$, $\varphi_t(\underline{x}, 0) = \psi(\underline{x})$.
   In three dimension the solution from (4.52) is

$$
\begin{aligned}
\varphi(\underline{x}, t) &= \frac{\partial}{\partial t}\{tM_t[\phi]\} + tM_t[\psi] \\
&= \frac{1}{4\pi}\frac{\partial}{\partial t}\left\{\frac{1}{t}\iint_{S_t}\phi(\xi, \eta)dS_t\right\} \\
&\quad + \frac{1}{4\pi t}\iint_{S_t}\psi(\xi, \eta)dS_t,
\end{aligned}
$$

where S_t is the surface of a sphere defined by $S \equiv (x - \xi)^2 + (y - \eta)^2 + (z - \zeta)^2 - t^2 \equiv 0$.
   We want to project from the spherical surface S_t onto the $\xi - \eta$ plane and integrate over the interior of the circle centered at (x, y) of radius t: $(x - \xi)^2 + (y - \eta)^2 \leq t^2$. Now $d\xi d\eta = dS_t n_\zeta$, where

$$n_\zeta = -\frac{\partial S}{\partial \zeta}\frac{1}{|\nabla_\xi S|}.$$

n_ζ is the ζ-component of the unit vector $\underline{n}$ normal to the surface S_t.
Then

$$dS_t = \frac{d\xi d\eta}{n_\zeta} = \frac{t\, d\xi d\eta}{\sqrt{t^2 - (x - \xi)^2 - (y - \eta)^2}},$$

since

$$|\nabla_\xi S| = 2[(x - \xi)^2 + (y - \eta)^2 + (z - \zeta)^2]^{\frac{1}{2}} = 2t$$

and

$$-\frac{\partial S}{\partial \zeta} = 2(z - \zeta) = 2\sqrt{t^2 - (x - \xi)^2 - (y - \eta)^2},$$

therefore

$$\varphi(\underline{x},t) = \frac{1}{4\pi}\frac{\partial}{\partial t}\left\{\frac{2}{t}\iint \frac{t\,\phi(\xi,\eta)d\xi d\eta}{\sqrt{t^2-(x-\xi)^2-(y-\eta)^2}}\right\}$$

$$+\frac{1}{4\pi t}\iint \frac{2t\,\psi(\xi,\eta)d\xi d\eta}{\sqrt{t^2-(x-\xi)^2-(y-\eta)^2}},$$

where the factor 2 arises since we must integrate over the top and bottom of the spherical surface S_t. The final form of the solution is

$$\varphi(\underline{x},t) = \frac{1}{2\pi}\frac{\partial}{\partial t}\iint_{r\leq t}\frac{\phi(\xi,\eta)}{\sqrt{t^2-(x-\xi)^2-(y-\eta)^2}}d\xi d\eta \quad (4.54)$$

$$+\frac{1}{2\pi}\iint_{r\leq t}\frac{\psi(\xi,\eta)}{\sqrt{t^2-(x-\xi)^2-(y-\eta)^2}}d\xi d\eta$$

and the integration is over $(x-\xi)^2+(y-\eta)^2\leq t^2$.

Suppose that initial data ϕ and ψ are nonzero in a neighborhood of the point P and zero otherwise. The solution (4.54) shows that $\varphi(\underline{x},t)\equiv 0$ until such time t as the circle centered at $\underline{x}$ with radius t intersects the relevant neighborhood of P. After that time $\varphi(\underline{x},t)\neq 0$, since the integration is carried out over the interior of the circle, not just on its boundary as in the three-dimensional case, and the neighborhood of P now always lies within the region of integration. Thus, while there is a sharp front to the two-dimensional wave, there is not a sharp tail.

As an example of (4.54) consider an infinitely long circular cylinder when the initial pressure inside the cylinder exceeds that outside the cylinder. When this excess pressure is released we want to find the time history of the pressure at the center of the cylinder. We assume the pressure $p(x,y,t)$ satisfies the wave equation and initial conditions

$$p_{tt} = p_{xx}+p_{yy}, \text{ with } p(r,0) = \begin{cases} p_1, & 0\leq r < r_0 \\ p_0, & r > r_0 \end{cases} \text{ and } \frac{\partial p}{\partial t}(r,0) = 0,$$

where $p_1 > p_0$ and $r^2 = x^2+y^2$.

By (4.54), the solution is

$$p(0,t) = \frac{1}{2\pi}\frac{\partial}{\partial t}\iint_{\rho \le t}\frac{p(\rho,0)}{\sqrt{t^2 - \rho^2}}\rho d\rho d\theta,$$

where $\rho^2 = \xi^2 + \eta^2$ since, at the centre, $x = y = 0$, and the element of area is $\rho d\rho d\theta$.

For $r_0 > t$, the wave from the surface has not yet reached the origin and then $p(\rho,0) = p_1$.

$$p(0,t) = \frac{p_1}{2\pi}\frac{\partial}{\partial t}\int_0^{ct}\int_0^{2\pi}\frac{\rho}{\sqrt{t^2 - \rho^2}}d\rho d\theta$$

$$= p_1\frac{\partial}{\partial t}\left[-\sqrt{t^2 - \rho^2}\Big|_{\rho=0}^{\rho=t}\right]$$

$$= p_1\frac{\partial}{\partial t}(t) = p_1,$$

as expected. The wave has not reached the origin and the initial pressure remains.

For $t > r_0$, the pressure from outside the cylinder, p_0, has had time to affect conditions at $r = 0$. So now the solution is

$$p(0,t) = \frac{1}{2\pi}\frac{\partial}{\partial t}\int_0^{2\pi}d\theta\left[\int_0^{r_0}\frac{p_1\rho}{\sqrt{t^2 - \rho^2}}d\rho + \int_{r_0}^{t}\frac{p_0\rho}{\sqrt{t^2 - \rho^2}}d\rho\right],$$

since for $0 \le t < r_0$, $p(r,0) = p_1$ and for $t > r_0$, $p(r,0) = p_0$. We need

$$\frac{\partial}{\partial t}\int_0^{r_0}\frac{\rho}{\sqrt{t^2 - \rho^2}}d\rho = \frac{\partial}{\partial t}\left[t\sqrt{t^2 - r_0^2}\right]$$

$$= 1 - t(t^2 - r_0^2)^{-\frac{1}{2}}$$

and

$$\frac{\partial}{\partial t}\int_{r_0}^{t}\frac{\rho}{\sqrt{t^2 - \rho^2}}d\rho = \frac{\partial}{\partial t}\sqrt{t^2 - r_0^2} = t(t^2 - r_0^2)^{-\frac{1}{2}}.$$

Then

$$p(0,t) = p_1[1 - t(t^2 - r_0^2)^{-\frac{1}{2}}] + p_0\, t(t^2 - r_0^2)^{-\frac{1}{2}}$$

$$= p_1 - (p_1 - p_0)t(t^2 - r_0^2)^{-\frac{1}{2}}$$

and

$$p(0,t) = p_1 - (p_1 - p_0)(1 - t^{-2}r_0^2)^{-\frac{1}{2}} \rightarrow p_0, \ t \rightarrow \infty.$$

For large time, the pressure at the center settles down to the external pressure p_0.

4.3.6. *Forced Vibrations by a Body Force*

Here we solve the wave equation with a specified body force term $g(x,t)$

$$\varphi_{tt} - c^2\varphi_{xx} = g(x,t), \tag{4.55}$$

subject to zero initial conditions, $\varphi(x,0) = \varphi_t(x,0) = 0$, $-\infty < x < \infty$, $t > 0$. If the initial conditions are nonzero, we can add the D'Alembert solution of the homogeneous equation with $g \equiv 0$ to that of (4.55).

One method of solution is to use a technique called *Duhamel's principle*, when the problem is converted into the standard D'Alembert problem with $g \equiv 0$, but $\varphi_t(x,0) \neq 0$.

Consider

$$\varphi(x,t) = \int_0^t v(x,t;\tau)d\tau, \tag{4.56}$$

where $v(x,t;\tau)$ satisfies the one-dimensional wave equation $v_{tt} - c^2 v_{xx} = 0$, with initial conditions at $t = \tau$: $v(x,\tau;\tau) = 0$, $v_t(x,\tau;\tau) = g(x,\tau)$, $-\infty < x < \infty$. Clearly from the D'Alembert solution (3.5)

$$v(x,t;\tau) = \frac{1}{2c}\int_{x-c(t-\tau)}^{x+c(t-\tau)} g(s,\tau)ds. \tag{4.57}$$

It remains to confirm that the representation (4.56) is consistent with the D'Alembert problem for $v(x,t;\tau)$. Differentiating

$\varphi(x,t)$ yields

$$\varphi_t(x,t) = v(x,t;t) + \int_0^t v_t(x,t;\tau)d\tau = \int_0^t v_t(x,t;\tau)d\tau,$$

$$\varphi_{tt}(x,t) = v_t(x,t;t) + \int_0^t v_{tt}(x,t;\tau)d\tau = g(x,t) + \int_0^t v_{tt}(x,t;\tau)d\tau.$$

Hence $\varphi(x,t)$ satisfies (4.55), and combining (4.56) and (4.57)

$$\varphi(x,t) = \frac{1}{2c}\int_0^t \int_{x-c(t-\tau)}^{x+c(t-\tau)} g(s,\tau)dsd\tau = \frac{1}{2c}\int\int_\Delta g(s,\tau)dsd\tau,$$

where Δ is the domain of dependence of the point (x,t), see Sec. 3.2.

Using the Duhamel principle, the solution for the equivalent forced wave equation in two dimensions,

$$\varphi_{tt} - c^2[\varphi_{xx} + \varphi_{yy}] = g(x,t), \tag{4.58}$$

subject to $\varphi(\underline{x},0) = \varphi_t(\underline{x},t) = 0$, is

$$\varphi(\underline{x},t) = \frac{1}{2\pi c}\iint_{r\leq ct} \frac{g(\xi,\eta)}{\sqrt{c^2(t-\tau)^2 - (x-\xi)^2 - (y-\eta)^2}}d\xi d\eta. \tag{4.59}$$

For the three-dimensional inhomogeneous wave equation

$$\varphi_{tt} - c^2(\varphi_{xx} + \varphi_{yy} + \varphi_{zz}) = g(x,y,z,t)$$

subject to $\varphi(\underline{x},0) = \varphi_t(\underline{x},t) = 0$, the solution is

$$\varphi(\underline{x},t) = \int_0^t (t-\tau)M_{c(t-\tau)}[g]d\tau,$$

$$= \frac{1}{4\pi c^2}\int_0^t \frac{d\tau}{t-\tau}\iint_{S_{c(t-\tau)}} g(\underline{\xi},\tau)dS, \tag{4.60}$$

on using the definition of $M_{ct}[g]$. Now $S_{c(t-\tau)}$ is the surface $r^2 = c^2(t-\tau)^2$, or $r = |\underline{\xi} - \underline{x}| = c(t-\tau)$, so that $t - \tau = \frac{r}{c}$ and $\tau = t - \frac{r}{c}$. Then the combination of integrals, i.e., the iterated integrals in (4.60) is simply $\iiint_{r \leq ct}$, i.e., the integral over the full sphere $r \leq ct$, and then

$$\varphi(\underline{x}, t) = \frac{1}{4\pi c^2} \iiint_{r \leq ct} \frac{1}{r} g(\underline{\xi}, t - \frac{r}{c}) d\underline{\xi},$$

since the volume element is $d\underline{\xi} = dS_{c(t-\tau)} c d\tau$. This is referred to as a retarded potential, due to the presence of $t - \frac{r}{c}$ in the argument of g. Only those points that can reach $(\underline{x}, t)$ and a ray with speed c from a past time $\tau, 0 \leq \tau \leq t$, contribute to the solution at $(\underline{x}, t)$. This is exactly the domain of dependence of the point $(\underline{x}, t)$.

Chapter 5

Linear Waves in Stratified Media

Constructing solutions that describe waves in stratified materials has proven difficult except in the *geometrical acoustics* or *high frequency* limit when the inhomogeneity is *slowly varying.* In this limit there have been various approaches that typically use perturbation solutions, see Keller [1962] and Seymour and Mortell [1975]. The methods used to construct these approximate solutions exploit the fact that the dimensionless parameter, ε, that is a measure of the stratification, is small. As explained in Sec. 4.1, typically ε is the ratio of the input wavelength to the length scale defined by the variation in the medium and is inversely proportional to the dimensionless frequency ω, so that when ε is small the frequency is high. There are several classical methods of solution, including a variation on the WKB or WKBJ method, usually derived for ordinary differential equations with variable coefficients. The name WKBJ derives from the names Wentzel, Kramer, Brillouin and Jeffreys, who independently discovered the method around 1925.

We outline several expansion approaches before giving a more formal perturbation method valid for small ε. There are some geometries (including spherical and cylindrical) where ε represents the ratio of the applied wavelength on the boundary to the radius of curvature of the boundary. As examples, we give solutions for high frequency pulsating spheres and cylinders.

A different approach is by approximating a slowly varying elastic medium by a series of thin elastic laminates, each with constant

56

material properties. The WKB result is obtained by considering the limit as the thickness of the layers tends to zero.

Of course when $\varepsilon = O(1)$ the stratification may not satisfy this restriction and another approach is required. With no small parameter to assist in finding an approximate solution, we need to construct exact solutions. Exact analytical solutions to wave propagation problems in continuously stratified materials are relatively rare. One way to achieve this is by choosing a stratification function belonging to a special class of inhomogeneities for which a closed form exact solution is available. This means that instead of solving an exact problem for a specific stratification approximately for small ε, an approximate problem is solved exactly for any ε. A general method of this kind was given in Varley and Seymour [1988] where it is shown that the solution to a linear partial differential with variable coefficients can be written in terms of the solution to an equivalent constant coefficient problem. The parameters defining the inhomogeneity allow $\varepsilon = O(1)$. The Webster horn equation is analyzed by this method and exact solutions are derived for periodic oscillations in a variety of shaped resonators. These illustrate the difference between *commensurate* and *incommensurate* eigenvalues. For a straight tube or homogeneous rod, the eigenvalues are all integer multiples of the fundamental so $\omega_n = n\omega_1$. However, in general for shaped resonators or an inhomogeneous rod $\omega_n \neq n\omega_1$ and the eigenvalues are called incommensurate. This difference is significant when considering the corresponding nonlinear problems in Chapter 12. The connection between exact solutions of the Webster horn equation and soliton solutions of the KdV equation is shown. The formula for N-soliton solutions is derived and an example of the evolution of a 2-soliton solution is given.

5.1. Webster Horn Equation

A well known variable coefficient wave equation is the Webster horn equation, Webster [1919], that describes a one-dimensional approximation for sound propagation in a rigid tube with a variable cross-sectional area $C(y)$,

$$w_{tt} - C(y)\frac{\partial}{\partial y}\left(\frac{1}{C(y)}w_y\right) = 0, \tag{5.1}$$

where y represents axial distance, $w(y,t)$ is the flux and the equation is in dimensionless form with $L = D$, the stratification length, see (4.2), and $C(0) = 1$. A typical boundary condition is $w(0,t) = Mh(\omega t)$ with $\omega = D/c_0\tau_p = D/\xi$, so here the dimensionless parameter ω occurs in the boundary condition, but not in the equation. For a review of the Webster horn equation see Eisner [1967].

Because the y dependence in Eq. (5.1) is in Sturm-Liouville or 'divergence' form, it is in a more convenient form for analysis than most of the second order equations derived in Chapter 2 [(2.6), (2.8), (2.19) and (2.30)] so we first consider changes of variable to put these latter equations in the form (5.1).

The example of an elastic panel with density and Young's modulus having the same stratification, (2.8), is already in the form (5.1) with $y = x$ and $C = E(x)$ for the σ equation and $C = 1/E$ for the u equation. For the more general equations for an elastic panel, (2.6), let

$$y = \int_0^x \sqrt{\frac{\rho(z)}{E(z)}}\,dz \quad \text{and} \quad C(y) = \sqrt{\rho(x)E(x)} = i(x), \tag{5.2}$$

where $i(x)$ is the impedance. Then σ satisfies (5.1) with $C = i$ and u satisfies (5.1) with $C = 1/i$.

Typically boundary conditions for σ and u are

$$\sigma(0,t) = 0 \text{ and } \sigma(1,t) = Mh(\omega t) \tag{5.3}$$

and

$$u(0,t) = 0 \text{ and } u(1,t) = Mh(\omega t). \tag{5.4}$$

Hence, the solution of (5.1) for σ with boundary conditions (5.3) and stratification $C = i(y)$ is identical to the solution for u with boundary conditions (5.4) for the stratification $C = 1/i(y)$.

It is often more convenient to separate the second-order equations (5.1) or (2.6) into two first-order equations. Then, in terms of (y, t), Eq. (2.5) for u and σ becomes

$$i(y)u_t = \sigma_y, \quad \text{and} \quad \sigma_t = i(y)u_y. \tag{5.5}$$

For a polytropic gas in a tube with variable cross section $s(x)$ the equation for condensation or pressure in (2.19) is already in the form (5.1) with $y = x$ and $C = 1/s(x)$, while the flux $v = su$ satisfies (5.1) with $C = s(x)$.

For shallow water waves over a variable topography, $z = -H(x)$, both u and η in (2.30) satisfy (5.1) with $y = \int_0^x \frac{1}{\sqrt{gH(z)}}dz$, and $C(y) = 1/\sqrt{gH(x)}$. The change of variable from x to y is often called the *Liouville transformation*.

5.2. Expansion Methods

Most examples considered in this section contain the parameter, $\varepsilon \ll 1$, a measure of the slowly varying stratification. One way the size of this parameter is exploited is through expansions in powers of ε. As explained in Sec. 4.1, for ε to appear in the stratification function $C(\varepsilon y)$ as opposed to in the boundary condition, length must be nondimensionalized with the applied wavelength. If the dimensional boundary condition is $w(0, t) = a\sin(\Omega nt)$ for integer n, pick $T = \Omega^{-1}$ and $L = \Omega^{-1}c_0$. The boundary condition in dimensionless variables is $w(0, t) = M\sin(nt)$, $M = a/c_0$, while $C = C(\varepsilon y)$, where $\bar{x} = \varepsilon y$, so $C'(y) = \varepsilon C'(\bar{x})$ and $C(0) = 1$.

As an opening example we consider time-periodic solutions of (5.1) for $w = \sigma(y, t)$ and let

$$\sigma(y, t) = \psi(y)e^{int}.$$

Now, $\psi(y)$ satisfies

$$\psi'' - \frac{C'(y)}{C(y)}\psi' + n^2\psi = 0. \tag{5.6}$$

The term involving ψ' is like a damping term in an oscillator and can be eliminated by introducing the transformation

$$\psi(y) = \exp\left[\frac{1}{2}\int^y \frac{C'(\varsigma)}{C(\varsigma)}d\varsigma\right]v(y) = [C(y)]^{1/2}v(y),$$

when the equation for $v(y)$ is

$$v'' + \left[n^2 + \frac{1}{2}\frac{C''}{C} - \frac{3}{4}\left(\frac{C'}{C}\right)^2\right]v = 0. \tag{5.7}$$

But C depends on εy, so then (5.7) becomes

$$v'' + n^2 v + \varepsilon^2\left[\frac{1}{2}\frac{C''}{C} - \frac{3}{4}\left(\frac{C'}{C}\right)^2\right]v = 0.$$

A regular perturbation scheme in ε, $v(y) = v_0 + \varepsilon v_1 + \cdots$ yields

$$v_0'' + n^2 v_0 = 0, \quad \text{i.e.,} \quad v_0(y) = b_n e^{\pm iny},$$

where b_n is a constant. Hence to first order

$$\sigma(y, t) = b_n C^{1/2}(\bar{x})e^{in(t \mp y)}, \quad \bar{x} = \varepsilon y.$$

For a general time varying solution we sum over n giving

$$\sigma(y, t) = C^{1/2}(\bar{x})h(t \pm y), \tag{5.8}$$

where $\sigma(0, t) = h(t)$. The solution $\sigma(y, t)$ has right and left traveling waves in which the amplitude is modulated by $C^{1/2}(\bar{x})$. This is the first term in the *geometrical acoustics* approximation.

For the more general equations for an elastic panel (2.6), the result (5.8) becomes, in dimensional form,

$$\sigma(x, t) = \left[\frac{\rho(x)E(x)}{\rho(0)E(0)}\right]^{1/4}h\left(t \pm \int_0^x \sqrt{\frac{\rho(z)}{E(z)}}dz\right),$$

using the transformation (5.2).

5.2.1. *Multiple Scale Expansion*

An alternative method to produce the result (5.8) is that of multiple scales, see Kervorkian and Cole [2011]. We again consider Eq. (5.6) for $\psi(y)$ with $C = C(\overline{x})$, but now assume that $\psi(y) = \Psi(y, \overline{x})$ where y and $\overline{x} = \varepsilon y$ are *different* independent variables. Then, since ψ now depends on two 'independent' variables, (5.6) becomes

$$\frac{d^2\Psi}{dy^2} - \varepsilon \frac{C'(\overline{x})}{C(\overline{x})} \frac{d\Psi}{dy} + n^2\Psi = 0. \tag{5.9}$$

This is the equation for an oscillator with frequency n and a "slow" damping term. The equation has two separate scales: the fast scale y, typically defined by the applied input, and the slow scale $\overline{x}$ defined by the stratification. To exploit this assume that $\Psi(y, \overline{x}) = \Psi_0(y, \overline{x}) + \varepsilon\Psi_1(y, \overline{x}) + \cdots$ where $\varepsilon \ll 1$. Then

$$\begin{aligned}
\frac{d\Psi}{dy} &= \frac{\partial\Psi_0}{\partial y} + \frac{\partial\Psi_0}{\partial\overline{x}}\frac{d\overline{x}}{dy} + \varepsilon\frac{\partial\Psi_1}{\partial y} + O(\varepsilon^2), \\
&= \frac{\partial\Psi_0}{\partial y} + \varepsilon\left[\frac{\partial\Psi_0}{\partial\overline{x}} + \frac{\partial\Psi_1}{\partial y}\right] + O(\varepsilon^2)
\end{aligned}$$

and

$$\frac{d^2\Psi}{dy^2} = \frac{\partial^2\Psi_0}{\partial y^2} + \varepsilon\left[2\frac{\partial^2\Psi_0}{\partial y\partial\overline{x}} + \frac{\partial^2\Psi_1}{\partial y^2}\right] + O(\varepsilon^2).$$

On substituting into (5.9), the sequence of equations at $O(1)$ and $O(\varepsilon)$ are

$$\frac{\partial^2\Psi_0}{\partial y^2} + n^2\Psi_0 = 0 \tag{5.10}$$

and

$$\frac{\partial^2\Psi_1}{\partial y^2} + n^2\Psi_1 = \frac{C'(\overline{x})}{C(\overline{x})}\frac{\partial\Psi_0}{\partial y} - 2\frac{\partial^2\Psi_0}{\partial y\partial\overline{x}}. \tag{5.11}$$

(5.10) implies that $\Psi_0(y, \overline{x}) = B_0(\overline{x})e^{\pm i\omega y}$, and then the right-hand side of (5.11) is

$$\left[B_0(\overline{x})\frac{C'(\overline{x})}{C(\overline{x})} - 2B_0'(\overline{x})\right]e^{\pm iny}.$$

Since a particular integral contains a term like $ye^{\pm iny}$, to ensure no growth term in y we set the coefficient in the square brackets to zero, giving

$$B_0(\overline{x}) = b_n C^{1/2}(\overline{x}).$$

It is typical in this method that the unknown coefficient $B_0(\overline{x})$ arising in the $O(1)$ solution is determined by a secularity condition at $O(\varepsilon)$. The full solution for $\sigma(y,t)$ is then

$$\sigma(x,t) = \psi(x)e^{int}e^{\pm iny} = b_n C^{1/2}(\overline{x})e^{in(t\mp y)}.$$

Summing over all frequencies produces the result (5.8) again.

5.2.2. *WKB Expansion*

Another way to find an approximate solution to Eq. (5.1) with stress input $w = \sigma(0,t) = h\left(\frac{t}{\tau_p}\right)$ is a variation of the WKB method [see Murray, 1974]. Here the variables y and t are scaled with the stratification length, $L = D$ and $T = Dc_0^{-1}$ when the (dimensionless) boundary condition becomes $w = \sigma(0,t) = h(\omega t)$ where $\omega = \frac{D}{c_0 \tau_p}$, and the parameter appears in the boundary condition, but not the equation. Then $\omega = 1/\varepsilon$ is the dimensionless frequency and in the high frequency limit $c_0 \tau_p \ll D$, so that as before $\varepsilon \ll 1$ or $\omega \gg 1$. In this case the expansion parameter is $\frac{1}{\omega}$.

In the WKB expansion we assume that the solution is periodic in time with frequency ω:

$$\sigma(y,t) = e^{i\omega(t-\theta(y))}\left[\sigma_0(y) + \frac{1}{\omega}\sigma_1(y) + \cdots\right], \qquad (5.12)$$

while the unknown phase is $\theta(y)$. Then

$$0 = \sigma_{tt} - C(y)\frac{\partial}{\partial y}\left(\frac{1}{C(y)}\sigma_y\right) = -\omega^2\left(\sigma_0 + \frac{1}{\omega}\sigma_1 + \cdots\right)e^{i\omega(t-\theta(y))}$$

$$+\omega^2\theta'^2\left(\sigma_0 + \frac{1}{\omega}\sigma_1 + \cdots\right)e^{i\omega(t-\theta(y))}$$

$$+2i\omega\theta'\left(\sigma_0' + \frac{1}{\omega}\sigma_1' + \cdots\right)e^{i\omega(t-\theta(y))}$$

$$- \left(\sigma_0'' + \frac{1}{\omega} \sigma_1'' + \cdots \right) e^{i\omega(t-\theta(y))}$$

$$+ \frac{C'(y)}{C(y)} \left[\left(-i\omega\theta' \left(\sigma_0 + \frac{1}{\omega} \sigma_1 + \cdots \right) + \left(\sigma_0' + \frac{1}{\omega} \sigma_1' + \cdots \right) \right) \right]$$

$$\times e^{i\omega(t-\theta(y))},$$

where prime denotes differentiation w.r.t the argument shown.

At $O(\omega^2)$

$$\theta'(y) = \pm 1, \tag{5.13}$$

so that $\theta(y) = \pm y$. Equation (5.13) is called the Eikonal equation. Then, on using (5.13), at $O(\omega)$

$$2\sigma_0' - \frac{C'}{C}\sigma_0 = 0, \text{ so that } \sigma_0(y) = \kappa C^{1/2}(y),$$

κ constant. But C is normalized so that $C(0) = 1$, hence $\kappa = 1$ and (5.12) has the form

$$\sigma(y,t) = e^{i\omega(t \mp y)}[C^{1/2}(y) + O(\omega^{-1})].$$

Summing over the frequency ω gives the general solution

$$\sigma(y,t) = C^{1/2}(y)g(t \mp y) + O(\omega^{-1}).$$

The boundary condition $\sigma(0,t) = h(\omega t)$ gives $g(t) = h(\omega t)$, so that

$$\sigma(y,t) = C^{1/2}(y)h(\omega(t-y)),$$

where the minus sign is chosen as the wave travels into $y > 0$. This is the first term in the WKB expansion and corresponds to (5.8). Note that from (5.5) the velocity $u(y,t)$ satisfies (5.1) with $C = i^{-1}$, so the equivalent solution for $w = u(y,t)$ (since $i(0) = 1/C(0) = 1$) is

$$u(y,t) = \frac{1}{\sqrt{i(y)}} h(\omega(t-y)),$$

where $i(y) = \sqrt{\rho(y)E(y)}$.

5.3. Geometric Acoustics from Laminates

There is a completely different approach that puts the previous solutions in context. This is by approximating a slowly varying elastic medium by a series of thin elastic laminates, each with constant material properties. Following Mortell and Seymour [1976], as a (linear) wave propagates through the laminates we consider the limit as the laminate thickness tends to zero and the number of laminates tends to infinity to construct the solution that describes propagation through a continuous, but slowly varying, medium. This calculation is in dimensional variables.

Consider a wave travelling from left to right emanating from the boundary $x = 0$. Initially the slabs are motionless and in equilibrium. To the right of $x = 0$ are multiple elastic laminates with different, but constant, properties. We use the notation introduced in Sec. 3.4. The material occupying $0 < x < x_1$ has density and Young's modulus (ρ_1, E_1), while the density and Young's modulus for $x_{r-1} < x < x_r$ (called region r) are (ρ_r, E_r). The laminates have thickness $l_1, l_2, \ldots$ etc. and the corresponding wave speeds and impedances are $c_r = \sqrt{\frac{E_r}{\rho_r}}$, $i_r = \sqrt{E_r \rho_r}$, $r = 1, 2, \ldots$, as shown in Fig. 5.1, and we work in dimensional variables.

The signal incident on $x = 0$ generates a transmitted velocity wave u_1 that leaves $x = 0$ at time $t = \alpha_1$ in the positive x direction

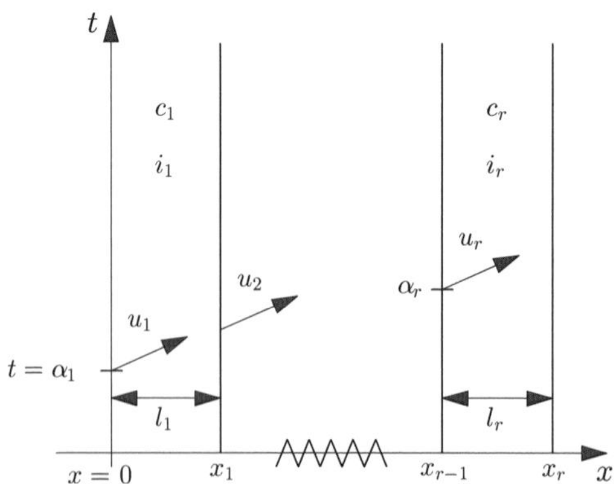

Fig. 5.1. $x - t$ Diagram showing slabs.

with $u_1(t) = h(t)$ specified on $x = 0$. Following just the primary transmitted signal through $r - 1$ laminates, it arrives at $x = x_{r-1} = \sum_{n=1}^{r-1} l_n$ at time $t = \alpha_r$. The signal emerges from the boundary $x = x_{r-1}$ as u_r at time $t = \alpha_r$. After passing through N laminates, we want to relate the amplitude and travel time of the transmitted signal $u_{N+1}(\alpha_{N+1})$ to the original signal at $x = 0$, i.e., $u_1(\alpha_1)$.

At the interface $x = x_r$ the transmission and reflection coefficients are $T_r = \frac{2i_r}{i_r + i_{r+1}}$ and $R_r = \frac{i_{r+1} - i_r}{i_r + i_{r+1}}$ respectively, see (3.16). We now make the assumption that the impedances are almost matching, so at any interface the reflected wave is always much smaller than the transmitted wave. Specifically this implies that $T_r \simeq 1$ and $R_r \ll 1$. This is consistent with the layered material modeling a slowly varying continuous medium. Then $u_{N+1} \simeq T_N u_N$ and by repeated application of this, we find $u_{N+1} \simeq K_N u_1$, where $K_N = \prod_{n=1}^{N} T_n$ and $\prod_{n=1}^{N}$ is the product symbol. Now, following Bremmer 1951, if $\Delta i_n = i_{n+1} - i_n$,

$$T_n = \frac{2i_n}{i_n + i_{n+1}} = 1 - \frac{\Delta i_n}{2i_n}.$$

Then

$$\prod_{n=1}^{N} T_r = \exp\left[\sum_{n=1}^{N} \ln(T_n)\right]$$

$$= \exp\left[-\sum_{n=1}^{N} \ln\left(1 + \frac{\Delta i_n}{2i_n}\right)\right]$$

$$= \exp\left[-\sum_{n=1}^{N} \frac{\Delta i_n}{2i_n} + O(\Delta i_n^2)\right].$$

Let $N \to \infty$, $l_n \to 0$ such that $x = \sum_{n=1}^{N} l_n$ remains fixed, i.e., the medium has a fixed length. Then

$$K_N = \prod_{n=1}^{N} T_n \to \exp\left[-\frac{1}{2}\int_{i(0)}^{i(x)} \frac{di}{i}\right] = \sqrt{\frac{i(0)}{i(x)}}$$

and

$$u_{N+1} = \sqrt{\frac{i(0)}{i(x)}}\, u_1. \tag{5.14}$$

To calculate the travel time, follow the characteristic leaving $x = 0$ at time $t = \alpha_1$ through N layers. In the rth layer, with sound speed c_r for $\sum_{n=1}^{r-1} l_n \ll x \ll \sum_{n=1}^{r} l_n$, the characteristic is

$$\alpha_r = t - c_r^{-1}\left[x - \sum_{n=1}^{r-1} l_n\right], \tag{5.15}$$

and satisfies the initial condition $t = \alpha_r$ on $x = \sum_{n=1}^{r-1} l_n$. But $t = \alpha_{r+1}$ on $x = \sum_{n=1}^{r} l_n$, and therefore (5.15) implies $\alpha_{r+1} - \alpha_r = c_r^{-1} l_r$. Summing over r gives the result

$$\alpha_{N+1} - \alpha_1 = \sum_{n=1}^{N}(\alpha_{n+1} - \alpha_n) = \sum_{n=1}^{N} \frac{l_n}{c_n}$$

and passing to the limit as $N \to \infty$, $l_n \to 0$ yields

$$\alpha_{N+1} = \alpha_1 + \int_0^x \frac{ds}{c(s)}.$$

Hence,

$$u_{N+1}(\alpha_{N+1}) = \sqrt{\frac{i(0)}{i(x)}}\, u_1(\alpha_1) = \sqrt{\frac{i(0)}{i(x)}}\, u_1\left(\alpha_{N+1} - \int_0^x \frac{ds}{c(s)}\right),$$

or

$$u(t,x) = \sqrt{\frac{i(0)}{i(x)}}\, h\left(t - \int_0^x \frac{ds}{c(s)}\right),$$

where $u_1(t) = h(t)$ is specified on $x = 0$. This is the first-term geometrical acoustic approximation, which follows from the assumption that all internal reflections are neglected.

If we now consider the transmitting material to consist of identical cells each containing J laminates; i.e., the material has periodic structure with the $(nJ + r)$th laminate being identical to the rth. Again, neglecting internal reflections, the relationship between the amplitudes of a signal entering and leaving a cell is given by $u_{J+1}(\alpha_{J+1}) \simeq (\prod_{n=1}^{J} T_n)u_1(\alpha_1)$. Noting that the $(J+1)$th laminate is identical to the first,

$$T_J = \frac{2i_J}{i_J + i_1} \text{ and } K_J = \prod_{n=1}^{J} T_n = \frac{2\sqrt{i_1 i_2}}{i_1 + i_2} \frac{2\sqrt{i_2 i_3}}{i_2 + i_3} \cdots \frac{2\sqrt{i_J i_1}}{i_J + i_1} \leq 1,$$

since the arithmetic mean of two numbers is greater than or equal to the geometric mean. Thus, a laminated material with a periodic structure attenuates a signal. Using this observation and defining the attenuation length l_r of a viscoelastic material by $K_j = \prod_{n=1}^{J} T_n = \exp(-l/l_r)$, where $l = \sum_{n=1}^{J} l_n$ is the cell width, as $N \to \infty$, $l_n \to 0$, (5.14) becomes

$$u_{N+1} = \exp(-x/l_r)u_1.$$

Hence, a laminated material with a periodic structure can be used as a discrete model for a viscoelastic material, see Mortell and Seymour [1976].

5.4. Linear Geometric Acoustics Expansion

In the preceding sections we have considered some simple methods that all produce the first-term geometrical acoustic solution for waves in a continuously stratified material. Here we give a regular perturbation scheme that can theoretically be continued for many terms, though in practice only the first two are used.

The method is valid when the applied wavelength is small compared with a natural length D defined by the material. In the examples considered so far, D is the length defined by the stratification, but the same methods can be used when, for example, D represents the radius of curvature of the wave-front or the relaxation length of a viscoelastic material. In all cases if $\xi = c_0 \tau_p$ is the wavelength of the input at the boundary, then $\varepsilon = \frac{\xi}{D} \ll 1$ in the geometric acoustics limit.

We begin with the example of a stratified material governed by a set of two linear equations for $u(y,t)$ and $\sigma(y,t)$ in the dimensionless form (5.5) containing the variable coefficient $i(y)$. Here we have picked $L = D$ (the stratification length) and $T = Dc_0^{-1}$ so that the boundary condition is $u(0,t) = Mh(\omega t)$, where M is the Mach number and the dimensionless frequency parameter is $\omega = \frac{D}{c_0 \tau_p} = \varepsilon^{-1} \gg 1$. Note that M plays no role here as the equations are linear.

The variables are normalized so that $i(y)$ is a function of the 'slow' variable y, while the boundary condition defines the 'fast' variable, the characteristic, $\alpha = \omega(t-y) = (t-y)/\varepsilon$ with $\alpha = \omega t$ on $y = 0$ and $u(0,\alpha) = Mh(\alpha)$. Initially the material is at rest in equilibrium. Then the solution to (5.5) can be represented in series form [see Seymour and Mortell, 1975] with each term separated into fast and slow factors

$$\sigma(y,t) = \sum_{n=0}^{\infty} \varepsilon^n g_n(\alpha) f_n(y), \quad u(y,t) = \sum_{n=0}^{\infty} \varepsilon^n g_n(\alpha) e_n(y), \qquad (5.16)$$

where we note that, in terms of the dependence on the fast variable, $\varepsilon^n \frac{\partial g_n(\alpha)}{\partial t} \curvearrowright \frac{\varepsilon^n}{\varepsilon} g_n'(\alpha) \curvearrowright O(\varepsilon^{n-1})$. Hence, to balance the terms when substituting (5.16) into Eqs. (5.5) we assume that

$$g_{n+1}'(\alpha) = g_n(\alpha) \qquad (5.17)$$

and then the 'slow' factors satisfy

$$i(y)e_n'(y) - f_n'(y) = 0, \qquad (5.18)$$
$$f_{n-1}'(y) - f_n(y) = i(y)e_n(y), \quad n \geq 0 \qquad (5.19)$$

with $\sigma_{-1}(y) = 0$. The first term is found from

$$i(y)e_0'(y) - f_0'(y) = 0, \quad f_0(y) + i(y)e_0(y) = 0$$

with $u_0(0) = 1$. Hence, (since $i(0) = 1$)

$$e_0(y) = \frac{1}{\sqrt{i(y)}}, \quad f_0(y) = -\sqrt{i(y)} \qquad (5.20)$$

and to first order

$$u(y,t) = \frac{M}{\sqrt{i(y)}} h(\alpha), \quad \text{and} \quad \sigma(y,t) = -\sqrt{i(y)} M h(\alpha), \qquad (5.21)$$

where $g_0(\alpha) = Mh(\alpha)$. In a regular geometric acoustics expansion (5.17)–(5.19) are solved recursively and at the next order (with $i(0) = 1$)

$$e_1' + \frac{i'}{2i}e_1 = \frac{1}{2i}\, f_0'' = -\frac{1}{4i}\frac{d}{dy}\left(\frac{i'}{\sqrt{i}}\right), \quad \text{and} \quad g_1'(\alpha) = h(\alpha), \quad (5.22)$$

giving, with $e_1(0) = 0$,

$$e_1 = -\frac{M}{\sqrt{i(y)}}\int_0^y \frac{1}{4\sqrt{i(r)}}\left(\frac{i'(r)}{\sqrt{i(r)}}\right)' dr, \quad g_1(\alpha) = M\int_0^\alpha h(p)dp.$$

$$(5.23)$$

Hence, from (5.16), (5.21) and (5.23), the first two terms for $u(y,t)$ are

$$u(y,t) = \frac{M}{\sqrt{i(y)}}\left[h(\alpha) - \varepsilon\int_0^\alpha\int_0^y \frac{h(p)}{4\sqrt{i(r)}}\left(\frac{i'(r)}{\sqrt{i(r)}}\right)' dr\, dp\right] + O(\varepsilon^2).$$

$$(5.24)$$

By comparison with the results of Sec. 5.3, the first term in (5.24) can be interpreted as the direct wave traveling to the right, while the second term is the sum of all double internal reflections traveling to the right, see Seymour and Mortell [1977].

5.5. Physical Examples

5.5.1. *Inhomogeneous Elastic Panel*

As described in Sec. 2.1, we consider the forced oscillations of a semi-infinite elastic panel, $x \geq 0$, initially at rest. We use the simplified model when the inhomogeneity scales both density and Young's modulus in the same way, so that $\rho(x) = E(x)$. Then the dimensionless equations of motion relating strain λ and the particle velocity u are (2.7)

$$\frac{\partial u}{\partial t} = \frac{\partial \lambda}{\partial x} + \frac{E'}{E}\lambda \quad \text{and} \quad \frac{\partial u}{\partial x} = \frac{\partial \lambda}{\partial t} \qquad (5.25)$$

with initial and boundary conditions

$$u(0,t) = Mh(\omega t) \quad \text{and} \quad u(x,0) = 0,\ \lambda(x,0) = 0, \qquad (5.26)$$

where $\omega = \frac{D}{c_0\tau_p}$.

In preparation for more complicated systems that do not fit neatly into the form (5.5), we make a change of variable before substituting the geometric acoustics expansion. In the high frequency limit we anticipate that the wave will oscillate on a scale $\omega \gg 1$, defined by the boundary condition, with an amplitude that changes slowly with x. Hence, we introduce the 'fast' characteristic coordinate $\alpha(x,t)$ with $\alpha_t = O(\omega)$ and consider $u(x,t) = U(x,\alpha)$ and $\lambda(x,t) = \Lambda(x,\alpha)$. We define $T(x,\alpha)$ as the arrival time of the characteristic $\alpha(x,t) =$ constant at the location x, but at this stage α is not defined precisely. This change of variable and associated notation will be useful when considering the equivalent nonlinear problems in Chapter 6. The change of variable implies that

$$u_x = U_x + U_\alpha \alpha_x = U_x - \alpha_t T_x U_\alpha, \qquad (5.27)$$

since $\frac{d\alpha}{dx} = \alpha_x + \alpha_t T_x = 0$. Then (5.25) becomes

$$U_\alpha + T_x \Lambda_\alpha = [\Lambda_x + \Lambda E'/E]/\alpha_t, \quad \Lambda_\alpha + T_x U_\alpha = U_x/\alpha_t \qquad (5.28)$$

with the characteristic condition

$$\det \left\| \begin{matrix} 1 & T_x \\ T_x & 1 \end{matrix} \right\| = 0,$$

so that the reciprocal of the wave speed, sometimes called the slowness, is $T_x = \pm 1$.

For a wave traveling to the right, define $\alpha = \omega(t - x) = (t - x)/\varepsilon$ with $\alpha = \omega t$ on $x = 0$. With $\varepsilon = \omega^{-1}$ (5.28) can be rewritten as

$$U_\alpha + \Lambda_\alpha = \varepsilon(\Lambda_x + \Lambda E'/E), \quad \Lambda_\alpha + U_\alpha = \varepsilon U_x. \qquad (5.29)$$

Hence, the compatibility condition or *transport equation* (equating the right-hand sides of (5.29)) yields

$$\Lambda_x + \Lambda E'/E = U_x. \qquad (5.30)$$

Equations (5.25) have been replaced by (5.30) and either one of (5.29). These new equations are a different form of the exact (linear) equations; no approximations have been made so far. The new equations are in a convenient form to construct a geometrical acoustics solution as an expansion in powers of ε of the form [see Seymour and Mortell, 1975; Mortell and Seymour, 2011],

$$U(x,\alpha) = \sum_{n=0}^{\infty} \varepsilon^n g_n(\alpha) u_n(x), \ \Lambda(x,\alpha) = \sum_{n=0}^{\infty} \varepsilon^n g_n(\alpha)\lambda_n(x), \quad (5.31)$$

where,

$$g'_{n+1}(\alpha) = g_n(\alpha), n \geq 0. \tag{5.32}$$

On substituting (5.31) into the second of (5.29) and (5.30), and using (5.32) we obtain

$$u_n + \lambda_n = u_{n-1}, \ \lambda'_n + \lambda_n E'/E = u'_n, \ n \geq 0 \tag{5.33}$$

with $g_0(\alpha) = Mh(\alpha)$, $u_0(0) = 1$ and $u_{-1} = \lambda_{-1} = 0$. Then, solving the equations for $n = 0$,

$$u_0 + \lambda_0 = 0, \ \lambda'_0 + \lambda_0 E'/E = u'_0,$$

yields (with $s(0) = 1$)

$$u_0(x) = \frac{1}{\sqrt{E(x)}}, \quad \lambda_0(x) = -\frac{1}{\sqrt{E(x)}}, \tag{5.34}$$

so that to $O(1)$

$$u(x,t) = \frac{M}{\sqrt{E(x)}}h(\alpha), \quad \text{and} \quad \lambda(x,t) = -\frac{M}{\sqrt{E(x)}}h(\alpha). \tag{5.35}$$

In a regular geometric acoustics expansion (5.32)–(5.33) are solved recursively, so that (solving the $n = 1$ equations)

$$u_1 = \frac{1}{\sqrt{E(x)}} \int_0^x \frac{1}{4\sqrt{E(r)}} \left(\frac{E'(r)}{\sqrt{E(r)}} \right)' dr, \ g_1(\alpha) = -\int_0^\alpha h(p)dp$$

and to $O(\varepsilon)$

$$
u(x,t) = \frac{M}{\sqrt{E(x)}} h(\alpha)
$$

$$
- \frac{\varepsilon}{\sqrt{E(x)}} M \int_0^x \frac{1}{4\sqrt{E(r)}} \left(\frac{E'(r)}{\sqrt{E(r)}} \right)' dr \int_0^\alpha h(p)dp
$$

$$
+ O(\varepsilon^2), \tag{5.36}
$$

in agreement with (5.24).

5.5.2. *Maxwell Solid*

The governing equations for the one-dimensional motion of a linear viscoelastic Maxwell panel are (2.3) and (2.10). If these are scaled with $T = 2\tau_r$, and $L = 2c_0\tau_r$ (the factor of 2 is included for convenience so that the decay is proportional to e^{-x}) where $c_0 = \sqrt{E_0/\rho_0}$, they can be written as

$$
u_t - \sigma_x = 0, \ \sigma_t = u_x - 2\sigma \tag{5.37}
$$

with boundary condition

$$
u(0,t) = Mh(\omega t), \tag{5.38}
$$

where $\omega = 2\tau_r/\tau_p = \frac{2(\text{relaxation time})}{\text{period of input}} \gg 1$ and $M = u_0/c_0$. For $x > 0$ the material is at rest at $t = 0$. In the geometric acoustics limit we anticipate that the wave will oscillate on a scale $\omega = \varepsilon^{-1}$, defined by the boundary condition, with an amplitude that changes with x. We define the 'fast' variable $\alpha = \omega(t - x)$ and for a right traveling wave consider $u(x,t) = U(x,\alpha)$ and $\sigma(x,t) = \Sigma(x,\alpha)$. Then (5.27) implies that (5.37) becomes

$$
U_\alpha + \Sigma_\alpha = \varepsilon\Sigma_x, \ \Sigma_\alpha + U_\alpha = \varepsilon U_x - 2\varepsilon\Sigma. \tag{5.39}
$$

The transport equation from the right-hand sides of (5.39) then yields

$$
U_x + U = 0, \tag{5.40}
$$

while to first order $U = -\Sigma$ so that the first order solution is

$$u = Mh(\alpha)e^{-x}, \tag{5.41}$$

on using the boundary condition (5.38).

Asymptotic expansions like linear geometric acoustics series usually have a limited range of validity that depends on the small parameter ε. This example is ideal to illustrate the various ranges, typically listed in terms of x and $\overline{x} = \varepsilon x$ as $\varepsilon \to 0$. The ranges and corresponding equations for $u(x,t)$ are best described in terms of (x,t) rather than (x,α) when (5.40) becomes

$$u_x + u_t + u = 0$$

which is valid for finite x as $\overline{x} \to 0$, so $x = O(1)$. For this range the signal is attenuated, but not dispersed. However, for finite $\overline{x}$ as $\varepsilon \to 0$, so $x = O(\varepsilon^{-1})$, $u(x,t)$ satisfies

$$u_x + u_t + u = k \int_0^\varepsilon (t-s)u(t-s,x)ds$$

for some constant k, and the signal is both attenuated and dispersed [see Seymour and Varley, 1970 for details]. To complete the picture, $u(x,t)$ is neither attenuated nor dispersed when $x = O(\varepsilon)$ when u satisfies

$$u_x + u_t = 0,$$

so that $u = Mh(t-x)$.

5.5.3. *Pulsating Sphere of Large Radius*

The linear geometric acoustics method can be applied to consider small-amplitude, high-frequency sound waves generated by a pulsating sphere. The nonlinear extension will be considered in Chapter 6. If the surface of the sphere of radius R_0 pulsates periodically with period τ_p, and c_0 is the ambient sound speed, then, in the geometric acoustics limit, $\varepsilon = \frac{c_0 \tau_p}{R_0} = \omega^{-1} \ll 1$; that is the applied wavelength is small compared with the radius of the sphere.

As in 4.3.1, we nondimensionalize by picking $L = R_0$ and $T = R_0/c_0$, where $c_0 = \sqrt{\gamma p_0/\rho_0}$, then the governing equations for linear motions are (4.24):

$$u_t + k^{-1}a_r = 0, \quad a_t + k(u_r + 2u/r) = 0, \quad r > 1, \tag{5.42}$$

where $u(r,t)$ is the particle velocity and $c(r,t) = 1 + a(r,t)$ is the (dimensionless) sound speed and $k = (\gamma - 1)/2$. The boundary condition on $R = 1$ is

$$u(1,t) = Mh(\omega t), \tag{5.43}$$

where $M = u_0/c_0$ and u_0 is the amplitude of the applied velocity. For $R > 1$ the gas is at rest at $t = 0$.

As in the previous example we introduce the characteristic coordinate $\alpha(r,t)$ and consider $u(r,t) = U(r,\alpha)$, $a(r,t) = A(r,\alpha)$. The change of variable implies that

$$u_r = U_r + U_\alpha \alpha_r = U_r - \alpha_t T_r U_\alpha. \tag{5.44}$$

Then (5.42) becomes

$$T_r A_\alpha - kU_\alpha = A_r/\alpha_t, \quad A_\alpha - kT_r U_\alpha = -k[U_r + 2U/r)]/\alpha_t \tag{5.45}$$

with the characteristic condition

$$\det \left\| \begin{array}{cc} k & -T_r \\ -kT_r & 1 \end{array} \right\| = 0,$$

so that $T_r = \pm 1$. For a wave traveling outward from $r = 1$ we define the 'fast' variable $\alpha = \omega(t - [r - 1])$ with $\alpha = \omega t = t/\varepsilon$ on $r = 1$. Equations (5.45) can be rewritten as

$$A_\alpha - kU_\alpha = \varepsilon A_r, \quad A_\alpha - kU_\alpha = -\varepsilon k[U_r + 2U/r)]. \tag{5.46}$$

Often we are only interested in the dependence of the one-term linear solution on the slow variable, here r. This can be found from the exact equations written in independent variables (r, α), here (5.46), by using the exact transport equation (found by equating the right-hand sides of (5.46)) and using the $O(1)$ part of one of the equations.

The transport equation from (5.46) yields

$$k^{-1}A_r + U_r + 2U/r = 0, \tag{5.47}$$

while $A = kU + O(\varepsilon)$, so that (5.47) becomes

$$U_r + U/r = 0. \tag{5.48}$$

Hence, using (5.43) the first order solution is

$$U(r,\alpha) = k^{-1}A(r,\alpha) = \frac{1}{r}Mh(\alpha) \tag{5.49}$$

which satisfies (5.48) and (5.43). $U(r,\alpha)$ in (5.49) is the approximate solution to (5.42) for $\varepsilon \ll 1$ corresponding to the first term in a geometric acoustics expansion.

For this particular example there is an exact solution to (5.42). To see this, define $w(r,t) = r^2 u(r,t)$ and eliminate A from (5.42) to give

$$w_{tt} = r^2 \frac{\partial}{\partial r}\left(\frac{1}{r^2}w_r\right).$$

Then (4.25) implies that, for an arbitrary function g,

$$w(r,t) = rg'(\omega(t - [r-1])) + g(\omega(t - [r-1])) \quad \text{and}$$
$$k^{-1}A(r,t) = r^{-1}g'(\omega(t - [r-1])),$$

or

$$U = \frac{g'(\alpha)}{r} + \frac{g(\alpha)}{r^2}, \quad A = k\frac{g'(\alpha)}{r}, \tag{5.50}$$

so that with $g'(\alpha) \backsim h(\alpha)$, the *exact* solution contains an extra term proportional to $1/r^2$. Note that for $r = O(1)$ both terms in u in (5.50) are of the same magnitude, but for $r \gg 1$ $u \backsim 1/r$.

5.5.4. *Pulsating Cylinder of Large Radius*

The analysis of the previous section can be extended to consider small-amplitude, high-frequency sound waves generated by a pulsating cylinder. The relevant equations are (4.37) with boundary condition (5.43). Using the same notation and change of variable as (5.44), the velocity and condensation, $u = U(r, \alpha)$ and $e = E(r, \alpha)$ satisfy

$$U_\alpha - E_\alpha = -\varepsilon E_r, \quad E_\alpha - U_\alpha = -\varepsilon(U_r + U/r). \tag{5.51}$$

Hence, the transport equation is

$$E_r + U_r + U/r = 0.$$

Then, since to first order $E = U$, this yields

$$2U_r + \frac{U}{r} = 0, \text{ or } U = \frac{1}{\sqrt{r}} M h(\alpha) = E. \tag{5.52}$$

The $1/\sqrt{r}$ dependence in (5.52) can also be deduced from the exact solution (4.41) by considering the solution near the wave-front $t - r = 0$ for large distances, see Eq. (7.32) in Whitham [1974].

5.5.5. *Sound Waves in a Tube with Variable Cross Section*

From (2.18) the dimensionless linear equations for sound waves in a tube with cross section $s(x)$ are

$$s e_t + (su)_x = 0, \quad u_t + e_x = 0, \tag{5.53}$$

where u and e are velocity and condensation. These are scaled with $L = D = \|s/\frac{ds}{dx}\|$ and $T = Dc_0^{-1}$, where $c_0 = \sqrt{\frac{\gamma p_0}{\rho_0}}$.

The boundary condition at $x = 0$ is the piston velocity, $u(0, t) = M h(\omega t)$ where $M = u_0/c_0 \ll 1$, $\omega = D/c_0\tau_p = \varepsilon^{-1}$ and τ_p is the period or pulse length. For $x > 0$ the gas is at rest at $t = 0$.

We again define the 'fast' variable $\alpha = \omega(t - x)$ and for a right travelling wave consider $u(x, t) = U(x, \alpha)$ and $e(x, t) = E(x, \alpha)$. This change of variable, see (5.27), implies that (5.53) can be written as

$$E_\alpha - U_\alpha = -\varepsilon(U_x + Us'/s), \quad U_\alpha - E_\alpha = -\varepsilon E_x. \tag{5.54}$$

Then to first order $E = U$, and the transport equation, obtained by adding the Eqs. (5.54), is

$$2U_x + Us'/s = 0,$$

so that the first order geometric acoustics solution is

$$u = e = Mh(\alpha)s^{-1/2}(x), \quad \alpha = \omega(t - x).$$

5.6. Exact Solutions

All of the approximate techniques introduced so far in Chapter 5 are valid for high frequency disturbances when $\varepsilon \ll 1$. This limits the form of the stratification to slowly varying functions. When this is not the case another option is to find exact solutions. In Seymour and Varley 1987 and Varley and Seymour 1988 a general algorithm was developed that produces many stratification functions $i(y)$ for which exact solutions to (5.5) exist. The motivation for this result came from looking for a condition on $i(y)$ that would result in the geometric acoustics series (5.16) **terminating** after a finite number of terms. Before outlining the full result we consider some simple examples that illustrate the process.

5.6.1. *Simple Examples*

First consider the anticipated form for a two-term series by comparison with (5.24). Assume each term separates as in (5.16) and the dependence on $t - y$ is as in (5.17), i.e., the first term is the derivative of the second, so we propose a two-term solution to (5.5) of the form

$$\sigma(y, t) = \sqrt{i(y)}F'(t - y) - l_1 F(t - y),$$

$$u(y, t) = \frac{-1}{\sqrt{i(y)}}F'(t - y) + k_1 F(t - y), \tag{5.55}$$

where $F(t - y)$ is an arbitrary function, (k_1, l_1) are arbitrary constants and $i(y)$ occurs as in (5.21). By direct substitution into (5.5), the expressions (5.55) are solutions to (5.5) if $i(y)$ satisfies the Riccati equation [see Boyce and DiPrima, 2012]

$$(\sqrt{i})' = k_1 i - l_1. \tag{5.56}$$

This equation has several sets of solution, including:

$$\sqrt{i(y)} = -\sqrt{\frac{l_1}{k_1}} \tanh(\sqrt{k_1 l_1}(y + \phi)) \quad \text{and}$$

$$\sqrt{i(y)} = -l_1(y + y_1), \quad k_1 = 0, \tag{5.57}$$

where k_1, l_1 and y_1 are arbitrary constants. The full range is shown later in (5.86) and (5.89). These functions are all monotonic solutions of (5.56) and form the basis of more complicated nonmonotonic functions.

The two-term solution (5.55) represents a wave traveling to the right. There is a similar solution corresponding to a wave traveling to the left in terms of $G(t + y)$:

$$\sigma(y, t) = \sqrt{i(y)}G'(t + y) + l_1 G(t + y),$$

$$u(y, t) = \frac{1}{\sqrt{i(y)}}G'(t + y) + k_1 G(t + y) \tag{5.58}$$

and of course, due to linearity, the two solutions (5.55), (5.58) can be added to form a solution containing left and right traveling waves.

If we now consider the constant coefficient system that is equivalent to (5.5) with $i(y) \equiv 1$, i.e.,

$$U_t = \Sigma_y \quad \text{and} \quad \Sigma_t = U_y \tag{5.59}$$

with solutions

$$\Sigma(y, t) = G(t + y) - F(t - y) \text{ and } U(y, t) = G(t + y) + F(t - y),$$
$$\tag{5.60}$$

it is not difficult to recognize that the sum of (5.55) and (5.58) can be represented in the form

$$\sigma(y, t) = \sqrt{i(y)}\frac{\partial \Sigma}{\partial y} + l_1 \Sigma, \quad u(y, t) = \frac{1}{\sqrt{i(y)}}\frac{\partial U}{\partial y} + k_1 U, \tag{5.61}$$

where U and Σ are given by (5.60). This points to a more general result that does not depend on the specific form (5.60). By direct substitution, it can be confirmed that (5.61) is a solution to (5.5) for any $i(y)$ satisfying (5.56) if U and Σ satisfy the constant coefficient Eqs. (5.59). The solution to the variable coefficient equations is then given in terms of the general solution to the equivalent constant coefficient equations, and this motivates the more general results described in Seymour and Varley [1987] and Varley and Seymour [1988].

A different approach that also produces exact solutions to (5.1) if $C(y) = i(y)$ satisfies the Riccati Eq. (5.56) is to consider periodic solutions of the form

$$w = \sigma(y, t) = \psi(y) \sin(\omega t), \tag{5.62}$$

when $\psi(y)$ satisfies

$$\frac{d}{dy}\left(\frac{1}{i}\psi'\right) + \frac{\omega^2}{i}\psi = 0. \tag{5.63}$$

Guided by the form of $\sigma(y, t)$ in (5.58), we let

$$\psi(y) = \sqrt{i(y)}F'(y) + l_1 F(y), \tag{5.64}$$

where

$$F'' + \omega^2 F = 0. \tag{5.65}$$

This form is suggested by assuming periodic solutions of the constant coefficient wave equation. Then, from (5.64),

$$\frac{1}{i}\psi' = -\frac{\omega^2}{i}F + \frac{1}{i}\left[(\sqrt{i})' + l_1\right]F'.$$

If $i(y)$ satisfies the Riccati Eq. (5.56), then

$$\frac{d}{dy}\left(\frac{1}{i}\psi'\right) = -\frac{\omega^2}{i}[\sqrt{i}F' + l_1 F] = -\frac{\omega^2}{i}\psi.$$

Thus ψ given by (5.64) is a solution of (5.63) provided (5.56) is satisfied and F satisfies (5.65).

Since σ satisfies (5.1) with $C = i$ and u satisfies (5.1) with $C = 1/i$, if $u(y, t) = \phi(y) \sin(\omega t)$, then the equation satisfied by ϕ is

$$\frac{1}{i}\frac{d}{dy}(i\phi') + \omega^2\phi = 0. \tag{5.66}$$

Then, noting (5.58),

$$\phi(y) = -\frac{1}{\sqrt{i(y)}}F'(y) + k_1 F(y),$$

satisfies (5.66) provided $i(y)$ satisfies (5.56) and F satisfies (5.65).

5.6.2. *General Result and Proof*

A generalization of (5.5) was considered in Varley and Seymour [1988] to include potential and diffusive equations of the form

$$i(y)u_t = \sigma_y \quad \text{and} \quad a\sigma_t + b\sigma = i(y)u_y, \tag{5.67}$$

where (a, b) are constants such that $(a, b) = (1, 0)$ for wave-like, $(-1, 0)$ for potential and $(0, 1)$ for diffusive solutions. For all of these equations, the basic result given in Varley and Seymour [1988] is that the solution to (5.67) for any constants (a, b) can be written in terms of the solution to the equivalent constant coefficient equations

$$U_t = \Sigma_y \quad \text{and} \quad a\Sigma_t + b\Sigma = U_y, \tag{5.68}$$

as series with $N + 1$ terms of the form

$$\sigma(y, t) = \sum_{n=0}^{N} f_n(y)\frac{\partial^{N-n}\Sigma}{\partial y^{N-n}}, \quad u(y, t) = \sum_{n=0}^{N} e_n(y)\frac{\partial^{N-n}U}{\partial y^{N-n}}, \tag{5.69}$$

where $f_n(y)$ and $e_n(y)$ satisfy the equations

$$(\sqrt{i(y)})' = i(y)e_1(y) - f_1(y) \tag{5.70}$$

$$e_n(y) + e'_{n-1}(y) = i^{-1}(y)f_n(y), \ 2 \leq n \leq N \tag{5.71}$$

$$f_n(y) + f'_{n-1}(y) = i(y)e_n(y), \ 2 \leq n \leq N \tag{5.72}$$

with $f_0 = e_0^{-1} = \sqrt{i(y)}$. We note that the constants (a, b) in (5.67) do not appear in (5.70)–(5.72). A condition for a terminating series is that the coefficients of the last terms are constant, hence, for $n = N$, $e_N(y) = k_N$ and $f_N(y) = l_N$. The result (5.61) is the two-term solution corresponding to $N = 1$ with $e_1 = k_1$ and $f_1 = l_1$.

It was shown in Varley and Seymour [1988] that the $2N - 1$ coupled nonlinear o.d.e.s (5.70)–(5.72) for $e_n(y)$, $f_n(y)$ and $i(y)$ can be solved *exactly* for any value of N. Since N is finite, this process produces exact solutions to (5.67). In practice only $N = 1, 2$ or 3 are used.

The proof of this result is complicated so we will give an outline for the case $N = 2$ when (5.70)–(5.72) becomes

$$f_1(y) + (\sqrt{i(y)})' = i(y)e_1(y), \tag{5.73}$$

$$k_2 + e_1'(y) = i^{-1}(y)l_2, \tag{5.74}$$

$$l_2 + f_1'(y) = i(y)k_2 \tag{5.75}$$

with $f_0 = e_0^{-1} = \sqrt{i(y)}$ and k_2 and l_2 are specified constants. The $N = 2$ version of (5.69) is

$$\sigma(y,t) = \sqrt{i}\Sigma_{yy} + f_1\Sigma_y + l_2\Sigma, \quad u(y,t) = \frac{1}{\sqrt{i}}U_{yy} + e_1 U_y + k_2 U, \tag{5.76}$$

where $\Sigma(y,t)$ and $U(y,t)$ satisfy (5.68). To confirm that (5.76) satisfies (5.67) if the coefficients satisfy (5.73)–(5.75) for any a, b, substitute (5.76) into (5.67) and consider the first of (5.67):

$$i(y)u_t - \sigma_y = ie_1 U_{yt} - ((\sqrt{i})' + f_1)\Sigma_{yy} + (ik_2 - l_2 + f_1')\Sigma_y \tag{5.77}$$

$$= [ie_1 - (\sqrt{i})' - f_1]\Sigma_{yy} + (ik_2 - l_2 - f_1')\Sigma_y, \tag{5.78}$$

using (5.68). Then, again using (5.68), from the second of (5.67)

$$a\sigma_t + b\sigma - i(y)u_y$$

$$= [f_1 + (\sqrt{i})' - ie_1]\{a\Sigma_{yt} + b\Sigma_y\} + a[l_2 - ie_1' - ik_2]\{a\Sigma_t + b\Sigma\}. \tag{5.79}$$

Setting the various coefficients to zero in (5.78) and (5.79) confirms that (5.67) are satisfied if the coefficients in (5.76) satisfy (5.73)–(5.75).

The next step is to solve (5.73)–(5.75) for $i(y)$, $f_1(y)$ and $e_1(y)$. These are three nonlinear first order o.d.e.s containing two arbitrary constants, k_2, l_2. First let $\theta(y)$ and $\phi(y)$ satisfy the homogeneous versions of (5.76), with $\theta(y) = u$ and $\psi(y) = \Sigma$,

$$0 = e_0(y)\theta'' + e_1(y)\theta' + k_2\theta, \tag{5.80}$$

$$0 = f_0(y)\phi'' + f_1(y)\phi' + l_2\phi, \tag{5.81}$$

where $f_0 = e_0^{-1} = \sqrt{i(y)}$.

Using (5.73)–(5.75) it can be shown that if θ satisfies (5.80) then θ' satisfies (5.81) and vice versa. This is the important result that allows us to find the solution to a nonlinear o.d.e. in terms of solutions of linear algebraic equations.

The proof comes from differentiating (5.80) and using (5.73)–(5.75) to give

$$0 = \sqrt{i(y)}\theta''' + f_1(y)\theta'' + l_2\theta'$$

which is (5.81) with $\phi = \theta'$.

Let the set $[\theta_1(y),\ \theta_2(y)]$ be two independent solutions of the second-order o.d.e. (5.80) and $[\phi_1(y),\ \phi_2(y)]$ be two independent solutions of (5.81). So θ_n' can be written as a linear combination of the ϕ_n, and ϕ_n' can be written as a linear combination of the θ_n. It then follows that we can write (using vector notation)

$$\underline{\theta}'' = \underline{A}\underline{\phi}' = \underline{B}\underline{\theta}$$

and diagonalizing $\underline{B}$ gives

$$\begin{bmatrix} \theta_1''(y) \\ \theta_2''(y) \end{bmatrix} = \begin{bmatrix} \lambda_1 & 0 \\ 0 & \lambda_2 \end{bmatrix} \begin{bmatrix} \theta_1(y) \\ \theta_2(y) \end{bmatrix} \tag{5.82}$$

for some real constants λ_1, λ_2. The constants need not be real, but this is the simplest case.

Then, using (5.82), (5.80) and (5.81) become (for $n = 1, 2$)

$$0 = e_0(y)\theta_n\lambda_n + e_1(y)\theta_n' + k_2\theta_n, \tag{5.83}$$

$$0 = f_0(y)\theta_n'\lambda_n + f_1(y)\theta_n\lambda_n + l_2\theta_n'. \tag{5.84}$$

Now observe that if $\theta_n'' = \lambda_n\theta_n$, for a real constant λ_n, then $z_n = \frac{\theta_n'}{\theta_n}$ satisfies the Riccati equation

$$z_n' + z_n^2 - \lambda_n = 0. \tag{5.85}$$

For $\lambda_n \neq 0$ (5.85) has four solutions

$$z_n(y) = \sqrt{\lambda_n}\tanh(\sqrt{\lambda_n}(y + y_n)), \quad \lambda_n > 0, \tag{5.86}$$

$$z_n(y) = \sqrt{\lambda_n}\coth(\sqrt{\lambda_n}(y + y_n)), \quad \lambda_n > 0, \tag{5.87}$$

$$z_n(y) = -\sqrt{-\lambda_n}\,\tan(\sqrt{-\lambda_n}(y+y_n)), \quad \lambda_n < 0, \qquad (5.88)$$

$$z_n(y) = -\sqrt{-\lambda_n}\,\cot(\sqrt{-\lambda_n}(y+y_n)), \quad \lambda_n < 0, \qquad (5.89)$$

where λ_n, y_n are arbitrary constants.

The $N = 1$ equivalents of (5.80) and (5.81) are

$$0 = e_0(y)z_1 + k_1, \quad f_0(y)\lambda_1 + l_1 z_1 = 0,$$

so that

$$\frac{1}{e_0(y)} = -\frac{z_1}{k_1} = f_0(y) = \sqrt{i(y)} = -\frac{l_1}{\lambda_1}z_1, \qquad (5.90)$$

implying that for consistency $\lambda_1 = k_1 l_1$. In (5.90) z_1 is any of the functions (5.86)–(5.89). Note that (5.56) and (5.85) are identical when $z_n = -k_1\sqrt{i}$ and then (5.86) corresponds to the first of (5.57), since $\lambda_1 = k_1 l_1$. The four functions (5.86)–(5.89) are used to produce a wide range of nonmonotonic functions $i(y)$.

To construct solutions to (5.73)–(5.75) for $i(y)$ we use (5.83)–(5.85). First, dividing by θ_n, (5.83) and (5.84) becomes

$$0 = e_0(y)\lambda_n + e_1(y)z_n + k_2, \qquad (5.91)$$

$$0 = f_0(y)z_n\lambda_n + f_1(y)\lambda_n + l_2 z_n. \qquad (5.92)$$

Now differentiate (5.91) to yield

$$0 = e_0'\lambda_n + e_1'z_n + e_1 z_n'$$

which, using (5.85), becomes

$$0 = e_0'\lambda_n + e_1'z_n + e_1(-z_n^2 + \lambda_n)$$

and then from (5.91), this becomes

$$0 = e_0'\lambda_n + e_1'z_n + z_n(k_2 + e_0\lambda_n) + e_1\lambda_n,$$
$$= z_n(k_2 + e_1') + \lambda_n(e_1 + e_0') + z_n\lambda_n e_0. \qquad (5.93)$$

Rearrange (5.92) as

$$0 = l_2 z_n + f_1\lambda_n + f_0 z_n\lambda_n \qquad (5.94)$$

and since z_n and λ_n are arbitrary, comparing coefficients in (5.93) and (5.94) gives:

$$\frac{k_2 + e_1'}{l_2} = \frac{e_1 + e_0'}{f_1} = \frac{e_0}{f_0}. \tag{5.95}$$

Conditions (5.95) are precisely (5.73)–(5.75) as long as $f_0 = e_0^{-1} = \sqrt{i(y)}$. To complete the proof we need to solve (5.73), or equivalently (5.91) and (5.92), for e_0 and f_0 in terms of z_1, z_2 for consistency with (5.73)–(5.75). This will also give us an expression for $i(y)$.

Equations (5.91) and (5.92) are linear algebraic equations for e_0 and e_1 or f_0 and f_1. To solve we substitute two different sets (z_n, λ_n) $n = 1, 2$ in both (5.91) and (5.92) to give two sets of linear algebraic equations:

$$0 = e_0\lambda_1 + e_1 z_1 + k_2, \ 0 = e_0\lambda_2 + e_1 z_2 + k_2$$

and

$$0 = f_0 z_1 \lambda_1 + f_1 \lambda_1 + l_2 z_1, \ 0 = f_0 z_2 \lambda_2 + f_1 \lambda_2 + l_2 z_2.$$

Solving for e_0 and f_0 gives

$$e_0 = \frac{k_2(z_1 - z_2)}{z_2 \lambda_1 - z_1 \lambda_2} = \frac{1}{\sqrt{i(y)}}, \tag{5.96}$$

$$f_0 = \frac{l_2(z_2 \lambda_1 - z_1 \lambda_2)}{\lambda_1 \lambda_2 (z_1 - z_2)} = \sqrt{i(y)}. \tag{5.97}$$

We can also solve for e_1 and f_1, but it is $f_0 = \sqrt{i(y)}$ that we are most interested in. Everything is consistent as long as $f_0 = e_0^{-1} = \sqrt{i(y)}$. Comparing (5.96) and (5.97): $\frac{l_2}{\lambda_1 \lambda_2} = \frac{1}{k_2}$, or $\lambda_1 \lambda_2 = k_2 l_2$, which is a constraint on the constants, and hence

$$\sqrt{i(y)} = \frac{z_2(y)\lambda_1 - z_1(y)\lambda_2}{k_2(z_1(y) - z_2(y))}, \tag{5.98}$$

where (z_n, λ_n) $n = 1, 2$ are any of (5.86)–(5.89). There are six constants, (y_n, λ_n) for $n = 1, 2$, k_2, l_2 with the one constraint $\lambda_1 \lambda_2 = k_2 l_2$.

So now, having picked the functions (z_1, z_2) and five constants to give a suitable form of $\sqrt{i(y)}$, e_0, e_1, f_0 and f_1 are known and the general solution can be written in the form (5.76) in terms of U and Σ that satisfy (5.68). By way of a comparison, for $N = 2$ the third order o.d.e. for $i(y)$ is, after eliminating $e_1(y)$ and $f_1(y)$ from (5.73)–(5.75),

$$(\sqrt{i(y)})''' + i'(3k_2 - l_2/i) - 2(ik_2 - l_2)i''/i' = 0. \qquad (5.99)$$

The solution will contains three integration constants plus k_2, l_2. Equation (5.98) is an exact solution of (5.99). Without the above analysis, looking for a terminating series leads to equations like (5.99).

The possible shapes of $i(y)$ for $N \geq 2$ are quite varied and include many nonmonotonic functions. Figure 5.2 shows a typical function using $N = 2$ with five arbitrary parameters that has a variation of over 10 units for $0 \leq y \leq 1$. The relevant parameters for Fig. 5.2 are $\lambda_1 = -9$, $\lambda_2 = -3$, $k_2 = 0.5$, $y_1 = -1.65$, $y_2 = 2$ using the form (5.88) in (5.98).

The condition $k_1 l_1 = 0$ yields two other sets of exact solutions when $i(y) = y^{2m}$ for positive and negative integer m; the details are given in Varley and Seymour [1988]. The $m = \pm 1$ cases are important, as for the Webster horn equation they describe cone-shaped

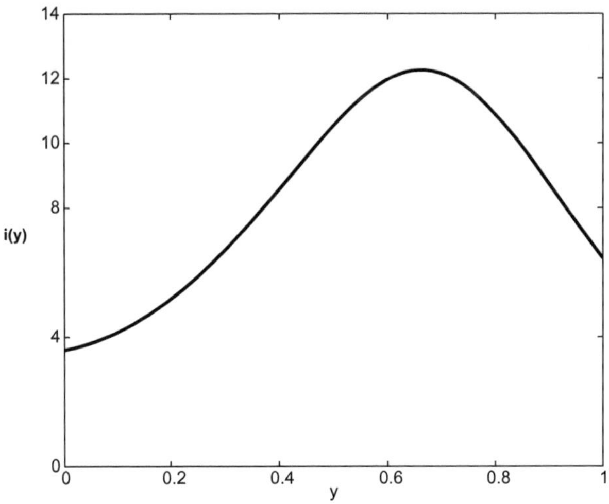

Fig. 5.2. $i(y)$ for $N = 2$.

'horns' with cross-sectional area increasing or decreasing like y^2. For $m = 1$ with $k_1 = 0$, $l_1 \neq 0$, (5.56) gives

$$i(y) = l_1^2(y + y_1)^2, \tag{5.100}$$

while $m = -1$ with $l_1 = 0$, $k_1 \neq 0$, gives

$$i(y) = \frac{1}{k_1^2(y + y_1)^2}, \tag{5.101}$$

see Mortell and Seymour [2004] and Keller [1977].

5.6.3. *Simple Shaped Horns*

Three exact solutions that exhibit resonance for the Webster horn equation with periodic forcing can be found for the shapes with cross-sectional areas $C(y) = e^{ky}$, $C(y) = (1+ay)^{\pm 2}$ and $C(y) = (1+ay)^{\pm 4}$. These illustrate properties that will be useful in later chapters.

5.6.3.1. $C(y) = e^{ky}$

For the first example, in (5.1) pick the obvious exponential horn shape $C(y) = e^{ky}$ to produce the constant coefficient equation

$$w_{tt} + kw_y - w_{yy} = 0. \tag{5.102}$$

This has time-periodic, wave-like, solutions of the form

$$w(y,t) = e^{\frac{1}{2}ky}e^{i\omega t}e^{\pm iD(\omega)y}, \ \ D(\omega) = \omega\sqrt{1 - \left(\frac{k}{2\omega}\right)^2}, \ \ \omega > k/2.$$

For $\omega \leq k/2$ there are no periodic solutions and all disturbances are damped out. Because of its simplicity, this shape is sometimes used as a test case for numerical and experimental investigations of acoustic resonance in shaped tubes, see for example Hiroyuki and Han [2015].

For $\omega > k/2$ periodic oscillatory motions with frequency ω exist when energy is introduced through the boundary condition at one end, such as for the conditions:

$$w(0,t) = 0 \text{ and } w(1,t) = M\sin(\omega t), \tag{5.103}$$

where M is a constant. Then a time-periodic solution can be constructed from the basic solutions of (5.102):

$$w = e^{\frac{1}{2}k(y-1)}[A_1 \sin(\omega t - Dy) + A_2 \cos(\omega t - Dy)$$
$$+ A_3 \sin(\omega t + D(\omega)y) + A_4 \cos(\omega t + Dy)], \qquad (5.104)$$

where the A_k are arbitrary constants and $D(\omega)$ is given above. The four constants A_k in (5.104) are determined by substituting (5.104) into the two boundary conditions (5.103). After some algebra we find, for $D(\omega) \neq n\pi$,

$$w(y,t) = e^{\frac{1}{2}k(y-1)} \frac{M}{2\sin D}(\cos(\omega t - Dy) - \cos(\omega t + Dy))$$

$$= e^{\frac{1}{2}k(y-1)} M \frac{\sin(yD)}{\sin D} \sin(\omega t).$$

This is an exact solution of (5.102) subject to (5.103) for the area shape $C(y) = e^{ky}$.

Note that the eigenvalues of the problem for the homogeneous boundary conditions, $w(0,t) = w(1,t) = 0$, are found from $\sin(D(\omega)) = 0$. They are

$$\omega_n = n\pi\sqrt{1 + \left(\frac{k}{2n\pi}\right)^2} > n\pi, \text{ integer } n.$$

This example illustrates very simply the difference between *commensurate* and *incommensurate* eigenvalues. When $k = 0$, corresponding to a straight tube or rod, the eigenvalues are all integer multiples of $\omega_1 = \pi$, so $\omega_n = n\pi = n\omega_1$. However, for $k \neq 0$, $\omega_n > n\pi$ and the eigenvalues are called incommensurate i.e., $\omega_n \neq n\omega_1$. This difference will be significant when considering the corresponding nonlinear problems in later chapters.

5.6.3.2. $C(y) = (1 + ay)^2$

The second example is for (5.1) in a cone-shaped horn or bar, when $C(y) = (1 + ay)^2$ for some constant a and the boundary conditions are again (5.103). This corresponds to the example given by (5.100) where $\sqrt{i(y)} = 1 + ay$, with $k_1 = 0$, $l_1 = -a$ in the Riccati Eq. (5.56). From (5.61) the $N = 1$ solution has the form

$$w(y, t) = (1 + ay)\frac{\partial \Sigma}{\partial y} - a\Sigma, \tag{5.105}$$

where $\Sigma(y, t)$ is given by (5.60). Direct substitution of (5.105) into the Webster horn Eq. (5.1) confirms it is the relevant solution for any Σ satisfying (5.68).

We now pick $\Sigma(y, t)$, which also satisfies (5.60), so that $w(y, t)$ satisfies the stress boundary conditions (5.103). To do this we assume $\Sigma(y, t)$ has the form

$$\Sigma(y, t) = A_1 \sin(\omega[t - y]) + A_2 \cos(\omega[t - y]) + A_3 \sin(\omega[t + y])$$
$$+ A_4 \cos(\omega[t + y]), \tag{5.106}$$

where the A_k are arbitrary constants. Then (5.106) is substituted in (5.105) to obtain a representation for $w(y, t)$ in terms of the four constants A_k. These are determined by substitution of $w(y, t)$ into the two boundary conditions (5.103). After some algebra, the *exact solution* to (5.1) is

$$w(y, t) = M\left[\frac{(a^2 + \omega^2 + ay\omega^2)\sin y\omega - a^2 y\omega \cos y\omega}{(a^2 + \omega^2 + a\omega^2)\sin \omega - a^2\omega \cos \omega}\right]\sin(\omega t)$$

or

$$w(y, t) = K(y, \omega, a)\frac{\sin(\omega y - \phi(y))}{\sin(\omega - \phi(1))}\sin(\omega t), \tag{5.107}$$

where

$$K(y, \omega, a) = M\frac{\sqrt{a^2 + \omega^2 (1 + ay)^2}}{\sqrt{a^2 + \omega^2 (1 + a)^2}} \quad \text{and}$$

$$\phi(y, \omega, a) = \tan^{-1}\left[\frac{a^2 y\omega}{a^2 + \omega^2[1 + ay]}\right].$$

Again, there is a resonance condition and (5.107) is valid only if $\sin(\omega - \phi(1)) \neq 0$. The solutions $\omega = \omega_n$ for which $\sin(\omega_n - \phi(1)) = 0$

give the resonant frequencies as solutions of

$$\tan \omega_n = \frac{a^2 \omega_n}{a^2 + \omega_n^2 [1 + a]}. \tag{5.108}$$

Note that as $a \to 0$ (representing a straight tube) (5.108) becomes $\tan \omega_n = 0$ with commensurate solutions $\omega_n = n\pi$, integer n, and $\phi(y, \omega, a) \to 0$. When $\sin(\omega - \phi(1)) = 0$ the resonant frequencies are incommensurate.

5.6.3.3. $C(y) = (1 + ay)^{-2}$

Here, w satisfies (5.1) with $C = i(y)$ given by (5.101) with $m = -1$ so that $\sqrt{i(y)} = (1 + ay)^{-1}$, with $l_1 = 0$, $k_1 = a$ in (5.56). From (5.61) the $N = 1$ general solution has the form

$$w(y, t) = (1 + ay)^{-1} \frac{\partial \Sigma}{\partial y},$$

so that the solution that satisfies the boundary conditions (5.103) is

$$w(y, t) = \frac{a + 1}{(ay + 1)} M \frac{\sin \omega t \sin \omega y}{\sin \omega}. \tag{5.109}$$

This solution is only valid for $\sin(\omega) \neq 0$. Hence the eigenvalues are $\omega_n = n\pi$ for integer n and are commensurate, even though there is stratification. This could have been anticipated from (5.7) by setting

$$\frac{1}{2} \frac{C''}{C} - \frac{3}{4} \left(\frac{C'}{C} \right)^2 = 0, \tag{5.110}$$

with solution $C(y) = (1 + ay)^{-2}$, since v then satisfies $v'' + \frac{\omega^2}{c_0^2} v = 0$.

Note that the solution (5.109) also applies to the velocity $u(y, t)$ that satisfies (5.1) with $C = 1/i(y)$ and $\sqrt{i(y)} = (1 + ay)$. Then $k_1 = 0$, $l_1 = 1$ and $u(y, t) = (1 + ay)^{-1} \frac{\partial U}{\partial y}$ where $U(y, t) = F(t - y) + G(t + y)$.

5.6.3.4. $C(y) = (1 + ay)^4$

The preceding two examples correspond to $N = 1$ solutions in the series (5.69). This example is for the stress, $w = \sigma(y, t)$ in (5.1) with boundary condition (5.3) when $i(y) = (1 + ay)^4$ for some constant

a and corresponds to $N = 2$ in (5.69). Motivated by (5.107) and (5.109), we let $w = \sigma(y,t) = \phi(y)\sin(\omega t)$ in (5.1) with $C = i$ when $\phi(y)$ satisfies

$$\frac{d}{dy}\left(\frac{1}{i(y)}\phi\right) + \frac{\omega^2}{i(y)}\phi = 0. \tag{5.111}$$

The general result for $N = 2$, *viz.*, (5.76), then implies that $\phi(y)$ has the form

$$\phi(y) = (1+ay)^2 F'' - k(1+ay)F' + lF, \tag{5.112}$$

where $F(y)$ satisfies $F'' + \omega^2 F = 0$ and k, l are constants. Substituting (5.112) into (5.111) and using $F'' = -\omega^2 F$ yields $l = ka$ and $l = -3a(2a - k)$ so that $k = 3a$, $l = 3a^2$ and

$$\phi(y) = (1+ay)^2 F'' - 3a(1+ay)F' + 3a^2 F,$$

where

$$F(y) = b_1 \cos(\omega y) + b_2 \sin(\omega y).$$

After some algebra, the *exact solution* to (5.1) is

$$\sigma(y,t) = K(y,\omega,a)\frac{\sin(\omega y - \phi(y))}{\sin(\omega - \phi(1))}\sin(\omega t), \tag{5.113}$$

where

$$K(y,\omega,a) = M\frac{\sqrt{\left(\omega^4(ay+1)^4 + 9a^4 + 3a^2\omega^2(ay+1)^2\right)}}{\sqrt{\left(\omega^4(a+1)^4 + 9a^4 + 3a^2\omega^2(a+1)^2\right)}},$$

$$\phi(y,\omega,a) = \tan^{-1}\left[\frac{3a^2 y\omega\left(\omega^2(1+ay) + 3a^2\right)}{\omega^4(ay+1)^2 + 9a^4 + 3a^2\omega^2\left(1 + ay - a^2 y^2\right)}\right].$$

Again, there is a resonance condition and (5.113) is valid only if $\sin(\omega - \phi(1)) \neq 0$. The solutions $\omega = \omega_n$ for which $\sin(\omega_n - \phi(1)) = 0$

give the resonant frequencies as solutions of

$$\tan \omega_n = \frac{3a^2\omega_n^3(1+a) + 9a^4\omega_n}{\omega_n^4(a+1)^2 + 3a^2\omega_n^2(1+a-a^2) + 9a^4}. \tag{5.114}$$

Note that as $a \to 0$ (representing a straight tube) (5.114) becomes $\tan \omega_n = 0$ with commensurate solutions $\omega_n = n\pi$, integer n, and $\phi(y, \omega, a) \to 0$. When $\sin(\omega - \phi(1)) \neq 0$ the resonant frequencies are incommensurate.

5.6.3.5.　$C(y) = (1 + ay)^{-4}$

There is a similar exact result when $i(y) = (1 + ay)^{-4}$ for some constant a when

$$\phi(y) = -\omega^2(1 + ay)^{-2}F - a(1 + ay)^{-3}F'.$$

The exact solution is again of the form (5.113) where

$$K(y, \omega, a) = \frac{\sqrt{(a^2 + \omega^2(1 + ay))^2 + a^4 y^2 \omega^2}}{\sqrt{(a^2 + \omega^2(1 + a))^2 + a^4\omega^2}},$$

$$\phi(y, \omega, a) = \tan^{-1}\left[\frac{a^2 y\omega}{(a^2 + \omega^2(1 + ay))}\right].$$

The solutions $\omega = \omega_n$ for which $\sin(\omega_n - \phi(1)) = 0$ give the resonant frequencies as solutions of

$$\tan \omega_n = \frac{a^2\omega_n}{(a^2 + \omega_n^2(1 + a))}.$$

5.6.4.　*General Solution with Cross Section:*　　*$s(x) = (1 + ax/D)^{\pm 2}$*

The results outlined in Sec. 5.6 allow us to find exact **general** solutions to linear problems for specific forms of inhomogeneity. In preparation for examples in future chapters, here we construct such solutions in the context of acoustic disturbances in a tube with cross-sectional area varying like $s(x) = (1 + ax/D)^{\pm 2}$; D is a length scale and a an $O(1)$ constant.

The dimensional equations are given in Sec. 2.2. If length and time are nondimensionalized in the manner described in Sec. 4.1 using $L = D$ and $T = Dc_0^{-1}$, then the linear dimensionless equations governing the motion of the gas are (5.53):

$$se_t + (su)_x = 0, \ u_t + e_x = 0 \qquad (5.115)$$

with dimensionless cross-sectional area $s(x) = (1 + ax)^{\pm 2}$.

The solutions (5.55), (5.58) can be rewritten in the context of acoustic waves in a tube. To do this, associate σ and u in (5.5) with su and $-e$ in (5.115) respectively. Together (5.55), (5.58) form a solution containing left and right traveling waves. The two special cases (5.100) and (5.101) can then be interpreted in terms of the tube cross sections $s(x) = (1 + ax)^{\pm 2}$. The case $k_1 = 0$, $l_1 = -a$ corresponds to $s(x) = (1 + ax)^2$, which describes a tube in the form of a cone, while $l_1 = 0$, $k_1 = -a$ corresponds to $s(x) = (1 + ax)^{-2}$. These cases produce the following two exact solutions in terms of the independent variables x, $\alpha = t - x$ and $\beta = t + x$. They can be verified by direct substitution in (5.115).

Case 1: $s(x) = (1 + ax)^2$

$$u = (1 + ax)^{-1}[F_1'(\alpha) + G_1'(\beta)] + a(1 + ax)^{-2}[F_1(\alpha) - G_1(\beta)], \qquad (5.116)$$

$$e = (1 + ax)^{-1}[F_1(\alpha) - G_1(\beta)]. \qquad (5.117)$$

Case 2: $s(x) = (1 + ax)^{-2}$

$$u = (1 + ax)[F_1'(\alpha) + G_1'(\beta)], \qquad (5.118)$$

$$e = (1 + ax)[F_1'(\alpha) - G_1'(\beta)] + a[F_1(\alpha) + G_1(\beta)]. \qquad (5.119)$$

$F_1(\alpha)$ and $G_1(\beta)$ are arbitrary functions.

The critical difference between Cases 1 and 2 was noted in the examples 5.6.3.2 and 5.6.3.3 that will be important in nonlinear theory: the resonant frequencies are incommensurate and commensurate respectively.

5.6.5. *Comparison of Exact and Approximate Solutions*

The forced oscillations of an inhomogeneous semi-infinite elastic panel initially at rest, is described in Sec. 5.5.1 in the high frequency, or geometric acoustics, limit when $\omega = \varepsilon^{-1} = \frac{D}{c_0 \tau_p} \gg 1$ and $D = |\frac{E(0)}{E'(0)}|$. Defining $E(x) = \frac{s(x)}{s_0}$, the two-term geometric acoustics solution, (5.36), is

$$u(x,t) = \sqrt{\frac{s_0}{s(x)}} h(\alpha) - \frac{\varepsilon}{4} \frac{1}{\sqrt{s(x)}} \int_0^x \sqrt{\frac{s(0)}{s(r)}} \left(\frac{s'(r)}{\sqrt{s(r)}} \right)' dr$$

$$\times \int_0^\alpha h(p)dp + O(\varepsilon^2), \tag{5.120}$$

where $\alpha = \omega\eta$, $\eta = t - x$, $s_0 = s(0)$ and the boundary condition on $x = 0$ is

$$u(0,t) = h(\omega t). \tag{5.121}$$

This approximate solution can now be compared with an exact solution using the results in Sec. 5.6.1, and to do this we follow an example in Mortell and Seymour [2005]. We assume $s(x)$ satisfies the Riccati Eq. (5.56)

$$(\sqrt{s})' = k_1 s - l_1, \tag{5.122}$$

where k_1, l_1 are arbitrary constants, and pick the stratification (5.87):

$$s(x) = \frac{l_1}{k_1} \coth^2 \left(l_1 \sqrt{\frac{k_1}{l_1}} (x + \phi) \right), \tag{5.123}$$

where ϕ is constant. Then, for the signalling problem on $x \geq 0$, $u(x,t)$ takes the form

$$u(x,t) = -\frac{1}{\sqrt{s(x)}} w'(\eta) + k_1 w(\eta), \tag{5.124}$$

in terms of the arbitrary unknown function $w(\eta)$.

The boundary condition (5.121) then implies that w satisfies

$$w'(t) - k_1 \sqrt{s_0} w(t) = -\sqrt{s_0} h(\omega t), \quad w(0) = 0 \tag{5.125}$$

with solution

$$w(t) = -\sqrt{s_0} \int_0^t e^{\gamma(t-y)} h(\omega y)\,dy, \tag{5.126}$$

where $\gamma = k_1\sqrt{s_0}$. Then, for $\eta \geq 0$, (5.124) implies that

$$u(x,t) = \sqrt{\frac{s_0}{s(x)}} h(\alpha) - k_1 J(x) \int_0^\eta e^{\gamma(\eta-y)} h(\omega y)\,dy, \tag{5.127}$$

where

$$J(x) = \sqrt{s_0} \left(1 - \sqrt{\frac{s_0}{s(x)}} \right). \tag{5.128}$$

Thus, from (5.127) the exact solution is

$$u(x,t) = \sqrt{\frac{s_0}{s(x)}} h(\alpha) - \varepsilon k_1 J(x) H(\alpha), \ \ H(\alpha) = \int_0^\alpha e^{\varepsilon\gamma(\alpha-y)} h(y)\,dy. \tag{5.129}$$

We are now in a position to compare the exact solution (5.129) and the two-term geometrical acoustics solution (5.120). When $\varepsilon \ll 1$ the exact solution (5.129) expands to give

$$u(x,t) = \sqrt{\frac{s_0}{s(x)}} h(\alpha) - \varepsilon k_1 J(x) \int_0^\alpha [1 + O(\varepsilon)] h(y)\,dy \tag{5.130}$$

for bounded α. Thus, for $\varepsilon \ll 1$ the second order geometric acoustic expansion differs from the exact solution at $O(\varepsilon^2)$. One reason for the difference is that the geometric acoustics expansion satisfies the boundary condition at $x = 0$ with the first-order solution, but not the second order, while the exact method uses the full solution to satisfy the boundary condition.

Comparing the solutions shows that there is an additional integral term in the exact solution that produces a 'tail' on the solution to a pulse input. We illustrate this with the input signal

$$h(\alpha) = \begin{cases} \sin(\pi\alpha) & \text{if } 0 \le \alpha \le 1 \\ 0 & \text{if } 1 \le \alpha, \end{cases} \tag{5.131}$$

so that

$$u(x,t) = \begin{cases} \sqrt{\frac{s_0}{s(x)}}\,\sin(\pi\alpha) - \varepsilon k_1 J(x) H_1(\alpha,\alpha), & \text{if } 0 \le \alpha \le 1 \\ -\varepsilon k J(x) H_1(\alpha,1), & \text{if } 1 \le \alpha, \end{cases}$$

where

$$H_1(\alpha,p) = \int_0^p e^{\varepsilon\gamma(\alpha-y)}\sin(\pi y)\,dy.$$

For h given by (5.131) this can be simplified to

$$u(x,t) = \begin{cases} \sqrt{\frac{s_0}{s(x)}}\,\sin(\pi\alpha) - \varepsilon\gamma J(x)\dfrac{(e^{\varepsilon\gamma\alpha}-\cos\pi\alpha)}{\pi\left(1+\frac{\gamma^2}{\pi^2}\right)\sqrt{s_0}} & \text{if } 0 \le \alpha \le 1 \\ -\varepsilon\gamma\dfrac{J(x)}{\pi(1+\frac{\gamma^2}{\pi^2})\sqrt{s_0}}(1+e^{-\varepsilon\gamma})e^{\varepsilon\gamma\alpha} & \text{if } 1 \le \alpha. \end{cases}$$
$$\tag{5.132}$$

We note that for fixed x and $\alpha \to \infty$, $u(x,t) \to 0$ since $\gamma < 0$, i.e., the tail of the exact solution decays exponentially behind the wavefront. In the geometrical acoustics limit the first term is

$$u(x,t) \sim \begin{cases} \sqrt{\frac{s_0}{s(x)}}\,\sin(\pi\alpha) & \text{if } 0 \le \alpha \le 1 \\ 0 & \text{if } 1 \le \alpha, \end{cases} \tag{5.133}$$

while the 'tail' in the second-order geometrical acoustics expansion for $1 \le \alpha$ is

$$-\varepsilon k_1 J(x) \int_0^1 h(y)\,dy \tag{5.134}$$

which is constant for fixed x. So the second-order expansion gives a 'tail' that remains constant at fixed x.

The exact linear solution (5.132) and asymptotic solution (5.133) are now compared for the stratification $s(x)$ for ε both finite and in the geometrical acoustics range. This stratification is shown in Fig. 5.3 where $s(x)$ is dashed for $\varepsilon = 0.39$ ($k = -0.1$, $l = -0.1$, $\phi = 4.5$) and solid for $\varepsilon = 2.3$ ($k = -0.1$, $l = -0.1$, $\phi = 0.85$).

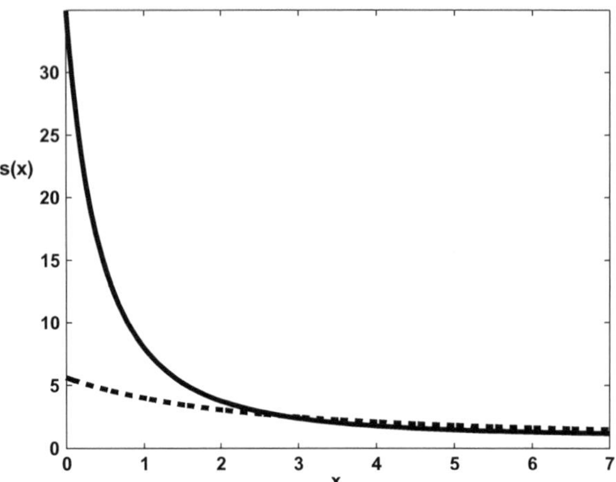

Fig. 5.3. $s(x)$ for $\varepsilon = 0.39$ (dashed), $\varepsilon = 2.3$, solid.

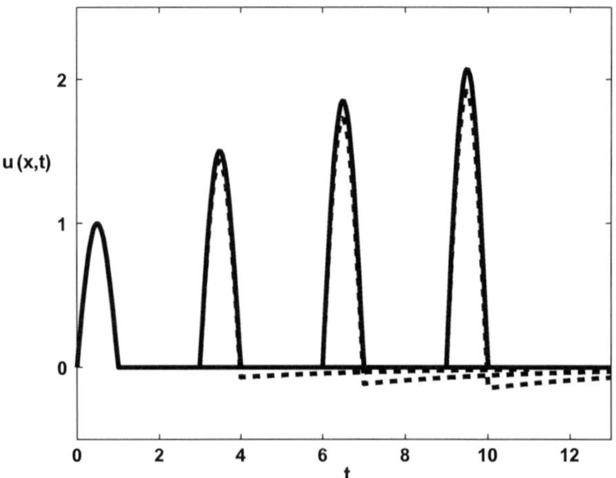

Fig. 5.4. $u(x,t)$ at $x = 0$, 3, 6, 9 for $\varepsilon = 0.39$.

Here, $\varepsilon = D^{-1} = |\frac{s'(0)}{s(0)}|$ is a measure of the maximum stratification. Figure 5.4, with $\varepsilon = 0.39$, within the geometrical acoustics range, and Fig. 5.5 with $\varepsilon = 2.3$, well outside this range, compare $u(x,t)$ for the exact solution (dashed) and the one-term geometrical acoustics solution (solid) at several values of $x = 0$, 3, 6, 9. The geometrical

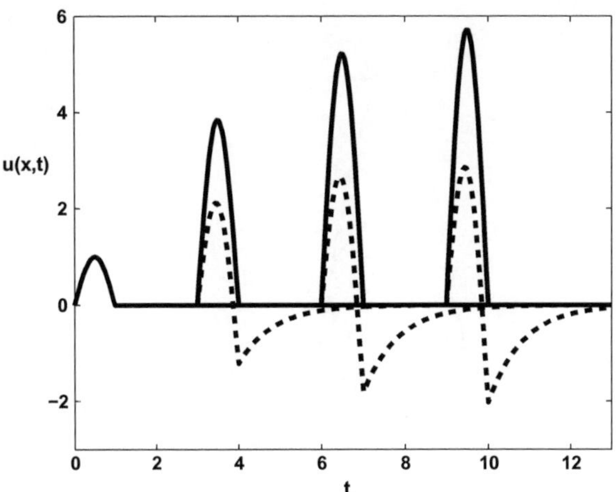

Fig. 5.5. $u(x, t)$ at $x = 0$, 3, 6, 9 for $\varepsilon = 2.3$.

acoustics solution is only valid for $\varepsilon \ll 1$ so it is not surprising that the comparison in Fig. 5.5 for $\varepsilon = 2.3$ is very poor. Note that the amplitude increases with x as $s(x)$ decreases sharply with x. The one-term geometrical acoustics solution overestimates the amplitude, has no tail and preserves the original pulse shape. In the exact solution, when $s(x)$ is a decreasing function, $J(x) \leq 0$, so that the tail is always negative.

Figures 5.6 and 5.7 compare the exact solution with the two-term geometrical acoustics solution (5.120). They show the exact solution (dashed), one-term geometrical acoustics solution (solid) and two-term geometrical acoustics solution (dotted) for the stratification (5.123) shown in Fig. 5.3 with $\varepsilon = 0.39$ and $\varepsilon = 2.3$. For $\varepsilon = 0.39$ there is good agreement between the exact and two-term expansion, but as ε increases the two-term expansion gives a very poor result, see Fig. 5.7 for $\varepsilon = 2.3$.

5.7. Connection to Solitons and the KdV Equation

It was shown by Varley and Seymour [1998] that there is a direct connection between the terminating series described in Sec. 5.6 that

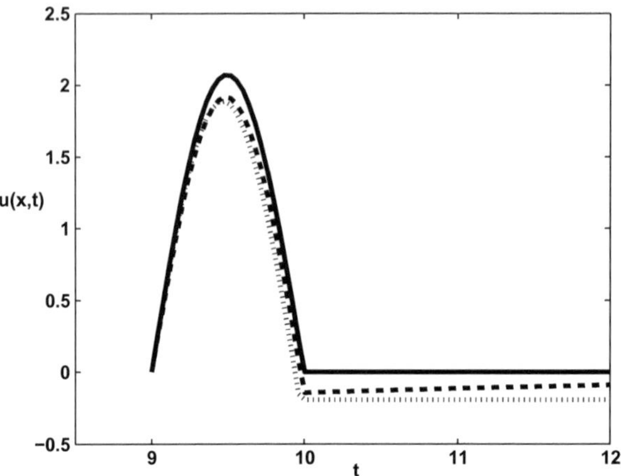

Fig. 5.6. $u(x,t)$ at $x = 9$ for $\varepsilon = 0.39$.

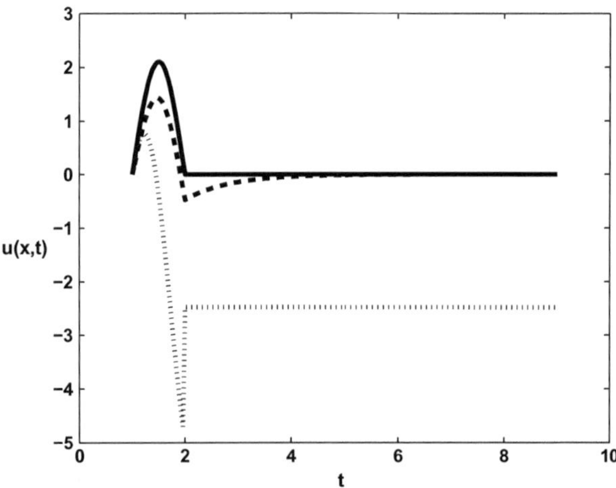

Fig. 5.7. $u(x,t)$ at $x = 1$ for $\varepsilon = 2.3$.

produce exact solutions to the Webster horn Eq. (5.1) and N-soliton solutions to the KdV equation. This gives a systematic procedure to produce exact representations of N-soliton solutions. Here we outline the process for $N = 1,\ 2$.

From (5.67), $u(y,t)$ satisfies the Webster horn equation in the form

$$\frac{\partial}{\partial y}\left(i(y)\frac{\partial u}{\partial y}\right) = i(y)\frac{\partial^2 u}{\partial t^2}.$$ (5.135)

To put this in the most convenient form to derive the N-soliton solutions, we define

$$\psi = u\psi_0 \quad \text{with } \psi_0 = \sqrt{i(y)}$$ (5.136)

then $\psi(y,t)$ satisfies a Schrödinger-type equation

$$\psi_{yy} - v(y)\psi = \psi_{tt},$$ (5.137)

where $v = \psi_0^{-1}\psi_0''$.

In Sec. 5.6.2, to construct exact solutions to the Webster horn Eq. (5.1) $u(y,t)$ was expanded in a series containing derivatives of the solution to the corresponding constant coefficient equation. The equivalent procedure here is to expand $\psi(y,t)$ in (5.137) as

$$\psi = \sum_{n=0}^{N} \gamma_{n+1}(y)\frac{\partial^n \Psi}{\partial y^n},$$ (5.138)

where $\Psi(y,t)$ satisfies the constant coefficient equation

$$\Psi_{yy} = \Psi_{tt}.$$ (5.139)

The series (5.138) terminates after $N+1$ terms when

$$\gamma_{N+1} = \text{constant}, = 1 \text{ say, and } \gamma_n = 0 \text{ for } n > N + 1.$$ (5.140)

Then, by substituting the series (5.138) into (5.137), using (5.139) and equating coefficients of $\frac{\partial^n \Psi}{\partial y^n}$, we obtain the system of N second-order nonlinear o.d.e.s

$$\gamma_1'' - v\gamma_1 = 0, \quad \gamma_n'' - v\gamma_n = -2\gamma_{n-1}', \ 2 \le n \le N$$ (5.141)

for the N functions $(\gamma_1(y),\ldots,\gamma_N(y))$ with the coefficient of $\frac{\partial^N \Psi}{\partial y^N}$ giving

$$v = 2\gamma_N'.$$ (5.142)

These are equivalent to the $2N$ 1st order nonlinear o.d.e.s (5.70)–(5.72) for $e_n(y)$ and $f_n(y)$.

For example, when $N = 1$

$$\psi = \Psi_y + \gamma_1 \Psi \tag{5.143}$$

and

$$v = 2\gamma_1' \quad \text{where } \gamma_1'' - 2\gamma_1\gamma_1' = 0. \tag{5.144}$$

This equation for γ_1 integrates once to give the Riccati Eq. (5.85), here written in the form

$$z_1' + z_1^2 = \kappa_1^2 \tag{5.145}$$

with $\gamma_1 = -z_1$. The two solutions of (5.145) of interest here, see (5.86) and (5.87), are

$$z_1 = \kappa_1 \tanh(\kappa_1(y + \delta_1)) \quad \text{and} \quad z_1 = \kappa_1 \coth(\kappa_1(y + \delta_1)), \tag{5.146}$$

where δ_1 and κ_1 are arbitrary constants.

When $N = 2$,

$$\psi = \Psi_{yy} + \gamma_2 \Psi_y + \gamma_1 \psi \tag{5.147}$$

and

$$v = 2\gamma_2', \ \gamma_1'' - 2\gamma_1\gamma_2' = 0 \quad \text{and} \quad \gamma_2'' - 2\gamma_2\gamma_2' = -2\gamma_1'. \tag{5.148}$$

It would seem that the two coupled second-order nonlinear equations given in (5.148) for $\gamma_1(y)$ and $\gamma_2(y)$ cannot be integrated except by using numerical procedures. (If γ_1 is eliminated these equations yield an horrendous fourth order nonlinear equation for γ_2.) However, Varley and Seymour [1988] showed that, for any choice of N, Eqs. (5.141) and (5.142) can be integrated in closed form. The procedure used depends on the remarkable fact that if $\theta(y)$ is any solution to the Nth order o.d.e.

$$\sum_{n=0}^{N} \gamma_{n+1}(y)\theta^{(n)}(y) = 0 \tag{5.149}$$

then so is $\theta''(y)$. This result is proven directly by differentiating (5.149) twice, and then using the recurrence relations (5.141) and (5.142) to eliminate the derivatives of the γ_n. Consequently, if $\underline{\theta} = (\theta_1, \theta_2, \ldots, \theta_N)$ is any set of linearly independent solutions to (5.149) then each component of $\underline{\theta}''$ can be expressed as a linear form in the components of $\underline{\theta}$. Hence, the components of $\underline{\theta}$ can always be chosen so that

$$\theta_i'' = \kappa_i^2 \theta_i, \quad i = 1, 2, \ldots, N, \tag{5.150}$$

where $\boldsymbol{\kappa} = (\kappa_1, \kappa_2, \ldots, \kappa_N)$ are distinct constants. For the cases that produce the $N-$ soliton solutions the $\boldsymbol{\kappa}$ are real positive constants. Thus every θ_i satisfying the second-order constant coefficient o.d.e. (5.150) also satisfies the Nth order variable coefficient o.d.e. (5.149), and each term in (5.149) is proportional to either θ_i or θ_i'. But defining

$$z_i = \theta_i'/\theta_i, \; i = 1, 2, \ldots, N, \tag{5.151}$$

then, according to (5.150), the z_i satisfy the Riccati Eq. (5.145) with (z_1, κ_1) replaced by (z_i, κ_i). Hence, the coefficients of the γ_n, i.e., $\theta^{(n)}(y)$, in (5.149) are either constant (depending on κ_i) or proportional to z_i. Using N different pairs (z_i, κ_i) produces N *algebraic* equations for the γ_n in the form

$$A_i[\underline{\gamma}]z_i + B_i[\underline{\gamma}] = C_i, \; i = 1, 2, \ldots, N, \tag{5.152}$$

where $\underline{\gamma} = (\gamma_1, \gamma_2, \ldots, \gamma_N)$. When N is odd

$$A_i[\underline{\gamma}] = \gamma_{N-1}\kappa_i^{N-3} + \gamma_{N-3}\kappa_i^{N-5} + \cdots + \gamma_4\kappa_i^2 + \gamma_2,$$
$$B_i[\underline{\gamma}] = \gamma_N\kappa_i^{N-1} + \gamma_{N-2}\kappa_i^{N-3} + \cdots + \gamma_3\kappa_i^2 + \gamma_1 \tag{5.153}$$
$$\text{and } C_i = -\kappa_i^{N-1}z_i \tag{5.154}$$

with a similar expression when N is even.

It then follows from (5.142) and (5.145) that if $\gamma_N = \Gamma_N(z_1, z_2, \ldots, z_N)$ then, since $z_n' + z_n^2 = \kappa_n^2$,

$$v(y) = 2\frac{\partial \Gamma_N}{\partial y} = 2\sum_{n=1}^{N}(\kappa_n^2 - z_n^2)\frac{\partial \Gamma_N}{\partial z_n}, = V(z_1, z_2, \ldots, z_N). \tag{5.155}$$

For example, when $N = 1$

$$\gamma_1 = -z_1 \quad \text{and} \quad v = 2\gamma_1' = -2(\kappa_1^2 - z_1^2), \tag{5.156}$$

so that when

$$z_1 = \kappa_1 \tanh(\kappa_1(y + \delta_1)), \ v = -2\kappa_1^2 \operatorname{sech}^2(\kappa_1(y + \delta_1)) \qquad (5.157)$$

and when

$$z_1 = \kappa_1 \coth(\kappa_1(y + \delta_1)), \ v = 2\kappa_1^2 \operatorname{csch}^2(\kappa_1(y + \delta_1)). \qquad (5.158)$$

When $N = 2$, (5.149) is

$$\gamma_1 \theta + \gamma_2 \theta' + \theta'' = 0$$

which becomes, dividing by θ and using this equation and (5.151),

$$\gamma_1 + \gamma_2 z_i + \kappa_i^2 = 0$$

for $i = 1, 2$. Then, since $v = 2\gamma_2'$, and using (5.145),

$$v = 2\frac{(\kappa_1^2 - \kappa_2^2)(\kappa_1^2 - z_1^2 + z_2^2 - \kappa_2^2)}{(z_1 - z_2)^2},$$

$$\gamma_2 = \frac{\kappa_2^2 - \kappa_1^2}{z_1 - z_2} \quad \text{and} \quad \gamma_1 = \frac{\kappa_1^2 z_2 - \kappa_2^2 z_1}{z_1 - z_2}. \qquad (5.159)$$

These functions, with

$$z_1 = \kappa_1 \tanh(\kappa_1(y + \delta_1)) \quad \text{or} \quad z_1 = \kappa_1 \coth(\kappa_1(y + \delta_1)) \qquad (5.160)$$

and with

$$z_2 = \kappa_2 \tanh(\kappa_2(y + \delta_2)) \quad \text{or} \quad z_2 = \kappa_2 \coth(\kappa_2(y + \delta_2)), \qquad (5.161)$$

are *exact* solutions to the nonlinear o.d.e.s (5.148).

There are several algorithms, all based on inverse scattering theory, that can be used to construct N-soliton solutions to the nonlinear KdV equation

$$\frac{\partial^3 v}{\partial y^3} - 6v\frac{\partial v}{\partial y} + \frac{\partial v}{\partial t} = 0, \qquad (5.162)$$

see Hirota [1972] and Miura [1976]. A soliton is a solitary, traveling wave pulse solution of certain nonlinear dispersive p.d.e.s that maintains its shape and speed. They also have the remarkable property that they emerge from interactions with other solitons with the speed and shape unchanged, see Drazin and Johnson [1989] for more details.

The functions $v(y)$ given by (5.155), (5.156) and (5.159) represent reflectionless potentials for the Schrödinger equation and are calculated from simple linear algebraic equations. Following Varley and Seymour [1998] it is possible to use these functions to construct N-soliton solutions, $v(y,t)$, to the KdV equation. To do this we first note that the z_i can be consistently regarded as functions of y and of an additional variable t satisfying the compatible equations

$$\frac{\partial z_i}{\partial y} + z_i^2 = \kappa_i^2 \text{ and } \frac{\partial z_i}{\partial t} + 4\kappa_i^2 \frac{\partial z_i}{\partial y} = 0, \quad i = 1, \ldots, N. \qquad (5.163)$$

By this we mean if $z_1(y) = \kappa_1 \tanh(\kappa_1(y+\delta_1))$ is a solution to (5.145), then $z_1(y,t) = \kappa_1 \tanh(\kappa_1(y - 4\kappa_1^2 t + \delta_1))$ is a solution to both of (5.163).

Varley and Seymour [1998] prove that $v(y,t)$ given by (5.155) also represents the N-soliton solution to the KdV equation and is simply

$$v(y,t) = V(z_1, z_2, \ldots, z_N), \qquad (5.164)$$

where

$$z_i(y,t) = \kappa_i \tanh(\kappa_i(y - 4\kappa_i^2 t + \delta_i)) \text{ or } \kappa_i \coth(\kappa_i(y - 4\kappa_i^2 t + \delta_i)). \qquad (5.165)$$

The details of the proof are given in Varley and Seymour [1998]. The proof hinges on showing that each component of $\underline{\gamma}$ satisfies the equation

$$\underline{\mathbf{E}} = \frac{\partial^3 \underline{\gamma}}{\partial y^3} - 3v \frac{\partial \underline{\gamma}}{\partial y} + \frac{\partial \underline{\gamma}}{\partial t} = 0 \qquad (5.166)$$

with this being a definition of $\underline{\mathbf{E}}$. Then with $v = 2\gamma_N'$, $v(y,t)$ satisfies (5.162). To prove the result (5.166) we first show that

$$A_i[\underline{\gamma}''' - 3v\underline{\gamma}']z_i + B_i[\underline{\gamma}''' - 3v\underline{\gamma}'] = 4\kappa_i^2(A_i[\underline{\gamma}']z_i + B_i[\underline{\gamma}']), \qquad (5.167)$$

where a dash denotes differentiation with respect to y. We indicate the steps that are needed to establish the relations (5.167) when N is odd — the equivalent manipulations when N is even are almost identical.

With A_i and B_i given by (5.153), differentiate (5.152) w.r.t. y and use (5.145) to show

$$A_i[\underline{\gamma}']z_i + B_i[\underline{\gamma}'] = -(\kappa_i^{N-1} + A_i[\underline{\gamma}])z_i' = -(\kappa_i^{N-1} + A_i[\underline{\gamma}])(\kappa_i^2 - z_i^2),$$
$$\tag{5.168}$$

then differentiate (5.168) twice and use (5.145) to show

$$A_i[\underline{\gamma}''']z_i + B_i[\underline{\gamma}'''] = -(\kappa_i^2 - z_i^2)(3A_i[\underline{\gamma}''] - 6z_i A_i[\underline{\gamma}']$$
$$+ (6z_i^2 - 2\kappa_i^2)(\kappa_i^{N-1} + A_i[\underline{\gamma}])). \tag{5.169}$$

Differentiating A_i and B_i in (5.153) w.r.t. y and using the recurrence relations (5.141) and (5.142) gives the identity

$$A_i[\underline{\gamma}''] = v(\kappa_i^{N-1} + A_i[\underline{\gamma}]) - 2B_i[\underline{\gamma}']. \tag{5.170}$$

It then follows from (5.168)–(5.170) that

$$A_i[\underline{\gamma}''']z_i + B_i[\underline{\gamma}'''] = -(\kappa_i^2 - z_i^2)(\kappa_i^{N-1} + A_i[\underline{\gamma}])(3v + 4\kappa_i^2)$$
$$= (A_i[\underline{\gamma}']z_i + B_i[\underline{\gamma}'])(3v + 4\kappa_i^2). \tag{5.171}$$

Rearranging (5.171) then gives (5.167).

Now let the γ_n and the z_n also depend on an additional variable t so that the dash denotes differentiation with respect to y at constant t. Using the first of Eqs. (5.168), condition (5.167) can be written

$$A_i\left[\frac{\partial^3 \underline{\gamma}}{\partial y^3} - 3v\frac{\partial \underline{\gamma}}{\partial y}\right]z_i + B_i\left[\frac{\partial^3 \underline{\gamma}}{\partial y^3} - 3v\frac{\partial \underline{\gamma}}{\partial y}\right] = -4\kappa_i^2(\kappa_i^{N-1} + A_i[\underline{\gamma}])\frac{\partial z_i}{\partial y}.$$
$$\tag{5.172}$$

Differentiating (5.152) with respect to t implies that

$$A_i\left[\frac{\partial \underline{\gamma}}{\partial t}\right]z_i + B_i\left[\frac{\partial \underline{\gamma}}{\partial t}\right] = -(\kappa_i^{N-1} + A_i[\underline{\gamma}])\frac{\partial z_i}{\partial t}. \tag{5.173}$$

Then, according to (5.166), (5.172) and (5.173)

$$A_i[\underline{\mathbf{E}}]z_i + B_i[\underline{\mathbf{E}}] = -(\kappa_i^{N-1} + A_i[\underline{\gamma}])\left(\frac{\partial z_i}{\partial t} + 4\kappa_i^2\frac{\partial z_i}{\partial y}\right). \tag{5.174}$$

In (5.174) the $z_i(y,t)$ satisfy the second of Eqs. (5.163).

When $z_i(y,t)$ is given by (5.165), Eqs. (5.163) are satisfied and the right-hand side of equation (5.174) is zero. It follows that for this choice of the z_i

$$A_i[\mathbf{E}]z_i + B_i[\mathbf{E}] = 0, \ 1 \leq i \leq N. \tag{5.175}$$

These are a set of linear algebraic equations for the E_n whose only solution is $E_n = 0, \ 1 \leq n \leq N$. It follows that each of the $\gamma_n, \ 1 \leq n \leq N$, satisfies the equation

$$\frac{\partial^3 \gamma_n}{\partial y^3} - 3v\frac{\partial \gamma_n}{\partial y} + \frac{\partial \gamma_n}{\partial t} = 0 \text{ with } v = 2\frac{\partial \gamma_N}{\partial y}. \tag{5.176}$$

In particular, γ_N satisfies the equation

$$\frac{\partial^3 \gamma_N}{\partial y^3} - 6\left(\frac{\partial \gamma_N}{\partial y}\right)^2 + \frac{\partial \gamma_N}{\partial t} = 0. \tag{5.177}$$

Differentiating this equation w.r.t. y then implies that $v(y,t)$ satisfies the KdV equation (5.162).

It is now straightforward to generate particular N-soliton solutions. When $N = 1$, from (5.156), $v = -2(\kappa_1^2 - z_1^2)$, then, either

$$z_1 = \kappa_1 \tanh(\kappa_1(y - 4\kappa_1^2 t + \delta_1)) \quad \text{and}$$
$$v(y,t) = -2\kappa_1^2 \operatorname{sech}^2(\kappa_1(y - 4\kappa_1^2 t + \delta_1)) \tag{5.178}$$

or

$$z_1 = \kappa_1 \coth(\kappa_1(y - 4\kappa_1^2 t + \delta_1)) \quad \text{and}$$
$$v(y,t) = 2\kappa_1^2 \operatorname{csch}^2(\kappa_1(y - 4\kappa_1^2 t + \delta_1)). \tag{5.179}$$

When $N = 2$, see (5.159),

$$v = 2\frac{(\kappa_1^2 - \kappa_2^2)(\kappa_1^2 - z_1^2 + z_2^2 - \kappa_2^2)}{(z_1 - z_2)^2}, \tag{5.180}$$

where $z_1(y,t)$ is given by (5.178) or (5.179) and where

$$z_2(y,t) = \kappa_2 \tanh(\kappa_2(y - 4\kappa_2^2 t + \delta_2)) \quad \text{or} \quad \kappa_2 \coth(\kappa_2(y - 4\kappa_2^2 t + \delta_2)). \tag{5.181}$$

An example of a two soliton solution to the KdV equation is

$$v(y,t) = -12\frac{3 + 4\cosh(2y - 8t) + \cosh(4y - 64t)}{(3\cosh(y - 28t) + \cosh(3y - 36t))^2}. \tag{5.182}$$

This corresponds to (5.180) with

$$(z_1(y,t), z_2(y,t)) = (\tanh(y - 4t), 2\coth(2(y - 16t))$$

and

$$(\kappa_1, \kappa_2) = (1, 2), \; (\delta_1, \delta_2) = (0, 0).$$

When $t = 0$,

$$v = -6\,\text{sech}^2 y.$$

Figure 5.8 shows the evolution of the 2-soliton solution given by (5.182) moving to the right at times $t = -2$ (thick solid line), -1 (dashed line), 0 (thin solid line), 1 (dotted line) and 2 (dash-dotted line). At $t = -2$ the soliton consists of two separate depressions. The smaller, slower moving one, has speed 4 and depth about 2, while the faster-moving one has speed 16 and a depth of about 8. At $t = 0$

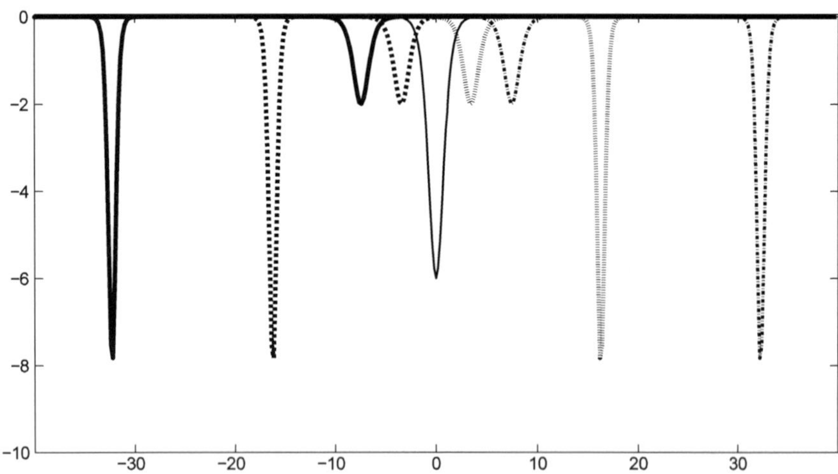

Fig. 5.8. $v(y,t)$ from (5.182) for $t = -2, -1, 0, 1, 2$.

the faster one catches the slower one and they combine to a single depression of depth about 6 at $y = 0$. After this "collision" they continue moving to the right, but now the faster one is ahead of the slower one. There is no steady state as they continue to spread apart.

Chapter 6

Kinematic and Simple Waves

The essential difference between waves propagating in linear and nonlinear hyperbolic systems is that the nonlinear characteristic curves, and hence the wave speeds, depend on the amplitude of the signal carried. As wavelets carrying different signal amplitudes travel at different speeds, this can produce a distortion of the wave profile. An initially continuous signal can 'break' and produce a discontinuous solution. Any nonlinear theory must account for this *amplitude dispersion* and possible shock formation.

Many of these properties are also present in kinematic waves that can be described with much simpler mathematics, so are reviewed here to introduce concepts like shocks and dissipation. Kinematic waves occur in many physical situations [see Whitham, 1974] and are typically described by a first-order equation of the form

$$u_t + c(u)u_x = b(u). \tag{6.1}$$

The solutions represent unidirectional waves traveling with speed $c(u)$, while $b(u)$ can describe some form of damping. In Sec. 6.1, we review several simple exact continuous and shocked solutions of (6.1).

To extend kinematic waves to waves traveling both left and right, (6.1) is generalized to a system typically of the form

$$u_t - c(e)e_x = 0, \quad e_t - c(e)u_x = 0. \tag{6.2}$$

This system describes nonlinear waves in homogeneous, nondissipative materials and admits exact (hence finite amplitude) solutions called simple waves. In Sec. 6.2, we review simple wave solutions to (6.2) and introduce the concept of the Riemann invariant.

108

6.1. Kinematic Waves, Shocks, Equal Area Rule

We saw in Sec. 3.1 that the equation $w_t + cw_x = 0$ with c constant can be interpreted as a directional derivative along the curve $\alpha(x,t) = \text{constant}$, given by $\frac{dx}{dt}\big|_\alpha = c$, with the solution $w = G(\alpha) = G(x - ct)$. This represents a wave traveling to the right with speed c and the shape is undistorted.

Now consider the case when $c = c(w)$, so that the speed depends on w, that satisfies

$$w_t + c(w)w_x = 0. \tag{6.3}$$

The directional derivative analogy still works and now along the curve $\alpha = \text{constant}$

$$\frac{dw}{dt}\bigg|_\alpha = 0 \text{ on } \frac{dx}{dt}\bigg|_\alpha = c(w). \tag{6.4}$$

Again, the first of (6.4) implies that $w = G(\alpha)$ for arbitrary G, but, since we are integrating along $\alpha = \text{constant}$, where $c = c(G(\alpha))$ is constant, the characteristic (from the second of (6.4)) is now

$$x = \alpha + c(w)t, \quad w = G(\alpha) \tag{6.5}$$

with the initial condition $\alpha = x$ on $t = 0$. Then $G(x)$ is the initial shape of w at $t = 0$. The solution (6.5) is implicit and this is typical of the solutions to quasilinear equations such as (6.3). The characteristics are still straight lines, but now the slope of each line depends on the magnitude of w carried by c. So different parts of the wave travel at different speeds leading to the possibility of the wave either spreading apart or concentrating. The phenomenon of the shape changing depending on the wave amplitude is called *amplitude dispersion*. For example, consider two initial points at $x = \alpha_1$ and α_2, $\alpha_2 < \alpha_1$ with $G(\alpha_2) < G(\alpha_1)$. At $t = 0$, they are a distance $\alpha_1 - \alpha_2$ apart, but after time t their separation has increased to $\alpha_1 + tG(\alpha_1) - \alpha_2 - tG(\alpha_2) > \alpha_1 - \alpha_2$, so the wave is more widely spread. Alternatively, if $G(\alpha_1) < G(\alpha_2)$ the separation will decrease and it is possible that at some time t_m the two straight-line characteristics carrying different amplitudes will

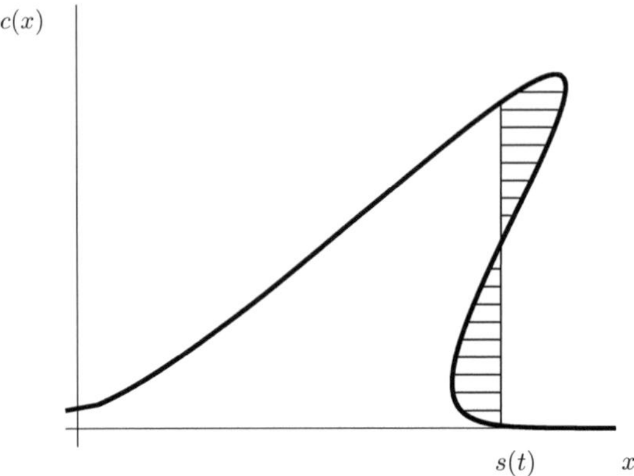

Fig. 6.1.   Multivalued solution.

intersect when $\alpha_1 + t_m G(\alpha_1) = \alpha_2 + t_m G(\alpha_2)$. To accommodate this latter phenomenon, we need to introduce the idea of 'weak' solutions that contain discontinuities or shocks. These are solutions that satisfy a discontinuity or shock condition that is compatible with the original equations whenever the continuous solution breaks down. If $G(x)$ is differentiable, we can predict when the solution (6.5) breaks down by calculating the change in the derivative w_x. A shock first forms when first $w_x \to \infty$. After that time, the solution predicts three values of w for a range of x at a particular time, see Fig. 6.1.

From (6.5),

$$w_x = G'(\alpha)\alpha_x, \quad 1 = \alpha_x + tc'(G)G'(\alpha)\alpha_x,$$

so that

$$\alpha_x = \frac{1}{1 + tc'(G)G'(\alpha)}, \quad w_x = \frac{G'(\alpha)}{1 + tc'(G)G'(\alpha)}.$$

This derivative becomes unbounded as $1 + tc'(G)G'(\alpha) \to 0$, so that after the time t_s, given by

$$t_s = -\frac{1}{c'(G)G'(\alpha)}, \tag{6.6}$$

the continuous initial shape $G(\alpha)$ will become multivalued.

While equations of the form (6.3) are useful to illustrate phenomena that arise in more complicated quasilinear systems of equations, they also arise in a variety of practical applications when written as conservation laws with solutions known as kinematic waves. Their properties were discussed by Lighthill and Whitham [1955] in the context of flood waves and traffic flow.

Equation (6.3) is typically written as a conservation law for $\rho(x, t)$:

$$\rho_t + q_x = 0, \quad q = Q(\rho), \tag{6.7}$$

where $Q(\rho)$ represents the flux of ρ and

$$q_x = Q'(\rho)\rho_x.$$

Then (6.7) has the same form as (6.3) with $\rho = w$ and the wave speed is given by $c(w) = Q'(w)$. The alternative form (6.7) is useful when considering shocks.

Once a shock has formed, it travels along a path $x = X(t)$ determined by the shapes of the waveform ahead and behind. Going back to first principles using the discrete or finite difference form of the derivatives in the conservation law (6.7) and taking the limit as $x \to X(t)$ from ahead of and behind the discontinuity, the shock condition is [see Whitham, 1974]

$$-\frac{dX}{dt}[\rho] + [q] = 0, \tag{6.8}$$

where $[\rho] = \rho_A - \rho_B$, and ρ_A and ρ_B are the values of ρ immediately ahead of and behind the shock. (6.8) is rewritten as

$$\frac{dX}{dt} = \frac{Q(\rho_A) - Q(\rho_B)}{\rho_A - \rho_B} \tag{6.9}$$

and, since ρ_A and ρ_B typically depend on t, (6.9) is an o.d.e. that traces the shock path.

We note that by multiplying (6.7) by $c'(w)$ it can be written for any Q as

$$c_t + cc_x = c_t + \frac{1}{2}(c^2)_x = 0. \tag{6.10}$$

The latter is a conservation law. This is the 'generic' form of the kinematic wave equation, valid for all Q. From (6.5) the general solution of (6.10), with initial condition $c(x,0) = F(x)$, outside of a shock is

$$x = \alpha + c(\alpha)t, \quad c = F(\alpha). \tag{6.11}$$

From (6.10), where $Q = c^2/2$, the shock condition (6.9) becomes

$$\frac{dX}{dt} = \frac{1}{2}\frac{c_A^2 - c_B^2}{c_A - c_B} = \frac{1}{2}(c_A + c_B). \tag{6.12}$$

However, while (6.12) is a possible weak solution, it is valid only when $Q(\rho)$ is quadratic in ρ or can be approximated by a quadratic, see Whitham [1974]. When this is the case, (6.12) is compatible with the "equal area rule"

$$\frac{1}{2}(F(\alpha_A) + F(\alpha_B))(\alpha_A - \alpha_B) = \int_{\alpha_B}^{\alpha_A} F(\alpha)d\alpha, \tag{6.13}$$

where α_A (ahead of the shock) and α_B (behind the shock) are the characteristics that meet at the shock; the initial condition is $c = F(x)$ at $t = 0$. The compatibility of (6.12) and (6.13) is shown in detail in Whitham [1974]; see also Kluwick [1981] and Landau [1945].

Analytical solutions for continuous initial shapes $F(x)$ that form shocks are limited and can be complicated, so we illustrate the use of the shock condition by considering a piecewise linear initial function. Then, while the general solution is implicit, for such initial conditions $c(x,t)$ can be calculated explicitly. Let us consider the solution to (6.10) with initial condition

$$c(x,0) = F(x) = \begin{cases} 0 & \text{if } x < 0 \\ x & \text{if } 0 \le x \le 1/2 \\ 1 - x & \text{if } 1/2 \le x \le 1 \\ 0 & \text{if } x > 1. \end{cases} \tag{6.14}$$

The initial shape is nonzero for $0 < x < 1$. Since for this equation the wave speed is the same as the wave height, we anticipate that the portions of the signal for $x < 0$ and $x > 1$ will remain fixed while the vertex will move to the right with speed $1/2$. Hence, when $t = 1$ the vertex will be at $x = 1$ and the solution contains a shock of

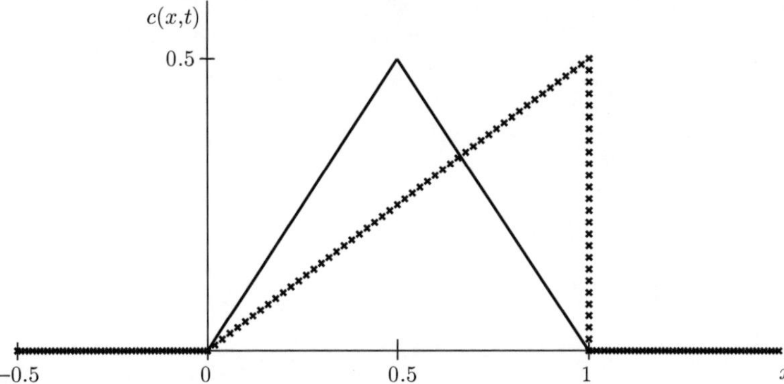

Fig. 6.2. $c(x, 0) = F(x)$, solid, and $c(x, 1)$, dashed.

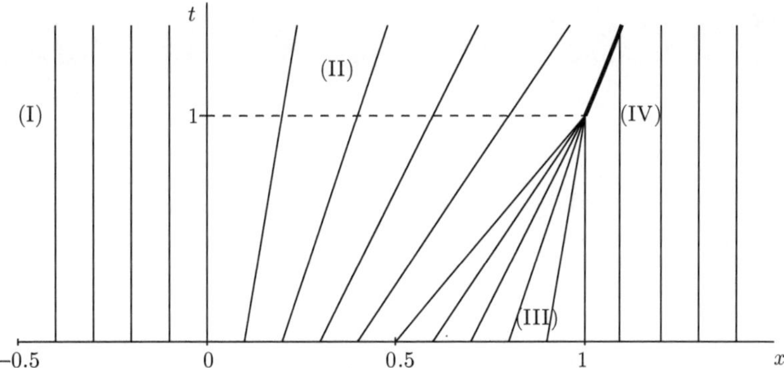

Fig. 6.3. Characteristic diagram.

strength $1/2$, as shown in Fig. 6.2. So for $x > 1$ and $t > 1$ there will be a shock region in the $x - t$ plane.

It is useful to divide the plane into four different regions corresponding to the different segments of $c(x, 0)$, see Fig. 6.3. In Regions I $(x < 0)$ and IV $(x > 1)$, where $c(x, 0) = F(x) = 0$, the characteristics are vertical in the $t - x$ plane and the solution is

$$c = 0, \quad x = \alpha, \quad \alpha < 0 \quad \text{and} \quad \alpha > 1.$$

In Region II, $0 \leq x \leq \frac{1}{2}$, $F(x) = x$ and

$$c = \alpha, \quad \frac{dx}{dt} = c, \quad x = \alpha + \alpha t, \quad 0 < \alpha < \frac{1}{2}. \tag{6.15}$$

Eliminating α

$$c(x,t) = \frac{x}{1+t} \quad \text{and} \quad t = \frac{x-\alpha}{\alpha}.$$

In Region III, $\frac{1}{2} \le x \le 1$,

$$c = 1 - \alpha, \quad \frac{dx}{dt} = c, \quad x = \alpha + (1-\alpha)t, \quad \frac{1}{2} < \alpha < 1,$$

so that the explicit solution is

$$c = 1 - \frac{x-t}{1-t} = \frac{1-x}{1-t}, \quad \text{and } t = \frac{x-\alpha}{1-\alpha}.$$

For $t > 1$ the solution is multivalued. This can be seen from the $x - t$ plane where the characteristics first intersect at $t = 1$ when $x = 1$. A shock with path $x = X(t)$, starting at $X(1) = 1$, must be fitted for $t > 1$ across Regions II and IV. From (6.12) the shock condition is

$$\frac{dX}{dt} = \frac{c_{\mathrm{II}} + c_{\mathrm{IV}}}{2}, \quad \text{where } c_{\mathrm{II}} = \left.\frac{x}{1+t}\right|_{x=X} \quad \text{and} \quad c_{\mathrm{IV}} = 0,$$

hence the shock path is calculated from

$$\frac{dX}{dt} = \frac{1}{2}\left(\frac{X}{1+t}\right), \quad X(1) = 1,$$

so

$$X(t) = \frac{1}{\sqrt{2}}(1+t)^{1/2}, \tag{6.16}$$

see Fig. 6.3. Since ahead of the shock $c = 0$, the shock strength $K(t)$ (the difference between c ahead and behind the shock) is simply the height of c at the shock in Region II:

$$K(t) = c(X(t), t) = \left.\frac{x}{1+t}\right|_{x=X} = \frac{1}{\sqrt{2}}\frac{(1+t)^{1/2}}{1+t}$$

$$= \frac{1}{\sqrt{2}}(1+t)^{-1/2}. \tag{6.17}$$

The shock strength decays like $t^{-1/2}$ as $t \to \infty$.

We can confirm the equivalence of the shock condition (6.12) and equal area rule by calculating the solution for the initial function

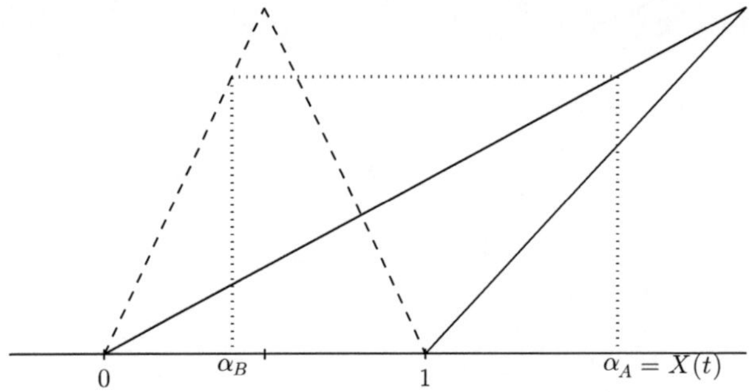

Fig. 6.4.   Shock diagram for $t > 1$.

(6.14) again, and using (6.13) for $t > 1$. Since $F(\alpha) = 0$ for $\alpha > 1$, the shock location and the characteristic ahead of the shock are always identical, so $X(t) = \alpha_A$, see Fig. 6.4. For $t > 1$, from (6.5) and (6.14), $X = \alpha_A > 1$, since $F(\alpha_A) = 0$, $F(\alpha_B) = \alpha_B$ for $0 < \alpha_B < 1/2$, and the shock location is given by

$$X(t) = \alpha_A = \alpha_B + F(\alpha_B)t = \alpha_B + \alpha_B t = \alpha_B(1 + t), \qquad (6.18)$$

so that

$$\alpha_A - \alpha_B = \alpha_B t. \qquad (6.19)$$

Then, since $F(\alpha) \equiv 0$ for $\alpha > 1$, (6.13) becomes

$$\frac{1}{2}(0 + \alpha_B)(\alpha_A - \alpha_B) = \int_{\alpha_B}^{\alpha_A} F(\alpha)d\alpha = \int_{\alpha_B}^{1} F(\alpha)d\alpha = A(\alpha_B),$$

$$(6.20)$$

where the area initially over $[\alpha_B, 1]$ is

$$A(\alpha_B) = \frac{1}{4} + \frac{1}{2}\left(\frac{1}{4} - \alpha_B^2\right). \qquad (6.21)$$

Hence, from (6.19)–(6.21),

$$\alpha_B^2 t = \frac{1}{2} + \left(\frac{1}{4} - \alpha_B^2\right),$$

or

$$\alpha_B = \frac{1}{\sqrt{2(1 + t)}}.$$

Then from (6.18)

$$X(t) = \alpha_B(1+t) = \frac{1}{\sqrt{2}}(1+t)^{1/2} \qquad (6.22)$$

which agrees with (6.16).

Equations (6.17) and (6.22) imply that the shock preserves area, since $A(t) = \frac{1}{2}X(t)c(X(t),t) = \frac{1}{4} = A(0) = A_0$, while the energy $E(t) = \int_0^{X(t)} \frac{1}{2}c^3(x,t)dx$ where, after the shock forms, $c(x,t) = \frac{x}{1+t}$, by (6.17). Hence, $E(t) = \frac{1}{8(1+t)^3}X^4(t) = \frac{1}{32(1+t)^3}(1+t)^2 = \frac{1}{32(1+t)}$ $\to 0$ as $t \to \infty$. We also note that from (6.17) and (6.22)

$$X(t) \sim \frac{1}{\sqrt{2}}t^{1/2} = \sqrt{2A_0 t}, \quad c(X(t),t) \sim \frac{1}{\sqrt{2t}} = \sqrt{\frac{2A_0}{t}} \text{ as } t \to \infty.$$
$$(6.23)$$

We note that the asymptotic results (6.23) apply for a general single hump initial shape with area A_0 and for an N-wave, see Whitham [1974], though the full details of the evolution cannot be calculated. It is the particular form of our initial conditions (6.14) that allows us to calculate the details of the evolution to the asymptotic state.

6.1.1. *Damped and Amplified Waves*

By adding a term proportional to $c(x,t)$, the solutions to (6.10) exhibit damping or amplification. We need to determine how the change in amplitude affects shock formation. Consider the problem

$$c_t + cc_x + \lambda c = 0, \quad 0 < t, \quad c(x,0) = F(x), \quad -\infty \le x \le \infty,$$
$$(6.24)$$

where the parameter $\lambda > 0$ or $\lambda < 0$. Along the characteristic curve $\alpha = $ constant, with $\alpha = x$ on $t = 0$,

$$\left.\frac{dc}{dt}\right|_\alpha = -\lambda c \quad \text{on} \quad \left.\frac{dx}{dt}\right|_\alpha = c, \qquad (6.25)$$

so that

$$c(x,t) = F(\alpha)e^{-\lambda t}, \quad x = \alpha + F(\alpha)(1 - e^{-\lambda t})\lambda^{-1}. \qquad (6.26)$$

The signal is distorted and either damped or amplified as it propagates depending on the sign of λ, but a shock can form if $c_x \to \infty$, where $c_x = F'(\alpha)\alpha_x e^{-\lambda t}$. From the second of (6.26)

$$\alpha_x = 1/D(\alpha, t), \quad D(\alpha, t) = 1 + F'(\alpha)(1 - e^{-\lambda t})\lambda^{-1}$$

and a shock will form when $\alpha_x \to \infty$, or when $D(\alpha, t) \to 0$.

First consider the damped case $\lambda > 0$. The signal amplitude decays, but is still distorted. The question is, can the damping dominate and prevent shock formation? To answer this, consider $D(\alpha, t)$. We need $D(\alpha, t) = 0$ for some α and t. Since for $\lambda > 0$, $0 < 1 - e^{-\lambda t} < 1$, a shock will only form if both $F'(\alpha) < 0$ and $|F'| > \lambda$, but no shock can form if $|F'| < \lambda$. Hence, there is a critical level of the initial value of $|F'|$ that determines whether a shock forms.

Alternatively, when $\lambda < 0$, the signal is amplified in time. Then $e^{-\lambda t} - 1 \to \infty$ as $t \to \infty$ and at any α when $F'(\alpha) < 0$, $D(\alpha, t) \to 0$. So, as long as $F'(\alpha) < 0$ for some α, a shock will always form.

Usually when the signal contains a shock this is a dissipative mechanism and the signal amplitude and energy decay with time. However, if $\lambda < 0$ and the signal is continuously amplified, the question we ask is: can the shock dissipation balance the amplification to produce a steady state as $t \to \infty$?

To resolve this we consider a simple symmetric piecewise linear initial signal:

$$c(x, 0) = F(x) = \begin{cases} 0 & \text{if } x < 0 \\ x & \text{if } 0 \le x \le 1/2 \\ 1 - x & \text{if } 1/2 \le x \le 3/2 \\ x - 2 & \text{if } 3/2 \le x \le 2 \\ 0 & \text{if } 2 < x. \end{cases}$$

as illustrated in Fig. 6.5. The portion on $[0, 1/2]$ is identical to (6.14) in the previous example, and hence a shock forms at $x = 1$ at $t = 1$. However, by symmetry and the equal area rule, since $F(\alpha_A) = -F(\alpha_B)$, this time there is a stationary shock at $x = 1$, after which the shock strength $K(t)$ is always $2c(1^-, t)$. Behind the shock c originates in the initial interval $[0, 1/2]$ where $F(x) = x$.

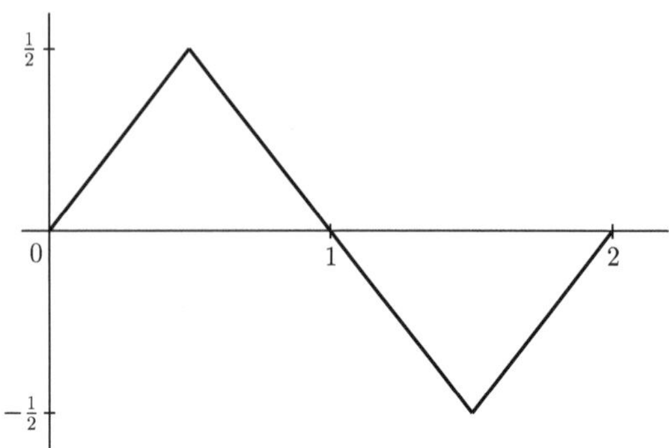

Fig. 6.5.   Initial signal $F(x)$.

Hence, $2c(1^-, t) = 2F(\alpha)e^{-\lambda t} = 2\alpha e^{-\lambda t}$, where, from the second of (6.26), $1 = \alpha + \alpha(1 - e^{-\lambda t})\lambda^{-1}$. Hence, eliminating α,

$$K(t) = \frac{2e^{-\lambda t}}{1 + (1 - e^{-\lambda t})\lambda^{-1}} \to -2\lambda > 0 \text{ as } t \to \infty.$$

Therefore, for any $\lambda < 0$ there is a steady state when the amplification and shock dissipation balance.

The steady state solution for $\lambda < 0$, $t \to \infty$, is

$$c(x, t) = \begin{cases} 0 & \text{if } x < 0 \\ -\lambda x & \text{if } 0 \leq x \leq 1 \\ -\lambda(x - 2) & \text{if } 1 \leq x \leq 2 \\ 0 & \text{if } 2 < x \end{cases} \tag{6.27}$$

as illustrated in Fig. 6.6. If we consider the steady state version of (6.24) for $\lambda < 0$, $c(c_x + \lambda) = 0$, the solutions are either $c = 0$ or given by (6.27). If $\lambda > 0$, $c = 0$ is the only possible steady state solution since a solution involving a jump and slope $-\lambda$ could not evolve, since the apex at $x = 1/2$ must move to the right.

6.2. Simple Waves: Riemann Invariants

We use the example of one-dimensional motion in a semi-infinite, homogeneous elastic panel or rod to introduce the concept of the simple wave. For a detailed description of simple wave solution in

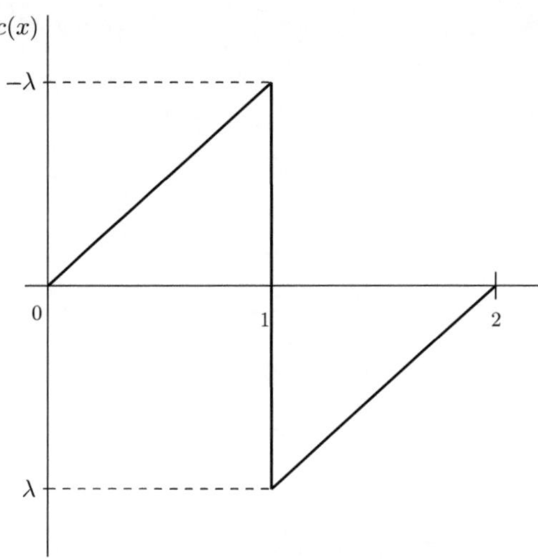

Fig. 6.6.   Steady state solution.

the context of gas dynamics, see Courant and Friedrichs [1999] or Kluwick [1981].

The governing equations in Lagrangian coordinates are (2.3) and (2.2), which in dimensionless variables (with $T = \tau_p$, $L = c_0\tau_p$ where $c_0 = \sqrt{E_0/\rho}$) are

$$u_t = \sigma_x, \quad \text{and} \quad u_x = \lambda_t, \tag{6.28}$$

$$\sigma(\lambda) = \Sigma(\lambda) = \lambda(1 + N\lambda + O(\lambda^2)), \tag{6.29}$$

where in the nonlinear stress–strain law (6.29) N is a dimensionless material constant and λ is the strain, $N > 0$ corresponds to a "hardening" material while $N < 0$ is a "softening" material. In terms of u and λ, (6.28) become

$$u_t = c^2(\lambda)\lambda_x, \text{ and } u_x = \lambda_t, \tag{6.30}$$

where $c^2(\lambda) = \Sigma'(\lambda)$. We define the nonlinear strain measure as

$$a(\lambda) = \int_0^\lambda c(y)dy, \tag{6.31}$$

then (6.30) become

$$u_t - c(\lambda)a_x = 0, \text{ and } a_t - c(\lambda)u_x = 0. \tag{6.32}$$

Using the notation $u(x,t) = U(x,\alpha)$, $a(x,t) = A(x,\alpha)$, where $T(x,\alpha)$ is the arrival time of the characteristic $\alpha(x,t) = $ constant at the location x, we make the change of variable (5.27)

$$u_x = U_x + U_\alpha \alpha_x = U_x - U_\alpha \alpha_t T_x, \tag{6.33}$$

since $\frac{d\alpha}{dx} = \alpha_x + \alpha_t T_x = 0$. Then (6.32) become

$$U_\alpha + cT_x A_\alpha = cA_x/\alpha_t, \quad A_\alpha + cT_x U_\alpha = cU_x/\alpha_t, \tag{6.34}$$

so that the characteristic condition defines the wave speeds

$$c(\lambda) = \pm\sqrt{\Sigma'(\lambda)} = 1 + N\lambda + \cdots$$

Defining the right (α) and left (β) traveling characteristics by

$$\left.\frac{dT}{dx}\right|_\alpha = \frac{1}{c(\lambda)} = 1 - N\lambda + O(\lambda^2), \tag{6.35}$$

$$\left.\frac{dT}{dx}\right|_\beta = -\frac{1}{c(\lambda)} = -[1 - N\lambda + O(\lambda^2)], \tag{6.36}$$

(6.34) become, on using (6.35) and (6.36) respectively,

$$U_\alpha + A_\alpha = cA_x/\alpha_t = cU_x/\alpha_t, \tag{6.37}$$
$$U_\beta - A_\beta = cA_x/\beta_t = -cU_x/\beta_t. \tag{6.38}$$

Now, assume that $A_x = U_x = 0$ and $U = U(\alpha,\beta)$ and $A(\alpha,\beta)$ depend on the characteristic variables, so that (6.37) and (6.38) imply

$$U + A = 2G(\beta) \quad \text{on} \quad \left.\frac{dT}{dx}\right|_\alpha = \frac{1}{c(\lambda)} \quad \text{and}$$

$$U - A = 2F(\alpha) \quad \text{on} \quad \left.\frac{dT}{dx}\right|_\beta = -\frac{1}{c(\lambda)}. \tag{6.39}$$

$U + A$ and $U - A$ are the **Riemann Invariants**. They are constant along $\beta(x,t) = $ constant and $\alpha(x,t) = $ constant respectively. Then

$$U = F(\alpha) + G(\beta), \quad A = G(\beta) - F(\alpha).$$

This is the nonlinear generalization of the linear general solution (3.2). The significant difference is that here α and β are unknown.

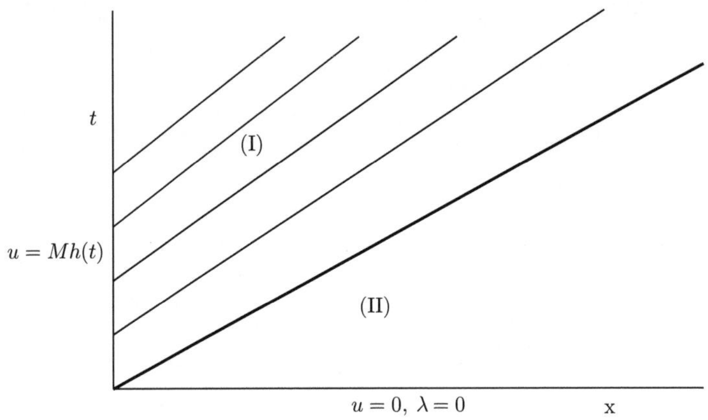

Fig. 6.7.   Simple wave characteristics.

Consider the initial-boundary value problem shown in Fig. 6.7 with

$$u(0,t) = Mh(t), \quad u(x,0) = 0, \quad \lambda(x,0) = 0, \quad x > 0, \qquad (6.40)$$

so that along $t = 0$, $x > 0$, $U = A = G = F = 0$. By (6.39), following the characteristics leaving $t = 0$, in all of Region II, $U = A = G = F = 0$. But the characteristics entering Region I from Region II carry the values $G(\beta) = 0$ and hence $U = -A = F(\alpha)$.

If we tag the characteristic leaving $x = 0$ by $\alpha = t$, then the first of (6.39) implies that

$$t = \frac{x}{c(\lambda)} + \alpha, \text{ and } U(\alpha) = -A(\alpha) = Mh(\alpha), \qquad (6.41)$$

since $c(\lambda)$ is constant on $\alpha = $ constant. (6.41) is an **exact** solution of the nonlinear equations (6.32) and is called a **simple wave** solution for a wave traveling to the right into an undisturbed region. It does not require the assumption of small amplitude.

The simple wave characteristics $\alpha(x,t) = $ constant can intersect and hence produce shocks when $\frac{\partial u}{\partial x} = U'(\alpha)\alpha_x \to \infty$. Differentiating the first of (6.41) w.r.t. x gives

$$0 = \frac{1}{c(\lambda)} - \frac{x}{c^2(\lambda)}c'(\lambda)\lambda'(\alpha)\alpha_x + \alpha_x. \qquad (6.42)$$

Hence,

$$\alpha_x = \frac{c(\lambda)}{xc'(\lambda)\lambda'(\alpha) - c^2(\lambda)} \tag{6.43}$$

and $\frac{\partial u}{\partial x} \to \infty$ when

$$xc'(\lambda)\lambda'(\alpha) - c^2(\lambda) \to 0. \tag{6.44}$$

From (6.31) and (6.41), $-Mh'(\alpha) = A'(\alpha) = A'(\lambda)\lambda'(\alpha) = c(\lambda)\lambda'(\alpha)$, hence (6.44) implies that a shock forms when

$$x = -\frac{c^3(\lambda)}{Mc'(\lambda)h'(\alpha)} \tag{6.45}$$

as long as $x > 0$.

For a small amplitude simple wave, when $M \ll 1$,

$$c(\lambda) = 1 + N\lambda + O(\lambda^2), \quad a(\lambda) = \lambda + O(\lambda^2)$$

and (6.41) becomes

$$t = x(1 - N\lambda + \cdots) + \alpha, \text{ and } a = -Mh(\alpha), \tag{6.46}$$

so that

$$U(\alpha) = Mh(\alpha), \quad \alpha = t - x - MNxh(\alpha). \tag{6.47}$$

The shock condition (6.44) is

$$xMNh'(\alpha) + 1 = 0 \tag{6.48}$$

which requires $Nh'(\alpha) < 0$ at a shock. Hence, for a pulse, for example like $h(t) = \sin t$, $0 \le t \le \pi$, if $N > 0$ the shock forms at the back of the pulse (where $h'(t) < 0$), while if $N < 0$ the shock forms at the front of the pulse.

After a shock has formed it propagates along a path $t = T(x)$, or $x = X(t)$, determined by the shapes of the waveform ahead and behind. In a similar way to (6.28), using the discrete or finite difference forms of the derivatives in the conservation laws (6.28) and taking the limit as $x \to X(t)$ from ahead of and behind the discontinuity, the shock conditions are

$$\frac{dT}{dx} = -\frac{[u]}{[\sigma]} = -\frac{[\lambda]}{[u]},\qquad(6.49)$$

where $[u] = u_A - u_B$, and u_A and u_B are the values of u immediately ahead of and behind the shock.

In the small amplitude, finite rate, limit, shocks are weak and produce a negligible reflected wave. Then, by similar arguments to those for a kinematic wave given in Sec. 6.1, $\frac{dT}{dx}$ for the α-wave is related to the characteristic wavelets ahead of the shock, α_A, and α_B, behind the shock, by

$$\frac{dT}{dx} = 1 + \frac{1}{2}NM(h(\alpha_A) + h(\alpha_B)).\qquad(6.50)$$

This is equivalent to (6.12) with $c = F(\alpha) = 1 - NMh(\alpha)$, which when substituted in (6.13), leaves (6.13) unchanged with $F(\alpha)$ replaced by $h(\alpha)$. The condition that α_A and α_B coalesce into the shock at the same time imply that

$$T - x = \alpha_A + MNxh(\alpha_A) = \alpha_B + MNxh(\alpha_B).\qquad(6.51)$$

The solution to the initial boundary-value problem for all wavelets that have not coalesced into the shock is given by (6.47). For a given initial signal function $h(\alpha)$, the shock trajectory is described by solutions to (6.50) and (6.51), which are consistent with the "equal area rule".

Chapter 7

Nonlinear Geometric Acoustics

In general it is not possible to construct exact solutions when either stratification or dissipation are included in nonlinear systems like (6.2). However, for small amplitude waves in the geometric acoustics limit, Whitham [1950], [1952] proposed a "rule" to correct linear theory to account for the observed nonlinear effects. This and the ray theory of geometric optics [Keller, 1962] is the foundation of *nonlinear geometrical acoustics*. In Sec. 7.1 we outline Whitham's rule and then show how it can also be derived by extending the results of Sec. 5.4 and independently by using the method of multiple scales.

While Whitham's rule extends the range of validity of the linear theory for small-amplitude but nonlinear disturbances and correctly predicts shock formation, it is not a systematic expansion procedure and hence is difficult to generalize or to obtain higher-order corrections. Varley and Cumberbatch [1966] recognized that linear geometric acoustics can be generalized to include nonlinear effects in a systematic way. They introduced the *theory of relatively undistorted waves* as an extension of an idea in Courant and Hilbert [1962] for linear waves. One advantage of the theory of relatively undistorted waves is that the solution can be determined by solving ordinary differential equations. The method was also used in Varley and Rogers [1967] and Seymour and Varley [1970] and similar expansions were subsequently proposed by Asano and Taniuti [1969]. All of these methods, which are limited to small-amplitude high frequency disturbances, agree to first order with the "rule" suggested by Whitham. In Sec. 7.2 we give four examples of the use of the method proposed

124

in Varley and Cumberbatch [1966]: longitudinal waves in an inhomogeneous elastic panel, shallow water surface waves over variable topography, and spherical and cylindrical waves in a polytropic gas.

The definition of a relatively undistorted wave for high-frequency disturbances is not restricted to small amplitudes. The simple wave solution can be of finite amplitude. Varley and Cumberbatch [1970] recognized how simple wave solutions can be used to produce approximate, finite amplitude, solutions when a high frequency mechanism is also present. A progressing wave is approximated locally by a simple wave that is considered to be "modulated" by the high frequency mechanism. The solutions are called *modulated simple waves*. The modulated simple-wave theory was used in Varley and Cumberbatch [1970] to describe large amplitude waves in a stratified atmosphere, and in Parker [1969], [1972] for pulses propagating through a relaxing gas. Probably the most significant results using the technique were obtained in Varley, Venkataraman and Cumberbatch [1971] for the behavior of a tsunami as it moves over a continental shelf toward a shoreline. There, since the equation of state is known explicitly, the governing equations are the simplest that describe finite-amplitude waves in an inhomogeneous medium. In Sec. 7.4 we outline how modulated simple waves are used to describe finite amplitude disturbances in an elastic panel and for shallow water waves.

7.1. Whitham's Nonlinearization Technique

In 1952, Whitham considered the flow pattern past an axisymmetric supersonic projectile. He noted that the case of two-dimensional supersonic flow past a projectile, solved by Friedrichs [1948], involved an exact solution, the simple wave, in which a shock had been inserted. He used this model, inter alia, to modify linear theory by means of the fundamental hypothesis: "linearized theory gives a valid first approximation to the flow *everywhere* provided that in it the approximate characteristics are replaced by the exact ones, or at least by a sufficiently good approximation to the exact ones". The "approximate characteristics" were the linear ones while the "good approximation" were nonlinear ones. The fundamental hypothesis obviated the need for an exact solution, and that put the focus on finding a sufficiently good nonlinear approximation to the exact characteristics.

While this method only produces a one-term approximation and might seem ad hoc, it is confirmed by the more general nonlinear geometric acoustics expansion. We give three examples of the non-linearization technique followed by examples of the full expansion.

7.1.1. *Steady Supersonic Projectile*

For supersonic flow past a thin projectile, the linear approximation is based on the equation

$$\phi_{yy} - B^2\phi_{xx} = 0, \quad B^2 = M^2 - 1,$$

where M is the constant Mach number and $\phi(x, y)$ the velocity potential. This approximate equation fails to predict the occurrence of shocks and fails at large distance from the body.

Van Dyke [1975] summarizes the steps leading to a uniform first approximation. Using the oblique coordinates $\xi = x - By$, $\eta = By$, the problem for a uniformly valid first approximation is defined by the kinematic wave equation (see Sec. 6.1)

$$u_\eta + \frac{\gamma + 1}{2}\frac{M^4}{B^2}uu_\xi = 0 \tag{7.1}$$

with

$$u(\xi, \eta = 0) = -\varepsilon T'(\xi)/B, \text{ and } u = 0 \text{ upstream.} \tag{7.2}$$

Here u is the velocity perturbation $\Delta u/U$ from the mainstream speed U.

Equation (7.1) describes a small amplitude simple wave, and the exact solution can be found from

$$\frac{du}{d\eta}\Big|_\alpha = 0 \text{ on } \frac{d\xi}{d\eta}\Big|_\alpha = \frac{\gamma + 1}{2}\frac{M^4}{B^2}u,$$

as

$$u = u(\xi), \quad \xi = \alpha + \frac{\gamma + 1}{2}\frac{M^4}{B^2}\eta u(\xi),$$

where $\alpha = \xi$ on $\eta = 0$. The boundary condition (7.2) gives

$$u(\xi) = -\varepsilon T'(\xi)/B, \text{ so that } \xi = \alpha - \varepsilon\frac{\gamma+1}{2}\frac{M^4}{B^3}\eta T'(\xi),$$

and

$$\alpha = x - By + \varepsilon\frac{\gamma+1}{2}\frac{M^4}{B^2}yT'(\xi). \tag{7.3}$$

Equation (7.3) is the "sufficiently good approximation" where the linear characteristic $\alpha = x - By$ has been corrected by a nonlinear term.

7.1.2. *High Frequency Waves in a Maxwell Solid*

The governing equations for the one-dimensional motion of a viscoelastic panel are (2.3) and (2.9). The linear problem was described in Sec. 5.5 for a Maxwell solid when the variables were scaled with $T = 2\tau_r$, and $L = 2c_0\tau_r$, where $c_0 = \sqrt{E_0/\rho_0}$. The equivalent nonlinear system can be written as

$$u_t - \sigma_x = 0, \ \lambda_t = u_x \tag{7.4}$$

with (2.9) replaced by

$$\sigma_t = \Gamma'(\lambda)\lambda_t - 2\sigma \tag{7.5}$$

and the boundary condition

$$u(0,t) = Mh(\omega t), \tag{7.6}$$

where $\omega = 2\tau_r/\tau_p \gg 1$ and $M = u_0/c_0$. As in previous examples, see (6.29), $\Gamma(\lambda) = \lambda + N\lambda^2 + O(\lambda^3)$. For $x > 0$ the material is at rest at $t = 0$ when the one-term linear solution (when $\Sigma(\lambda) = \lambda$) is

$$u = -\lambda = Mh(\alpha)e^{-x}, \ \alpha = \omega(t - x). \tag{7.7}$$

The lower-order term proportional to σ in (7.5) does not affect the characteristic and hence the two-term nonlinear characteristic equation for a wave traveling to the right is given by

(6.35): $\frac{dT}{dx}|_\alpha = 1 - N\lambda + O(\lambda^2)$. Substituting (7.7) into (6.35) and integrating produces the two-term nonlinear characteristic

$$\alpha = \omega(t - x - Mh(\alpha)(1 - e^{-x})). \tag{7.8}$$

7.1.3. *Spherical Wave for Large Radius*

The equivalent linear problem was described in Sec. 5.5 and the nonlinear one is given in 7.2.4. The governing nonlinear equations are (4.23), while the linear equations are (4.24) and the boundary condition is given by $u(1, t) = Mh(\alpha)$. When $\omega \gg 1$ the one-term linear result is (5.49), *viz.*,

$$u(r, t) = U(r, \alpha) = \frac{1}{r} Mh(\alpha), \text{ with } \alpha = \omega(t - [r - 1]). \tag{7.9}$$

Then the two-term nonlinear characteristic is given by (7.75), i.e.,

$$\alpha = \omega\left(t - [r - 1] + \frac{\gamma + 1}{2} Mh(\alpha)\ln(r)\right). \tag{7.10}$$

This latter is the sufficiently good approximation to the exact result.

7.1.4. *Cylindrical Wave for Large Radius*

For the cylindrical wave, in the high frequency limit the linear solution is given by (5.52)

$$u(r, t) = U(r, \alpha) = \frac{1}{\sqrt{r}} Mh(\alpha) \quad \text{with } \alpha = \omega(t - [r - 1]). \tag{7.11}$$

The nonlinear two-term characteristic is given in (7.82),

$$\alpha = \omega(t - [r - 1] + (\gamma + 1)Mh(\alpha)[\sqrt{r} - 1]). \tag{7.12}$$

Again, the nonlinear approximation (7.12) to the exact characteristic $\alpha(r, t)$ is inserted in the linear approximation (7.11). Unlike the spherical case, the exact linear solution (4.41) is not directly amenable and so the asymptotic form (7.11) must be used.

The details of (7.10) and (7.12) are given in Secs. 7.2.4 and 7.2.5.

7.2. Regular Expansion

In the preceding section Whitham's nonlinearization technique is used to produce a one-term solution. Here we give a regular perturbation scheme that can theoretically be continued for many terms, though in practice at most the first two are used. The method is valid when the applied wavelength $\xi = c_0\tau_p$ is small compared with a natural length D, where D is typically the radius of curvature of the wave-front, the relaxation length of a viscoelastic material, or the length defined by the stratification. In all cases $\varepsilon = \frac{\xi}{D} \ll 1$.

7.2.1. *Elastic Panel*

For the first illustration we consider the same example as in Sec. 5.5, the forced oscillations of a semi-infinite elastic panel when the inhomogeneity scales both density and Young's modulus in the same way, so that $\rho(x) = E(x)$. The variables are nondimensionalized in the same way as in Sec. 2.1 so that $L = D = \left\|E/\frac{dE}{dx}\right\|$. Then the dimensionless equations of motion relating stress σ, strain λ and particle velocity u are

$$E(x)u_t = \sigma_x \text{ and } u_x = \lambda_t, \tag{7.13}$$
$$\sigma(\lambda, x) = E(x)\lambda(1 + N\lambda + O(\lambda^2))$$

and $E(0) = 1$. The initial and boundary conditions are

$$u(0, t) = Mh(\omega t) \text{ and } u(x, 0) = \lambda(x, 0) = 0, \ x > 0, \tag{7.14}$$

where $\omega = \frac{D}{c_0\tau_p} = \varepsilon^{-1} \gg 1$, the Mach number $M \ll 1$ and $|h| = O(1)$.

In all examples we make a change of variable (5.27) before substituting the expansion. Let $\alpha(x, t)$ be the 'fast' characteristic coordinate with $\alpha_t = O(\omega)$ and consider $u(x, t) = U(x, \alpha)$ and $\lambda(x, t) = \Lambda(x, \alpha)$. $T(x, \alpha)$ is the arrival time of the characteristic $\alpha(x, t) = $ constant at the location x with $\alpha = \omega t$ on $x = 0$. Since $\frac{d\alpha}{dx} = \alpha_x + \alpha_t T_x = 0$, the change of variable implies that

$$u_x = U_x + U_\alpha \alpha_x = U_x - \alpha_t T_x U_\alpha.$$

Then (7.13) becomes

$$U_\alpha + T_x \Lambda_\alpha [1 + 2N\lambda + \cdots] = [\Lambda_x + \Lambda(1 + \cdots)E'/E]/\alpha_t,$$

$$\Lambda_\alpha + T_x U_\alpha = U_x/\alpha_t \tag{7.15}$$

with the characteristic condition

$$\det \left\| \begin{matrix} 1 & [1 + 2N\lambda + \cdots]T_x \\ T_x & 1 \end{matrix} \right\| = 0,$$

so that the reciprocal of the wave speed, sometimes called the slowness, for a wave travelling to the right, is

$$T_x = 1 - N\lambda + \cdots . \tag{7.16}$$

The compatibility condition from (7.15) yields

$$T_x[\Lambda_x + \Lambda(1 + \cdots)E'/E] = U_x \tag{7.17}$$

which, together with the second of (7.15), form the new equations for U and Λ. The solution is represented in series form with x being the 'slow' variable and α the 'fast' variable:

$$U(x,\alpha) = M \sum_{n=0}^{\infty} \varepsilon^n U_n(x,\alpha), \ \ \Lambda(x,\alpha) = M \sum_{n=0}^{\infty} \varepsilon^n \Lambda_n(x,\alpha). \tag{7.18}$$

Since the nonlinear characteristic is unknown, we add a series representing the arrival time $T(x,\alpha)$ of the characteristic $\alpha(x,t) = $ constant at the location x:

$$T(x,\alpha) = \frac{\alpha}{\omega} + x + \sum_{n=0}^{\infty} \varepsilon^n M T_n(x,\alpha) \tag{7.19}$$

with $T_n(0,\alpha) = 0$ since $\alpha = \omega t$ on $x = 0$. Substituting the series (7.18) and (7.19) into (7.17) and the second of (7.15) respectively produces the $O(\varepsilon)$ equations

$$\Lambda_{0x} + \Lambda_0 E'/E = U_{0x}, \ \ \frac{\partial}{\partial \alpha}[U_0 + \Lambda_0] = 0$$

which yield

$$U_0 = -\Lambda_0 \quad \text{and} \quad 2\Lambda_{0x} + \Lambda_0 E'/E = 0. \tag{7.20}$$

With the boundary condition (7.14), $U_0(0,\alpha) = -\Lambda_0(0,\alpha) = h(\alpha)$, and, on integration of the second of (7.20), the zeroth order solution is, since $E(0) = 1$,

$$U = \frac{Mh(\alpha)}{E^{1/2}(x)}, \quad \Lambda = -\frac{Mh(\alpha)}{E^{1/2}(x)}. \tag{7.21}$$

The first correction to the characteristic, from (7.19) and (7.16), is

$$MT_{0x} = -N\Lambda_0 = NM\frac{h(\alpha)}{E^{1/2}(x)}.$$

This yields

$$T_0 = Nh(\alpha) \int_0^x \frac{dy}{E^{1/2}(y)},$$

so that the corrected nonlinear characteristic is

$$\alpha = \omega \left[t - x - MNh(\alpha) \int_0^x \frac{dy}{E^{1/2}(y)} \right]. \tag{7.22}$$

which, together with (7.21), forms the first-order nonlinear geometric acoustics solution.

Mortell and Seymour [2005] extended this solution outside the geometric acoustics regime for the special stratification (5.123) as used in Sec. 5.6.5 by substituting the **exact** linear solution into the corrected characteristic rather than the first term geometrical acoustics solution. This idea was used by Whitham [1974] in the context of spherical waves in a sphere of large radius, see Sec. 7.2.4.

With $E(x) = \frac{s(x)}{s_0}$, the full linear solution is (5.130) for $u(x,t)$ and

$$\lambda(x,t) = -\sqrt{\frac{s_0}{s(x)}}h(\alpha) - \frac{\varepsilon K(x)}{s(x)}H(\alpha) \tag{7.23}$$

for the strain $\lambda(x,t)$, where

$$H(\alpha) = \int_0^\alpha e^{\varepsilon\gamma(\alpha-y)} h(y)dy \quad \text{and} \quad K(x) = s_0\left(k_1\sqrt{s(x)} - \frac{l}{\sqrt{s_0}}\right),$$

see Mortell and Seymour [2005]. From (7.16) the first nonlinear correction to the right travelling characteristic is given implicitly by

$$\frac{\alpha}{\omega} = t - x + N\int_0^x \lambda(y,\alpha)dy, \tag{7.24}$$

where, now $\lambda(y,\alpha)$ is given by (7.23). Then (7.22) is replaced by

$$\frac{\alpha}{\omega} = t - x - Nh(\alpha)\int_0^x \sqrt{\frac{s_0}{s(y)}}dy - \varepsilon N \int_0^x \frac{K(y)}{s(y)}dy H(\alpha) \tag{7.25}$$

and $u(x,t)$ and $\lambda(x,t)$ are now given implicitly by (5.130) and (7.23) together with (7.25). In these expressions N is a measure of the nonlinearity and can be positive or negative. The nonlinear solutions combine amplitude changes due to the stratification, measured by the size of ε, and distortion due to the nonlinearity, measured by the size of N. Figure 7.1 shows the nonlinear distortion at $x = 6.5$ for the stratification (5.123) with the same boundary condition (5.131) and the same stratification parameters as Fig. 5.5 except $N = 0.02$. The dashed curve is the nonlinear solution derived from the exact linear solution, the solid curve is the one-term nonlinear geometric acoustics

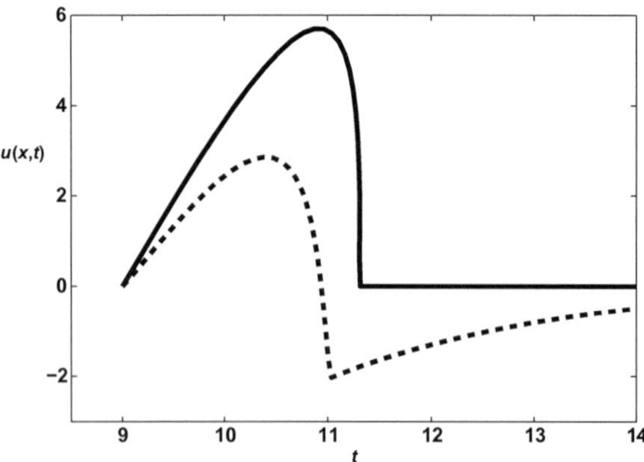

Fig. 7.1. $u(6.5,t)$ for $\varepsilon = 2.3$, $N = 0.02$.

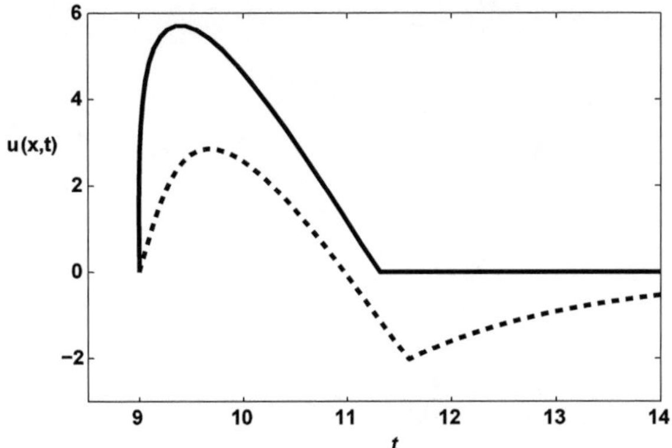

Fig. 7.2. $u(6.5, t)$ for $\varepsilon = 2.3$, $N = -0.02$.

solution. A shock is just forming at the back of the pulse, before the tail. For a material with $N < 0$ (like a gas) the shock will form at the front of the pulse, as illustrated in Fig. 7.2 with $N = -0.02$. Note that in this case the nonlinear geometric acoustics solution produces a similar distortion but overestimates the amplitude significantly and has no tail. The geometric acoustics solution preserves the pulse shape of the input, while the nonlinearization of the exact linear solution becomes negative before tending to zero. Solutions become multivalued and a shock forms when $\frac{\partial t}{\partial \alpha} = 0$. From (7.25)

$$t = t(x, \alpha) = \varepsilon \alpha + x + N h(\alpha) A(x) + \varepsilon N H(\alpha) B(x), \qquad (7.26)$$

where

$$A(x) = \int_0^x \sqrt{\frac{s_0}{s(y)}} dy, \quad B(x) = \int_0^x \frac{K(y)}{s(y)} dy. \qquad (7.27)$$

Defining $D(x, \alpha)$ by

$$D(x, \alpha) = \omega t_\alpha = 1 + \varepsilon^{-1} N h'(\alpha) A(x) + N(h(\alpha) + \varepsilon \gamma H(\alpha)) B(x) \qquad (7.28)$$

with $D(0, \alpha) = 1$, a shock forms when first

$$D(x, \alpha) = 0.$$

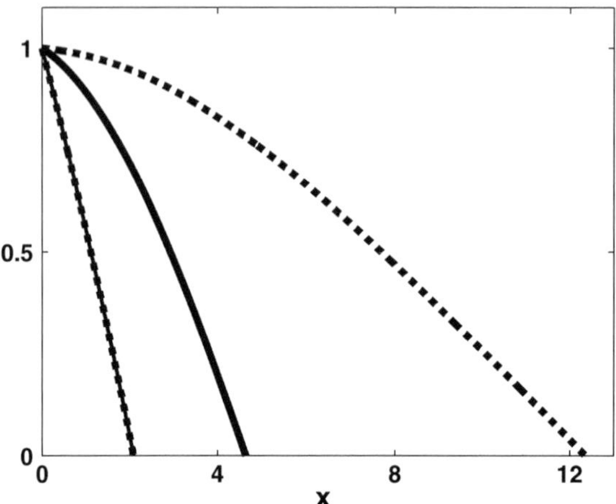

Fig. 7.3. $D(x, 1)$ (Dashed) and $D_G(x, 1)$ for the stratification (5.123).

Using the nonlinear geometrical acoustics solution, (7.28) is replaced by

$$D_G(x, \alpha) = 1 + \varepsilon^{-1} N h'(\alpha) A(x). \qquad (7.29)$$

To investigate shock formation predictions using the exact and geometrical acoustic approximation, we examine $D(x, \alpha)$ and $D_G(x, \alpha)$ for high and low frequencies. Since for $N > 0$ a shock forms at the back of the pulse, this will correspond to $\alpha = 1$. Figure 7.3 compares $D(x, 1)$ (dashed) and $D_G(x, 1)$ (solid) for the stratification function (5.123) shown in Fig. 7.3 with $N = 0.05$. There is a considerable difference between the exact and approximate solutions for the low frequency (large ε) case. A shock forms for $x \backsim 12$ for the low frequency ($\varepsilon = 2.3$) exact solution (dashed) but the geometrical acoustics approximation (solid) predicts a shock at $x \backsim 5$. The two superposed almost identical lower curves are for the high frequency case ($\varepsilon = 0.39$), showing shock formation at a value of $x \backsim 2$.

7.2.2. *Maxwell Solid*

The governing equations for the one-dimensional motion of a viscoelastic panel were derived in 7.1.2 as

$$u_t - \sigma_x = 0, \quad \lambda_t = u_x, \quad \sigma_t = \Gamma'(\lambda)\lambda_t - 2\sigma, \qquad (7.30)$$

where $\Gamma(\lambda) = \lambda + N\lambda^2 + O(\lambda^2)$ and u, σ and λ are particle velocity, stress and strain. The boundary condition is

$$u(0, t) = Mh(\omega t), \tag{7.31}$$

where $M \ll 1$, and $u = \lambda = 0$ at $t = 0$ for $x > 0$.

Again, we make the change of variable (5.27). Then (7.30) become

$$U_\alpha + T_x \Sigma_\alpha = \Sigma_x/\alpha_t, \ \ \Lambda_\alpha + T_x U_\alpha = U_x/\alpha_t, \ \ \Sigma_\alpha - \Gamma'(\Lambda)\Lambda_\alpha = -2\Sigma/\alpha_t \tag{7.32}$$

with the characteristic condition $T_x = \pm 1/\sqrt{\Gamma'(\Lambda)}$, so that the reciprocal of the wave speed for a wave traveling to the right is $T_x = 1 - N\lambda + \cdots$.

For a wave traveling to the right, $T_x = 1/C$, $C = \sqrt{\Gamma'(\Lambda)}$ and Eqs. (7.32) become

$$U_\alpha + \Sigma_\alpha/C = \Sigma_x/\alpha_t, \ \ \Lambda_\alpha + U_\alpha/C = U_x/\alpha_t,$$

$$\Sigma_\alpha - C^2 \Lambda_\alpha = -2\Sigma/\alpha_t. \tag{7.33}$$

To calculate the transport equation along a characteristic, first eliminate U_α from the first two of (7.33) to give

$$\Sigma_\alpha/C - C\Lambda_\alpha = \Sigma_x/\alpha_t - CU_x/\alpha_t, \ \ \Sigma_\alpha - C^2 \Lambda_\alpha = -2\Sigma/\alpha_t \tag{7.34}$$

which implies that

$$\Sigma_x - CU_x = -2\Sigma/C. \tag{7.35}$$

This transport equation, together with any two of (7.33), form the new equations for U, Σ and Λ. The solution is represented in series form with x being the 'slow' variable and α the 'fast' variable:

$$U(x, \alpha) = M \sum_{n=0}^{\infty} \varepsilon^n U_n(x, \alpha), \quad \Sigma(x, \alpha) = M \sum_{n=0}^{\infty} \varepsilon^n \Sigma_n(x, \alpha),$$

$$\Lambda(x, \alpha) = M \sum_{n=0}^{\infty} \varepsilon^n \Lambda_n(x, \alpha). \tag{7.36}$$

We add a series representing the arrival time $T(x, \alpha)$ of the characteristic $\alpha(x, t) = $ constant at the location x:

$$T(x, \alpha) = \frac{\alpha}{\omega} + x + \sum_{n=0}^{\infty} \varepsilon^n M T_n(x, \alpha) \qquad (7.37)$$

with $T_n(0, \alpha) = 0$ since $\alpha = \omega t$ on $x = 0$. Substituting the series (7.36) and (7.37) into (7.32) produces the $O(\varepsilon)$ equations

$$\Sigma_{0x} - U_{0x} = -2\Sigma_0, \quad \frac{\partial}{\partial \alpha}[U_0 + \Sigma_0] = 0, \quad \frac{\partial}{\partial \alpha}[\Sigma_0 - \Lambda_0] = 0 \qquad (7.38)$$

which yield

$$U_0 = -\Lambda_0 = -\Sigma_0, \quad \text{and} \quad \Sigma_{0x} + \Sigma_0 = 0. \qquad (7.39)$$

The boundary condition (7.31) implies $U_0(0, \alpha) = -\Lambda_0(0, \alpha) = h(\alpha)$ and on integration the zeroth-order solution is, since $E(0) = 1$,

$$U_0 = -\Lambda_0 = -\Sigma_0 = Mh(\alpha)e^{-x}, \quad T = \frac{\alpha}{\omega} + x. \qquad (7.40)$$

The first correction to the characteristic is found from (7.19) and (7.16)

$$MT_{0x} = -N\Lambda_0 = NMh(\alpha)e^{-x}. \qquad (7.41)$$

This yields

$$T_0 = Nh(\alpha) \int_0^x e^{-y} dy, \qquad (7.42)$$

so that the corrected nonlinear characteristic is

$$\alpha = \omega[t - x - MNh(\alpha)(1 - e^{-x})], \qquad (7.43)$$

in agreement with (7.8).

7.2.3. *Shallow Water Waves Over Variable Bottom*

Another example of the use of the nonlinear geometric acoustics expansion is the one-dimensional motion of a surface wave in shallow water propagating over a variable bottom $z = -H(x)$, with $H(0) = H_0$. The ambient wave speed is $c_0 = \sqrt{gH_0}$. If τ_p is the period of the surface wave, the water is "shallow" if the surface wavelength $(c_0\tau_p)$ is large compared with H_0, $H_0 \ll c_0\tau_p$. see Sec. 2.3.

We will consider a slowly varying bottom topography (the high frequency limit) so that if D is the length scale of the variation in $H(x)$ (typically $D = \min|\frac{H(x)}{H'(x)}|$), then $c_0\tau_p \ll D$. The water surface height above $z = 0$ is $\eta(x,t)$, while the bottom is at $z = -H(x)$, then the governing shallow water equations in Eulerian coordinates are

$$u_t + uu_x + g\eta_x = 0, \; h_t + (hu)_x = 0, \tag{7.44}$$

where $u(x,t)$ is the particle velocity, and $h(x,t) = H(x) + \eta(x,t)$ is the total water depth.

A typical boundary condition generating a surface wave traveling into $x > 0$ is

$$\eta(0,t) = af\left(\frac{t}{\tau_p}\right), \tag{7.45}$$

where a is the wave amplitude and $f = O(1)$ is dimensionless.

Velocity and water depth are nondimensionalized w.r.t. (c_0, H_0), and (u, η) are considered as functions of length and time respectively $(Dx, Dc_0^{-1}t)$. Picking $L = D$ the dimensionless equations are:

$$u_t + uu_x + \eta_x = 0, \; \eta_t + ([H + \eta]u)_x = 0 \tag{7.46}$$

with the boundary condition

$$\eta(0,t) = Mf(\omega t), \;\; |f| = O(1), \tag{7.47}$$

where $M = a/H_0 \ll 1$ and $\omega = D/c_0\tau_p \gg 1$. The boundary condition defines the 'fast' variable ωt on $x = 0$ and so we 'tag' the characteristic $\alpha(x,t) = $ constant leaving $x = 0$ at time t by $\alpha(0,t) = \omega t$. Using the notation $u(x,t) = U(x,\alpha)$ and $\eta(x,t) = E(x,\alpha)$, we make the change of variable (6.33). Then (7.46) become

$$U_\alpha(1 - UT_x) - T_x E_\alpha = -(UU_x + E_x)/\alpha_t \tag{7.48}$$

$$E_\alpha(1 - UT_x) - T_x(H + E)U_\alpha = -(U(H + E))_x/\alpha_t. \tag{7.49}$$

We note that $\alpha_t = O(\omega)$ so that the right-hand side of (7.48) and (7.49) are $O(\omega^{-1})$. The characteristic condition is

$$\begin{vmatrix} (1 - UT_x) & -T_x \\ -T_x(H + E) & (1 - UT_x) \end{vmatrix} = 0, \tag{7.50}$$

giving the nonlinear wave speed c as

$$c = 1/T_x = U \pm \sqrt{H + E}. \tag{7.51}$$

For a wave traveling to the right, with $\varepsilon = \omega^{-1}$,

$$T_x = (U + \sqrt{H + E})^{-1} = \frac{1}{\sqrt{H(x)}}\left[1 - \frac{U}{\sqrt{H(x)}} - \frac{E}{2H} + O(\varepsilon^2)\right]. \tag{7.52}$$

The compatibility condition from (7.48) and (7.49) yields

$$(UT_x - 1)(UU_x + E_x) = T_x(U(H + E))_x \tag{7.53}$$

and hence U, E and T are determined from (7.53), (7.52) and one of (7.48) and (7.49).

The solution to (7.46) can be represented in series form with x being the 'slow' variable and α the 'fast' variable:

$$u(x, t) = U(x, \alpha) = M \sum_{n=0}^{\infty} \varepsilon^n U_n(x, \alpha),$$

$$\eta(x, t) = E(x, \alpha) = M \sum_{n=0}^{\infty} \varepsilon^n E_n(x, \alpha), \tag{7.54}$$

where $\varepsilon = \omega^{-1}$.

Since the nonlinear characteristic is unknown, we add a series representing the arrival time $T(x, \alpha)$ of the characteristic $\alpha(x, t) = $ constant at the location x:

$$T(x, \alpha) = \frac{\alpha}{\omega} + \int_0^x \frac{dy}{\sqrt{H(y)}} + M \sum_{n=0}^{\infty} \varepsilon^n T_n(x, \alpha), \tag{7.55}$$

using (7.52) and with $T_n(0, \alpha) = 0$. Substituting the series (7.54) and (7.55) into (7.53) and (7.48) respectively produces the $O(\varepsilon)$ equations

$$\frac{\partial}{\partial \alpha}\left[U_0 - \frac{1}{\sqrt{H(x)}}E_0\right] = 0, \quad \sqrt{H(x)}E_{0x} + (U_0 H)_x = 0,$$

which yield

$$U_0 = \frac{1}{\sqrt{H(x)}}E_0 \quad \text{and} \quad 2\sqrt{H(x)}E_{0x} + E_0\left(\sqrt{H(x)}\right)_x = 0. \quad (7.56)$$

With the boundary condition (7.47), $E_0(0, \alpha) = f(\alpha)$, and $H(0) = 1$, on integration (7.56) give the zeroth-order solution

$$E_0 = \frac{f(\alpha)}{H^{1/4}(x)}, \quad U_0 = \frac{f(\alpha)}{H^{3/4}(x)}, \quad T = \frac{\alpha}{\omega} + \int_0^x \frac{dy}{\sqrt{H(y)}}. \quad (7.57)$$

Equations (7.57) are the first terms in the linear geometric acoustics solution. The first correction to the characteristic from (7.52) is

$$T_x = \frac{1}{\sqrt{H(x)}} - \varepsilon\frac{U_0}{H(x)} - \varepsilon\frac{E_0}{2H^{3/2}(x)} + O(\varepsilon^2). \quad (7.58)$$

Using (7.57) this yields

$$T_x = \frac{1}{\sqrt{H(x)}} - \frac{3}{2}\frac{\varepsilon f(\alpha)}{H^{7/4}(x)} + O(\varepsilon^2), \quad (7.59)$$

so that

$$T_1 = -\frac{3}{2}f(\alpha)\int_0^x \frac{dy}{H^{7/4}(y)} \quad (7.60)$$

and the corrected characteristic is

$$T(x, \alpha) = \frac{\alpha}{\omega} + \int_0^x \frac{dy}{\sqrt{H(y)}} - \frac{3}{2}Mf(\alpha)\int_0^x \frac{dy}{H^{7/4}(y)} + O(\varepsilon^2). \quad (7.61)$$

The representations for E_0 and U_0 in (7.57) together with (7.61) give the first term in the nonlinear geometric acoustics solution.

7.2.4. *Pulsating Sphere of Large Radius*

We consider the extension of the problem considered in Sec. 5.5 of small-amplitude, high-frequency sound waves generated by a pulsating sphere to include nonlinear effects. The surface of the sphere of radius R_0 pulsates periodically with period τ_p, and c_0 is the ambient sound speed so that in the geometric acoustics limit, $\varepsilon = \frac{c_0 \tau_p}{R_0} = \omega^{-1} \ll 1$. As in Sec. 4.3.1, we nondimensionalize by picking $L = R_0$ and $T = R_0/c_0$, where $c_0 = \sqrt{\gamma p_0/\rho_0}$, then the governing equations are (4.23):

$$u_t + uu_r + k^{-1}cc_r = 0, \quad c_t + uc_r + kc(u_r + 2u/r) = 0, \quad r > 1,$$

$$(7.62)$$

where $u(r,t)$ is the particle velocity and $c(r,t) = 1 + a(r,t)$ is the (dimensionless) sound speed and $k = (\gamma - 1)/2$. The boundary condition on $R = 1$ is

$$u(1,t) = Mh(\omega t), \tag{7.63}$$

where $M = u_0/c_0$ and u_0 is the amplitude of the applied velocity. For $r > 1$ the gas is at rest at $t = 0$.

As in the previous example we introduce the characteristic coordinate $\alpha(r,t)$ and consider $u(r,t) = U(r,\alpha)$, $a(r,t) = A(r,\alpha)$. The change of variable implies that

$$u_r = U_r + U_\alpha \alpha_r = U_r - \alpha_t T_r U_\alpha. \tag{7.64}$$

Then (7.62) become

$$U_\alpha(1 - UT_r) - T_r k^{-1}(1 + A)A_\alpha$$

$$= -(UU_r + k^{-1}(1 + A)A_r)\frac{1}{\alpha_t}, \tag{7.65}$$

$$A_\alpha(1 - UT_r) - kT_r(1 + A)U_\alpha$$

$$= -\left(UA_r + k(1 + A)\left[U_r + 2\frac{U}{r}\right]\right)\frac{1}{\alpha_t}. \tag{7.66}$$

We note that $\alpha_t = O(\omega)$ so that the right-hand side of (7.65) and (7.66) are $O(\omega^{-1})$. The characteristic condition is determined by rewriting (7.65) and (7.66) as

$$(1 - UT_r)U_\alpha - T_r k^{-1}(1 + A)A_\alpha = O(\omega^{-1}) \qquad (7.67)$$

$$kT_r(1 + A)U_\alpha - (1 - UT_r)A_\alpha = O(\omega^{-1}), \qquad (7.68)$$

giving

$$\begin{vmatrix} (1 - UT_r) & -T_r k^{-1}(1 + A) \\ kT_r(1 + A) & -(1 - UT_r) \end{vmatrix} = 0, \ \text{i.e.,} \ 1/T_r = U \pm (1 + A). \qquad (7.69)$$

For a wave traveling to the right,

$$T_r = (U + (1 + A))^{-1}. \qquad (7.70)$$

The transport equation (equating the right-hand sides from (7.65) and (7.66)) yields

$$k^{-1}A_r + U_r + 2(1 - U(1 + U + A)^{-1})U/r = 0 \qquad (7.71)$$

which, with one of (7.65) and (7.66), form the exact new equations for U and A. Substituting series like (7.18) and (7.19) for $U(x, \alpha)$, $A(x, \alpha)$ and $T(x, \alpha)$ in powers of $\varepsilon = \omega^{-1}$ into (7.65) and (7.71) yields at $O(\varepsilon)$

$$U_0 = k^{-1}A_0, \ \ k^{-1}A_{0r} + U_{0r} + 2U_0/r = 0. \qquad (7.72)$$

Equation (7.72) with (7.63) implies that

$$U_{0r} + \frac{U_0}{r} = 0, \ \text{or} \ U_0 = k^{-1}A_0 = \frac{1}{r}Mh(\alpha), \qquad (7.73)$$

confirming (5.49). The first-order nonlinear geometrical acoustics approximation is completed by calculating the characteristic. Substituting (7.73) into (7.70) gives

$$T_{0r} = 1 - U_0 - A_0 = 1 - (1 + k)U_0 \qquad (7.74)$$

which integrates to give

$$T = \frac{\alpha}{\omega} + r - 1 - (1 + k)Mh(\alpha)\ln(r) \qquad (7.75)$$

with $\alpha = \omega T$ on $r = 1$. Note that using the full linear solution (5.50) in (7.74) gives

$$\alpha = \omega \left[t - r + 1 + (1 + k)Mh(\alpha)\ln(r) - r^{-1} \int^{\alpha} Mh(\alpha) \right] \quad (7.76)$$

which agrees with Whitham [1974] with $h = g'$.

7.2.5. *Pulsating Cylinder of Large Radius*

The dimensionless governing equations in Eulerian coordinates are (4.36)

$$u_t + uu_r + (1 + e)^{\gamma-2}e_r = 0, \;\; e_t + ue_r + (1 + e)u_r + u/r = 0, \quad (7.77)$$

where $u(r,t)$ is the particle velocity, and $e(r,t)$ the condensation. The boundary condition is (7.63), while for $r > 1$ the gas is at rest at $t = 0$. Again using the notation $u(r,t) = U(r,\alpha)$ and the change of variable (7.64), (7.77) become

$$U_\alpha(1 - UT_r) - T_r(1 + E)^{\gamma-2}E_\alpha = -(UU_r + (1 + E)^{\gamma-2}E_r)/\alpha_t \quad (7.78)$$

$$E_\alpha(1 - UT_r) - T_r(1 + E)U_\alpha = -(UE_r + (1 + E)U_r + U/r)/\alpha_t. \quad (7.79)$$

The characteristic condition is determined by rewriting (7.78) and (7.79) as

$$(1 - UT_r)U_\alpha - T_r(1 + E)^{\gamma-2}E_\alpha = O(\omega^{-1}), \quad (7.80)$$

$$T_r(1 + E)U_\alpha - (1 - UT_r)E_\alpha = O(\omega^{-1}), \quad (7.81)$$

giving

$$\begin{vmatrix} (1 - UT_r) & -T_r(1 + E)^{\gamma-2} \\ T_r(1 + E) & -(1 - UT_r) \end{vmatrix} = 0$$

or

$$1/T_r = U \pm (1 + E)^{(\gamma-1)/2} = U \pm C,$$

where $C = (1 + E)^{(\gamma-1)/2}$. For a wave traveling to the right, $T_r = (U + C)^{-1}$, and, as in Sec. 5.5, to complete the first order nonlinear solution we use the linear solution (5.52) $U = \frac{1}{\sqrt{r}}Mh(\alpha) = E$ to correct the characteristic

$$T_r = 1 - U - \frac{\gamma - 1}{2}E = 1 - \frac{(\gamma + 1)}{2\sqrt{r}}Mh(\alpha).$$

Integrating gives the first-order nonlinear characteristic

$$T = \frac{\alpha}{\omega} + r - 1 - (\gamma + 1)Mh(\alpha)(\sqrt{r} - 1), \qquad (7.82)$$

where $T = \frac{\alpha}{\omega}$ on $r = 1$.

7.2.6. *Flow with Cross Section:* $s(x) = (1 + ax/D)^{-2}$

In this example there is an exact solution to the linear equation, (5.118) and (5.119) as given in Sec. 5.6.4 with $a = D^{-1}$. Here we examine the corresponding nonlinear problem in the geometric acoustics limit and make use of the known linear solution.

Length and time are nondimensionalized in the manner described in Sec. 4.1 using $L = D$ and $T = Dc_0^{-1}$. Then the equations governing the motion of a polytropic contained in a tube with the dimensionless cross-sectional area $s(x) = (1+ax)^{-2}$ ($x > 0$) and excited by a piston at $x = 0$ are approximated by (2.17) with $c_0 = 1$.

The boundary condition is

$$u(0, t) = Mh(\omega t), \qquad (7.83)$$

where $M = u_0/c_0 \ll 1$, $\omega = D/(c_0 T_p) \gg 1$ and $u = e = 0$ at $t = 0$ for $x > 0$.

Again we make the change of variable (5.27). Then, with $u(x, t) = U(x, \alpha)$ and $e(x, t) = E(x, \alpha)$, Eqs. (2.17) become

$$sE_\alpha - T_x[(sU)_\alpha + (sUE)_\alpha] = -[(sU)_x + (sUE)_x]/\alpha_t, \qquad (7.84)$$

$$U_\alpha - T_x[E_\alpha + UU_\alpha + EE_\alpha(\gamma - 2)] = [E_x + UU_x + EE_x(\gamma - 2)]/\alpha_t. \qquad (7.85)$$

To first order (on noting that $\alpha_t = O(\omega^{-1})$), (7.84) and (7.85) become

$$E_\alpha(1 - UT_x) - U_\alpha(1 + E)T_x = 0,$$
$$U_\alpha(1 - UT_x) - E_\alpha(1 + (\gamma - 2)E)T_x = 0$$

with the characteristic condition

$$1 - UT_x = \pm T_x[1 + E(\gamma - 1)/2 + O(E^2)]. \tag{7.86}$$

Then the reciprocal of the wave speed for a wave traveling to the right is

$$T_x = 1 - U - E(\gamma - 1)/2 + O(E^2) + \cdots . \tag{7.87}$$

From Case 2 in Sec. 5.6.4 with $\alpha = \omega t$ on $x = 0$, and the boundary condition (7.83), the *exact* linear solution is

$$U = (1 + ax)Mh(\alpha), \quad E = (1 + ax)Mh(\alpha) + Ma \int^\alpha h(s)\, ds. \tag{7.88}$$

The solution (7.88) is, in this special case, the two-term linear geometrical acoustics solution, see Sec. 5.4. Then the corrected nonlinear characteristic for the α-wave is, by integrating (7.87) using (7.88), (with $N = -\frac{\gamma+1}{2}$)

$$t = \frac{\alpha}{\omega} + x + (2a)^{-1}N[(1 + ax)^2 - 1]Mh(\alpha)$$
$$+ (N + 1)xaM \int_0^\alpha h(s)\, ds. \tag{7.89}$$

A shock forms when $\frac{\partial u}{\partial x} \to \infty$, i.e., when $|\alpha_x| \to \infty$ or $|x_\alpha| \to 0$, hence from differentiating (7.89), the shock condition is

$$0 = 1 + \omega(2a)^{-1}N[(1 + ax)^2 - 1]Mh'(\alpha) + \omega(N + 1)xaMh(\alpha). \tag{7.90}$$

As a check, taking the limit as $a \to 0$, the shock condition (7.90) becomes

$$0 = 1 + \omega x N M h'(\alpha)$$

consistent with (6.48).

7.3. Multiple Scale Expansion: Elastic Panel

Here we illustrate how the method of multiple scales can be used as a systematic procedure to calculate a nonlinear geometric acoustics expansion. The example we use is the forced oscillations of a semi-infinite elastic panel, $x \geq 0$, initially at rest where the inhomogeneity is introduced through a nonuniform Young's modulus, $E(x/D)$. $D = \|\frac{E}{E'}\|$ is a measure of the variation in the material properties and has the dimension of length. The density of the panel ρ is a constant. The elastic material of the panel is described by the nonlinear stress–strain law

$$\sigma(\lambda, x) = E(x/D)(\lambda + N\lambda^2 + O(\lambda^3)), \qquad (7.91)$$

where $\sigma(\lambda, x)$ is the Piola–Kirchoff stress, $E(0) = E_0$ is a constant. Here the strain $\lambda(x,t) = \frac{\partial \xi}{\partial x}$, where $\xi(x,t)$ is the particle displacement, x is the Lagrangian position of a particle, and N is a dimensionless material constant, the ratio of second to first order elastic constant. Then the equations of motion relating λ and the particle velocity u are, as in (6.28),

$$\rho u_t = \sigma_x \text{ and } u_x = \lambda_t. \qquad (7.92)$$

The initial and boundary conditions are

$$u(0,t) = u_0 h(t/\tau_p) \text{ and } u(x,0) = \lambda(x,0) = 0, \qquad (7.93)$$

where u_0 is the constant input amplitude so that $|h| = O(1)$ and τ_p is the duration or period of the input. The linear wave speed at $x = 0$ is $c_0 = \sqrt{\frac{E_0}{\rho}}$ and the applied wavelength is $\xi = c_0 \tau_p$.

To use the method of multiple scales we need to nondimensionalize in a way that produces a slow variable in the coefficients and not in the boundary condition. As described in Sec. 4.1 we do this by picking $L = \xi$ and $T = \tau_p$, then there is no parameter in the boundary condition, but the stratification function is scaled as $E(\varepsilon x)$, where $\varepsilon = \xi/D \ll 1$ and E is considered a slowly varying function of $y = \varepsilon x$.

u is scaled with c_0 and σ with E_0, then Eqs. (7.91)–(7.93) become

$$\sigma(\lambda, x) = E(y)(\lambda + N\lambda^2 + O(\lambda^3)) \tag{7.94}$$

$$u_t = \sigma_x \quad \text{and} \quad u_x = \lambda_t \tag{7.95}$$

and

$$u(0, t) = Mh(t), \tag{7.96}$$

where the Mach number $M = u_0/c_0 \ll 1$. We anticipate that for a balance between nonlinearity and stratification, $M = O(\varepsilon)$, so we take $M = \varepsilon$.

In the method of multiple scales, we consider $y = \varepsilon x$ as a new, *additional*, independent variable, so write $E = E(y)$ and $u = u(t, x, y)$, $\sigma = \sigma(t, x, y)$. Also, from Sec. 5.3 we know that the linear characteristic variables are

$$\alpha = t - \int_0^x \frac{dz}{c(\varepsilon z)} \quad \text{and} \quad \beta = t + \int_0^x \frac{dz}{c(\varepsilon z)},$$

where $c^2 = E(y)$, so we assume a series solution of the form

$$u(x, t; \varepsilon) = \varepsilon U_1(\alpha, \beta, y) + \varepsilon^2 U_2(\alpha, \beta, y) + \cdots ,$$

$$\lambda(x, t) = \varepsilon \Lambda_1(\alpha, \beta, y) + \varepsilon^2 \Lambda_2(\alpha, \beta, y) + \cdots \tag{7.97}$$

Then

$$\frac{\partial}{\partial t}\big|_x = \frac{\partial}{\partial \alpha} + \frac{\partial}{\partial \beta} \quad \text{and} \quad \frac{\partial}{\partial x}\big|_t = -\frac{1}{c}\left(\frac{\partial}{\partial \alpha} - \frac{\partial}{\partial \beta}\right) + \varepsilon\frac{\partial}{\partial y},$$

so that from (7.95) and (7.97) at $O(\varepsilon)$ we get

$$\left(\frac{\partial}{\partial \alpha} + \frac{\partial}{\partial \beta}\right) U_1 = \frac{E}{c}\left(\frac{\partial}{\partial \beta} - \frac{\partial}{\partial \alpha}\right) \Lambda_1 = c\left(\frac{\partial}{\partial \beta} - \frac{\partial}{\partial \alpha}\right) \Lambda_1,$$

$$\tag{7.98}$$

$$c\left(\frac{\partial}{\partial \alpha} + \frac{\partial}{\partial \beta}\right) \Lambda_1 = \left(\frac{\partial}{\partial \beta} - \frac{\partial}{\partial \alpha}\right) U_1. \tag{7.99}$$

Then (7.98) and (7.99) imply

$$\frac{\partial}{\partial \beta}(U_1 - c\Lambda_1) = 0, \quad \frac{\partial}{\partial \alpha}(U_1 + c\Lambda_1) = 0$$

and hence

$$U_1 - c\Lambda_1 = 2cH(\alpha, y), \ \ U_1 + c\Lambda_1 = 2cG(\beta, y).$$

For right travelling waves $G \equiv 0$, and then

$$U_1 = -c\Lambda_1 = cH(\alpha, y). \tag{7.100}$$

At $O(\varepsilon^2)$,

$$\left(\frac{\partial}{\partial\alpha} + \frac{\partial}{\partial\beta}\right) U_2 - c\left(\frac{\partial}{\partial\beta} - \frac{\partial}{\partial\alpha}\right)\Lambda_2$$

$$= Nc\left(\frac{\partial}{\partial\beta} - \frac{\partial}{\partial\alpha}\right)\Lambda_1^2 + \frac{\partial}{\partial y}(E\Lambda_1), \tag{7.101}$$

$$\left(\frac{\partial}{\partial\beta} - \frac{\partial}{\partial\alpha}\right) U_2 - c\left(\frac{\partial}{\partial\alpha} + \frac{\partial}{\partial\beta}\right)\Lambda_2 = -c\frac{\partial U_1}{\partial y}. \tag{7.102}$$

Then (7.101) and (7.102) together with (7.100) imply

$$2\frac{\partial}{\partial\beta}(U_2 - c\Lambda_2) = -Nc\frac{\partial}{\partial\alpha}(H^2) - \frac{\partial}{\partial y}(EH) - c\frac{\partial}{\partial y}(cH), \tag{7.103}$$

$$2\frac{\partial}{\partial\alpha}(U_2 + c\Lambda_2) = -Nc\frac{\partial}{\partial\alpha}(H^2) - \frac{\partial}{\partial y}(EH) - c\frac{\partial}{\partial y}(cH). \tag{7.104}$$

The right-hand side of (7.103) is independent of β, and integration w.r.t. β produces a β-secular term, so we set the right-hand side to zero:

$$Nc\frac{\partial}{\partial\alpha}(H^2) + \frac{\partial}{\partial y}(EH) + c\frac{\partial}{\partial y}(cH) = 0$$

or

$$\frac{N}{c}HH_\alpha + H_y + \frac{3}{2}\frac{c'}{c}H = 0. \tag{7.105}$$

Now define the curve $\gamma(\alpha, y) = $ constant by

$$\frac{d}{dy}\Big|_\gamma = \frac{\partial}{\partial y} + \frac{\partial\alpha}{\partial y}\frac{\partial}{\partial\alpha}$$

and picking $\frac{\partial \alpha}{\partial y}|_\gamma = \frac{N}{c}H$ with the notation $H(\alpha, y) = F(\gamma, y)$, then (7.105) becomes

$$F_y + \frac{3}{2}\frac{c'}{c}F = 0. \tag{7.106}$$

Hence, integrating (7.106),

$$F(\gamma, y) = c^{-3/2}G(\gamma),$$
$$U_1 = -c\Lambda_1 = cH(\alpha, y) = c^{-1/2}G(\gamma). \tag{7.107}$$

At the boundary, $y = 0$, pick $\alpha = \gamma = t$, then the boundary condition $u(0,t) = \varepsilon h(t)$ implies $U_1(\alpha, 0) = h(\gamma)$, so that $G(\gamma) = h(\gamma)$. Now integrate

$$\frac{\partial \alpha}{\partial y}\Big|_\gamma = \frac{N}{c}H = Nc^{-5/2}h(\gamma),$$

to give

$$\alpha = \gamma + Nh(\gamma)\int_0^y c^{-5/2}dz.$$

Hence, the corrected nonlinear characteristic is

$$\begin{aligned}
t &= \gamma + \int_0^x \frac{dr}{c(\varepsilon r)} + \varepsilon Nh(\gamma)\int_0^x c(\varepsilon r)^{-5/2}dr \\
&= \gamma + \int_0^x \frac{dr}{\sqrt{E(\varepsilon r)}} + \varepsilon Nh(\gamma)\int_0^x E(\varepsilon r)^{-5/4}dr, \tag{7.108}
\end{aligned}$$

where $E(\varepsilon x)$ is the dimensionless Young's modulus and $c^2 = E$. The first term in the nonlinear geometric acoustics expansion for the velocity $u(x,t)$ is given by (7.107) and (7.108).

7.3.1. *Nonlinear Characteristic from Laminates*

We derived the first term in the linear theory of geometric acoustics in Sec. 5.3 by approximating a slowly varying elastic medium by a series of thin elastic laminates, each with constant material properties, then taking the limit as the width of the laminates tends to zero and the number of laminates tends to infinity. We also know that the amplitude variation in the signal is given to a first approximation by

linear theory, so all that remains to be calculated is the travel time
or nonlinear characteristic in each laminate.

For small strains $|\lambda| \ll 1$, we assume that the sound speed in the
laminate $x_{r-1} \ll x \ll x_r$ may be approximated by

$$c(\lambda_r) = c_r(1 + N_r\,\lambda_r + \cdots), \qquad (7.109)$$

where $c_r = \sqrt{\frac{E_r}{\rho_r}}$ and N_r is the ratio of second- to first-order elastic
constant. In this section, we use dimensional variables.

We now assume that in each laminate the reflection from the
interface is negligible, and thus the motion is effectively a simple
wave. For the laminate $x_{r-1} \ll x \ll x_r$

$$\frac{dx}{dt}\big|_{\alpha_r} = c_r(1 + N_r\lambda_r + \cdots)$$

and $\lambda_r = c_r^{-1}g_r(\alpha_r)$. This follows from $u = -g_r(\alpha_r) = -a$ for a wave
to the right, $f \equiv 0$. Then $a(\lambda) = \int_0^\lambda c(y)dy$, so that $a_r(\lambda_r) = c_r\lambda_r$,
since c_r is constant in a laminate. Integration yields

$$c_r(t - \alpha_r) = [1 + N_r c_r^{-1}g_r(\alpha_r) + \cdots](x - x_{r-1}),$$

where $\alpha_r = t$ on $x = x_{r-1}$, i.e., α_r is the time the wave leaves the
boundary at $x = x_{r-1}$. The arrival time of this wave at the right-hand
boundary $x = x_r$ is $t = \alpha_{r+1}$, and then

$$\alpha_{r+1} - \alpha_r = l_r c_r^{-1} + l_r N_r c_r^{-2}g_r(\alpha_r),$$

where $l_r = x_r - x_{r-1}$. Then

$$\alpha_{N+1} - \alpha_1 = \sum_{r=1}^{N}(\alpha_{r+1} - \alpha_r) = \sum_{r=1}^{N} l_r c_r^{-1} + g(\alpha_1)\sum_{r=1}^{N} l_r N_r c_r^{-2}K_{r-1},$$

since $g_{r+1}(\alpha_{r+1}) = K_r g(\alpha_1)$, where $K_r = \prod_{j=1}^{r} T_j = \prod_{j=1}^{r} \frac{2i_j}{i_j+i_{j+1}}$,
see Sec. 5.3.

As $N \to \infty$, $l_n \to 0$ such that $x = \sum_{n=1}^{N} l_n$ remains fixed, then
$K_N = \prod_{j=1}^{N} \frac{2i_j}{i_j+i_{j+1}} \to \sqrt{\frac{i(0)}{i(x)}}$ and $u_{N+1} = \sqrt{\frac{i(0)}{i(x)}}u_1$, see (5.14), where

$u_1 = f(t)$ on $x = 0$. Hence $u(x,t) \to \sqrt{\frac{i(0)}{i(x)}} f(\alpha)$. Also,

$$\alpha = t - \sum_{r=1}^{N} l_r c_r^{-1} + g(\alpha) \sum_{r=1}^{N} l_r N_r c_r^{-2} K_{r-1},$$

where $\alpha_1 \equiv \alpha$ and $\alpha_{N+1} \equiv t$, i.e.,

$$\alpha \to t - \int_0^x \frac{ds}{c(s)} - f(\alpha) \int_0^x \frac{N(s)}{c^2(s)} \sqrt{\frac{i(0)}{i(s)}} ds, \quad f(\alpha) = -g(\alpha).$$

$$(7.110)$$

The first term in the nonlinear geometric acoustics expansion for an inhomogeneous material is given by (7.110). We refer to the solution (7.110) as a modulated, small amplitude, simple wave, since the simple wave is recovered when the material properties are uniform. The generalizations to a uniform and an inhomogeneous viscoelastic medium are given in Mortell and Seymour [1976].

7.4. Large Amplitude Modulated Simple Waves

Simple wave solutions can be used to approximate finite amplitude waves when a high frequency mechanism is present. The wave is approximated locally by a simple wave that is modulated by the high frequency mechanism. Here we consider two examples: longitudinal waves in an elastic panel and surface gravity waves.

7.4.1. *Inhomogeneous Elastic Panel*

This first example is formulated to keep the calculation simple as the inhomogeneity appears through one dimensionless stratification function, $E(x)$, that scales both density and Young's modulus, as outlined in Sec. 7.2. Without this simplification the algebra is complicated and analytic integrations difficult.

As in Sec. 7.2, the variables are nondimensionalized so that $L = D = \|E/E'\|$. Then the dimensionless equations of motion relating stress σ, strain λ and particle velocity u are

$$E(x)u_t = \sigma_x \quad \text{and} \quad u_x = \lambda_t, \tag{7.111}$$

$$\sigma(\lambda, x) = E(x)(\lambda + N\lambda^2 + O(\lambda^3)) = E(x)Y(\lambda) \tag{7.112}$$

with $Y(\lambda) = \lambda + N\lambda^2 + O(\lambda^3)$ and $E(0) = 1$. The initial and boundary conditions are

$$u(0, t) = Mh(\omega t) \text{ and } u(x, 0) = \lambda(x, 0) = 0, \; x > 0,$$

where $\omega = \frac{D}{c_0 \tau_p} = \varepsilon^{-1} \gg 1$, and Mach number M can be $O(1)$.

Using the notation $u(x, t) = U(x, \alpha)$, $\lambda(x, t) = \Lambda(x, \alpha)$, where $T(x, \alpha)$ is the arrival time of the characteristic $\alpha(x, t) = \text{constant}$ at the location x, we make the change of variable (5.27). Then, defining $p = 1/\alpha_t$ as the incremental arrival time, (7.111) become

$$U_\alpha + T_x Y' \Lambda_\alpha = pY'\Lambda_x + pYE'/E, \; \Lambda_\alpha + T_x U_\alpha = pU_x. \tag{7.113}$$

The characteristic condition is

$$\begin{vmatrix} 1 & Y'T_x \\ T_x & 1 \end{vmatrix} = 0, \tag{7.114}$$

determining the nonlinear wave speed $c(\Lambda)$ as

$$c(\Lambda) = 1/T_x = \pm\sqrt{Y'(\Lambda)}. \tag{7.115}$$

For a wave traveling to the right,

$$T_x = \frac{1}{\sqrt{Y'(\Lambda)}}. \tag{7.116}$$

Defining the right (α) and left (β) traveling characteristics by

$$\frac{dT}{dx}\Big|_\alpha = \frac{1}{c(\Lambda)}, \; \frac{dT}{dx}\Big|_\beta = -\frac{1}{c(\Lambda)} \tag{7.117}$$

and the new stress measure $A(\Lambda) = \int_0^\Lambda c(e)de$, the Eqs. (7.113) become

$$U_\alpha + A_\alpha = pcA_x + pYE'/E, \; U_\alpha + A_\alpha = pcU_x \tag{7.118}$$

for the α wave, and

$$U_\beta - A_\beta = pcA_x + pYE'/E, \; U_\beta - A_\beta = -pcU_x \tag{7.119}$$

for the β wave. Now assume that $E(x) \equiv$ constant, so that $E'(x) = 0$, and U and A depend on only one characteristic variable, so $A_x = U_x = 0$, then (7.118) and (7.119) imply that

$$U + A = 2G(\beta) \text{ on } \frac{dT}{dx}\big|_\alpha = \frac{1}{c(\Lambda)}, \quad \text{and}$$

$$U - A = 2F(\alpha) \text{ on } \frac{dT}{dx}\big|_\beta = -\frac{1}{c(\Lambda)}.$$

$U + A$ and $U - A$ are the **Riemann Invariants**. They are constant along $\alpha(x,t) = $ constant and $\beta(x,t) = $ constant respectively. Then

$$U = F(\alpha) + G(\beta), \quad A = G(\beta) - F(\alpha) \tag{7.120}$$

with no restriction on the amplitude. Consider the initial-boundary value problem with

$$u(0,t) = Mh(t), \quad u(x,0) = \lambda(x,0) = 0, \tag{7.121}$$

so that along $t = 0$, $x > 0$, $U = A = G = F = 0$. This implies that in the region $x > 0$, $t > 0$ the left-traveling signal carries zero signal, i.e., $G(\beta) = 0$, and hence from (7.120) $U = -A = F(\alpha)$. If we tag the characteristic leaving $x = 0$ by $\alpha = t$, then integration of (7.117), with the conditions (7.121), implies that

$$t = \frac{x}{c(\Lambda)} + \alpha, \text{ and then } U(\alpha) = -A(\alpha) = Mh(\alpha), \tag{7.122}$$

since c is a function of α only. (7.122) is the **exact simple wave** solution for a wave traveling to the right into an undisturbed region. It does not require the assumption of small amplitude.

To examine a finite amplitude wave when $E'(x)$ is **small,** we now consider just the α wave traveling to the right and consider U and A to be functions of x and α in terms of new dependent variables $F(x, \alpha)$ and $G(x, \alpha)$. Using (7.120) with, for example $U = F(x, \alpha) + G(x, \alpha)$, we rewrite (7.118) as

$$2G_\alpha - pYE'/E = pc(G_x - F_x), \quad 2G_\alpha = pc(F_x + G_x)$$

or, subtracting and adding these, respectively,

$$F_x = \frac{Y}{2c}\frac{E'}{E},\tag{7.123}$$

$$G_\alpha = 4^{-1}p[2cG_x + YE'/E],\tag{7.124}$$

where the compatibility condition for p is, on using the definition of A and (7.120),

$$p_x = (t_x)_\alpha = \left(\frac{1}{\sqrt{Y'(\Lambda)}}\right)_\alpha = \frac{1}{2}\frac{Y''(\Lambda)}{[Y'(\Lambda)]^2}(F_\alpha - G_\alpha)\tag{7.125}$$

and ahead of the wave $F = G = 0$, since $U = A = 0$ there. These equations are exact.

In the geometric acoustics approximation, since the stratification is small to first order, we neglect the internal reflections, leaving the left traveling Riemann invariant unchanged. Hence, $F(x,\alpha)$ is considered the 'fast' Riemann Invariant, since it changes on the same scale as the boundary condition (7.121), while $G(x,\alpha)$, the backward traveling wave, is the 'slow' Riemann Invariant that changes on the scale of $E'(x)$. For the finite amplitude case, we rewrite (7.123), using (7.116), as

$$F_x = \frac{1}{2}\frac{Y(\Lambda)}{\sqrt{Y'(\Lambda)}}\frac{E'}{E}, \text{ so that } \frac{\sqrt{Y'(\Lambda)}}{Y(\Lambda)}\frac{dF}{dE} = \frac{1}{2E}.\tag{7.126}$$

Then, since $\frac{dA}{d\Lambda} = \sqrt{Y'(\Lambda)}$,

$$\frac{1}{2E} = \frac{\sqrt{Y'(\Lambda)}}{Y(\Lambda)}\frac{dF}{dA}\frac{dA}{d\Lambda}\frac{d\Lambda}{dY}\frac{dY}{dE} = \frac{1}{Y}\frac{dF}{dA}\frac{dY}{dE}.\tag{7.127}$$

Equation (7.127) is exact, but since G changes slowly and $G \backsim 0$, using (7.120) we approximate A by $-F$, and (7.127) by

$$\frac{1}{2E} \backsim -\frac{1}{Y}\frac{dY}{dE}.$$

Hence, we have

$$Y(\Lambda) \backsim \phi(\alpha)[E(x)]^{-1/2},\tag{7.128}$$

where $\phi(\alpha) = Y(\Lambda(0,\alpha))$ and $E(0) = 1$. Then from (7.112) (with $\Lambda = 1$)

$$\sigma(\Lambda, x) = E(x)Y(\Lambda) \backsim \phi(\alpha)[E(x)]^{1/2}. \tag{7.129}$$

We now check the results (7.123) and (7.129) in the small amplitude limit: $A \backsim \Lambda \backsim -F$:

$$F_x = \frac{Y}{2c}\frac{E'}{E} = \frac{\Lambda + N\Lambda^2 + \cdots}{2(1 + N\Lambda + \cdots)}\frac{E'}{E} \backsim \frac{-F}{2}\frac{E'}{E}. \tag{7.130}$$

Integrating (7.130) with $F(0,\alpha) = Mh(\alpha)$ gives

$$F(x,\alpha) = Mh(\alpha)[E(x)]^{-1/2}. \tag{7.131}$$

Also, using (7.94), (7.125), and noting that $G_\alpha \backsim 0$ and $T_x(0,\alpha) = 1$,

$$\begin{aligned}
p_x &= NF_\alpha = NMh'(\alpha)[E(x)]^{-1/2}, \\
T_x &= 1 + NMh(\alpha)[E(x)]^{-1/2}, \\
T &= \alpha + x + NMh(\alpha)\int_0^x [E(z)]^{-1/2}dz.
\end{aligned} \tag{7.132}$$

Hence, the small amplitude result, consistent with (7.110) is given by (7.131) and (7.132). Also (7.128) and (7.112) imply that $Y(\Lambda(x,\alpha)) \backsim \Lambda(x,\alpha) \backsim A(x,\alpha) \backsim -F(x,\alpha) \backsim [E(x)]^{-1/2}$, consistent with (7.131).

7.4.2. *Surface Gravity Waves*

To consider long wave, large amplitude, surface gravity waves propagating over a variable bottom $z = -H(x)$, with $H(0) = H_0$ we follow the notation of Varley, Venkataraman and Cumberbatch [1971] and use the new dependent variable, $a^2(x,t) = h(x,t)$, where $h(x,t) = H(x) + \eta(x,t)$ is the total water depth, so that the surface elevation $\eta(x,t) = a^2(x,t) - H(x)$. The ambient wave speed is $c_0 = \sqrt{gH_0}$.

Following Sec. 7.2.3 if τ_p is the period of the surface wave, the water is "shallow" if the surface wavelength $(c_0\tau_p)$ is large compared with H_0, $H_0 \ll c_0\tau_p$. We will consider slowly varying bottom topography (the high frequency limit) so that if D is the length scale of the variation in $H(x)$ (typically $D = \min |\frac{H(x)}{H'(x)}|$), then $c_0 \tau_p \ll D$. The variables are nondimensionalized so that $L = D$ and the dimensionless equations of motion (7.46) become

$$u_t + \left(a^2 + \frac{1}{2}u^2\right)_x = H'(x), \quad (a^2)_t + (ua^2)_x = 0. \qquad (7.133)$$

Again using the notation $u(x,t) = U(x,\alpha)$ and $a(x,t) = A(x,\alpha)$, (7.133) become

$$2^{-1}U_\alpha(1 - UT_x) - AT_x A_\alpha = 2^{-1}p(H' - 2AA_x - UU_x), \qquad (7.134)$$
$$A_\alpha(1 - UT_x) - 2^{-1}AT_x U_\alpha = -p(UA_x + 2^{-1}AU_x), \qquad (7.135)$$

where $p = t_a$. Defining the right (α) and left (β) traveling characteristics by

$$\frac{dT}{dx}|_\alpha = \frac{1}{U+A}, \quad \frac{dT}{dx}|_\beta = \frac{1}{U-A}, \qquad (7.136)$$

for the α wave (7.134) and (7.135) become

$$A_\alpha - \frac{1}{2}U_\alpha = \frac{(U+A)}{2A}p(2AA_x + UU_x - H') \qquad (7.137)$$

$$A_\alpha - \frac{1}{2}U_\alpha = -\frac{(U+A)}{A}p(UA_x + 2^{-1}AU_x), \qquad (7.138)$$

while for the β wave (7.134) and (7.135) become

$$A_\beta + \frac{1}{2}U_\beta = \frac{(U-A)}{2A}p(2AA_x + UU_x - H') \qquad (7.139)$$

$$A_\beta + \frac{1}{2}U_\beta = \frac{(U-A)}{A}p(UA_x + 2^{-1}AU_x). \qquad (7.140)$$

If we now assume that $H' = 0$ and that U and A depend only on one characteristic variable, so $A_x = U_x = 0$, then (7.137) and (7.139) imply

$$A - U/2 = G(\beta) \text{ on } \frac{dT}{dx}\Big|_\alpha = \frac{1}{U + A},$$
$$A + U/2 = F(\alpha) \text{ on } \frac{dT}{dx}\Big|_\beta = \frac{1}{U - A}.$$

$A - U/2$ and $A + U/2$ are the **Riemann Invariants**. They are constant along $\alpha(x,t) = \text{constant}$ and $\beta(x,t) = \text{constant}$ respectively. Then

$$U = F(\alpha) - G(\beta), \ \ 2A = G(\beta) + F(\alpha). \tag{7.141}$$

We now consider just the α wave, traveling to the right, in the high frequency case. We consider U and A to be functions of x and α and, using (7.141), rewrite (7.134) and (7.135) in terms of new dependent variables $F(x, \alpha)$ and $G(x, \alpha)$ when

$$\frac{dT}{dx}\Big|_\alpha = \frac{1}{U + A} = \frac{2}{3F - G}.$$

Adding (7.137) and (7.138) and using (7.141) yields

$$2G_\alpha = 2A_\alpha - U_\alpha = \frac{(U + A)}{2A}p(2AA_x + UU_x - H' - 2UA_x - AU_x),$$

$$= \frac{(U + A)}{2A}p((U - A)(U_x - 2A_x) - H'),$$

$$= \frac{(3F - G)}{2(F + G)}p\left(\left(-\frac{1}{2}(3G - F)\right)(-2G_x) - H'\right),$$

so that

$$G_\alpha = 4^{-1}p(3F - G)(F + G)^{-1}[(3F - G)G_x - H'(x)].$$

Subtracting (7.137) and (7.138) and using (7.141) yields

$$0 = \frac{(U + A)}{2A}p(2AA_x + UU_x - H') + \frac{(U + A)}{A}p(UA_x + 2^{-1}AU_x)$$

$$H' = (U + A)(2A + U)_x = \frac{1}{2}(3F - G)2F_x.$$

Hence,

$$(3F - G)F_x = H'(x). \tag{7.142}$$

The compatibility condition for p is

$$p_x = (t_x)_\alpha = \left(\frac{2}{3F - G}\right)_\alpha = -2(3F - G)^{-2}(3F_\alpha - G_\alpha). \tag{7.143}$$

We consider the α-wave propagating to the right, where ahead of the wave $u = \eta = 0$, when $a^2 = H$ and then $F = G = A = H^{1/2}$ by (7.141). The boundary condition at $x = 0$ is $u(0, t) = Mf(t)$, so that

$$F(0, \alpha) = H(0) + u(0, \alpha) = 1 + Mf(\alpha), \tag{7.144}$$

where M is the Mach number of the input.

In the geometric acoustics approximation, since the stratification is small we neglect (to first order) the internal reflections, leaving the left traveling Riemann invariant $G(x, \alpha)$ unchanged. Hence, the right traveling invariant $F(x, \alpha)$ is considered the 'fast' Riemann Invariant, since it changes on the same scale as the boundary condition (7.47), while $G(x, \alpha)$ is the 'slow' Riemann Invariant that changes on the scale of $H'(x)$. Since $G = H^{1/2}$ ahead of the wave, and we assume G changes slowly behind the wave front, take $G \sim H^{1/2}$ and approximate (7.142) by

$$(3F - H^{1/2})\frac{dF}{dx} = H'(x)$$

or

$$(3F - n)\frac{dF}{dn} = 2n, \tag{7.145}$$

where $n = H^{1/2}$. This is an homogeneous equation, so let $V = F/n$ and (7.145) becomes

$$(3V - 1)\left(n\frac{dV}{dn} + V\right) = 2$$

which integrates to give

$$(V + 2/3)^{3/2}(V - 1) = \phi(\alpha)n^{-5/2} = \phi(\alpha)H^{-5/4}$$

or

$$\left(F + \frac{2}{3}H^{1/2}\right)^{3/2} (F - H^{1/2}) = \phi(\alpha), \qquad (7.146)$$

where, by (7.144), $\phi(\alpha) = (F + \frac{2}{3}H^{1/2})^{3/2}(F - H^{1/2}) = (5/3 + Mf(\alpha))^{3/2} f(\alpha)$ is the value of the left-hand side at $x = 0$. Differentiating (7.146) w.r.t α,

$$F_\alpha = \frac{2\phi'(\alpha)}{5\left(F - \frac{H^{1/2}}{3}\right)\left(F + \frac{2H^{1/2}}{3}\right)^{1/2}},$$

so that from (7.143)

$$p_x = -2(3F - H^{1/2})^{-2}3F_\alpha$$

$$= \frac{-4\phi'(\alpha)}{15\left(F - \frac{H^{1/2}}{3}\right)^3\left(F + \frac{2H^{1/2}}{3}\right)^{1/2}} \qquad (7.147)$$

$$= \frac{-4Mf'(\alpha)(1 + Mf(\alpha))(5/3 + Mf(\alpha))^{1/2}}{9\left(F - \frac{H^{1/2}}{3}\right)^3\left(F + \frac{2H^{1/2}}{3}\right)^{1/2}}, \qquad (7.148)$$

since $\phi'(\alpha) = 5/3Mf'(\alpha)(1 + Mf(\alpha))(5/3 + Mf(\alpha))^{1/2}$.

To this approximation

$$u = F - H^{1/2},$$

$$\eta = a^2 - H = \frac{1}{4}(F + H^{1/2}) - H$$

$$= \frac{1}{4}(F - H^{1/2})(F + 3H^{1/2}), \qquad (7.149)$$

where $F(\alpha, x)$ is given implicitly by (7.146) in terms of $H(x)$ and the prescribed boundary function $f(\alpha)$.

Using (7.146)–(7.149), Varley, Venkataraman and Cumberbatch [1971] showed that the variations of the amplitudes of the flow variables, such as maximum surface elevation, can be calculated as functions of the undisturbed depth $H(x)$. For example, they show that the variation of maximum surface wave elevation $\eta_M = 0.78h_M$, where

h_M is the value of h at the maximum. The ratio 0.78 of elevation to undisturbed depth also occurs in solitary waves, see Lamb [1945], and is the ratio at which the wave crests of swell become unstable and break, see Bascom [1964].

Note that in the small amplitude limit, $M \ll 1$, when $u = F - H^{1/2} = O(M)$, (7.146) yields

$$u(x,\alpha) \left(H^{1/2} + \frac{2}{3} H^{1/2} \right)^{3/2} = (5/3 + Mf(\alpha))^{3/2} Mf(\alpha), \quad (7.150)$$

so that, from (7.149),

$$u(x,\alpha) = Mf(\alpha)H^{-3/4}, \quad \eta(x,\alpha) = Mf(\alpha)H^{-1/4}. \quad (7.151)$$

Since in this limit $\phi'(\alpha) \backsim (\frac{5}{3})^{3/2} Mf'(\alpha)$, (7.148) becomes

$$p_x \backsim \frac{-4Mf'(\alpha)(5/3)}{15H^{3/2} \left(\frac{2}{3}\right)^3 H^{1/4}} = -\frac{3}{2} Mf'(\alpha) H^{-7/4}.$$

Noting that $p(0,\alpha) = \frac{1}{\omega}$, this integrates to

$$p(x,\alpha) = \frac{1}{\omega} - \frac{3}{2} Mf'(\alpha) \int_0^x H^{-7/4} ds \quad (7.152)$$

which then integrates (using the linear result) to

$$t = \int_0^x \frac{dy}{\sqrt{H(y)}} + \frac{\alpha}{\omega} - \frac{3}{2} Mf(\alpha) \int_0^x H^{-7/4} ds. \quad (7.153)$$

(7.151) and (7.153) agree with (7.57) and (7.61).

Chapter 8

Bounded Media

Here we investigate small amplitude nonlinear waves in bounded media whose different components interact either in the body of the medium or at its boundaries. When only one component of a disturbance is excited, as in Chapter 6, it is generated by the passage of a simple wave. When two oppositely traveling components are excited we need to consider possible interactions. One class of disturbances is that for which, even though the components do not interact in the body of the medium, nonlinearity (amplitude dispersion) may distort the signal carried by the component waves, so that shocks may form and ultimately dissipate the disturbance. Such phenomena are often associated with the cumulative effect of locally small nonlinearities.

8.1. Noninteracting Simple Waves

To illustrate the phenomena of a nonlinear standing wave we consider the small amplitude one-dimensional motion in a finite homogeneous elastic panel or rod. Following Mortell and Varley [1970] we show that a general disturbance, in the small amplitude limit, without restricting the level of the strain rate, can be represented by the passage of two distorting simple waves which interact only at the boundaries.

The governing equations in Lagrangian coordinates are (2.3) and (2.2), which in dimensionless variables (with $T = l/c_0$, $L = l$ where

160

$c_0 = \sqrt{E_0/\rho}$) are

$$u_t = \sigma_x = c^2(\lambda)\lambda_x \quad \text{and} \quad u_x = \lambda_t, \ 0 < x < 1, \tag{8.1}$$
$$\sigma(\lambda) = \Sigma(\lambda) = \lambda(1 + N\lambda + O(\lambda^2)),$$

where $c^2(\lambda) = \Sigma'(\lambda) = 1 + 2N\lambda + O(\lambda^2)$ and $\lambda \ll 1$. N is an $O(1)$ dimensionless material constant. We define the nonlinear strain measure as

$$a(\lambda) = \int_0^\lambda c(y)dy = \lambda(1 + \frac{1}{2}N\lambda + O(\lambda^2)), \tag{8.2}$$

then (8.1) become

$$u_t - c(\lambda)a_x = 0, \ \text{and} \ a_t - c(\lambda)u_x = 0.$$

The right (α) and left (β) traveling characteristics are defined by

$$\frac{dT}{dx}\big|_\alpha = \frac{1}{c(\lambda)}, \ \frac{dT}{dx}\big|_\beta = -\frac{1}{c(\lambda)}, \tag{8.3}$$

while

$$u = G(\beta) + F(\alpha), \ a = G(\beta) - F(\alpha) \tag{8.4}$$

and we define the incremental arrival time as

$$p(\alpha, x) = t_\alpha(\alpha, x). \tag{8.5}$$

Differentiating the first of (8.3)

$$p_x + c'(a)c^{-2}a_\alpha = 0 \tag{8.6}$$

but (8.4) implies

$$-F'(\alpha) = a_\alpha\big|_\beta = a_\alpha\big|_x + x_\alpha a_x, \tag{8.7}$$

so that

$$t_\alpha\big|_\beta = p + x_\alpha t_x = p + \frac{1}{c}x_\alpha, \tag{8.8}$$
$$t_\alpha\big|_\beta = t_x\big|_\beta x_\alpha\big|_\beta = -\frac{1}{c}x_\alpha, \tag{8.9}$$

hence, $x_\alpha = -\frac{1}{2}cp$. Then (8.7) implies

$$a_\alpha|_x = -F'(\alpha) - x_\alpha a_x = -F'(\alpha) + \frac{1}{2}cpa_x \qquad (8.10)$$

and (8.6) can be written

$$c^2 p_x + c'(a)\left[-F'(\alpha) + \frac{1}{2}cpa_x\right] = 0,$$

or

$$c^{1/2}p_x + c^{-1/2}c'(a)\frac{1}{2}pa_x = c^{-3/2}c'(a)F'(\alpha)$$

which becomes

$$(c^{1/2}p)_x = c^{-3/2}c'(a)F'(\alpha). \qquad (8.11)$$

Then, (8.7) and (8.10) become

$$a_\alpha|_\beta = a_\alpha|_x - \frac{1}{2}cpa_x = -F'(\alpha). \qquad (8.12)$$

Integrating (8.11) with $p(0) = 1$

$$p(\alpha, x) = \frac{c^{1/2}(0)}{c^{1/2}(x)} + \frac{1}{c^{1/2}(x)}F'(\alpha)\int_0^x \frac{c'(a)}{c^{3/2}}dx \qquad (8.13)$$

$$= \frac{c^{1/2}(0)}{c^{1/2}(x)} + \frac{1}{c^{1/2}(x)}F'(\alpha)\int_0^x \frac{c'(\lambda)}{c^{5/2}}dx \qquad (8.14)$$

since $\frac{da}{d\lambda} = c(\lambda)$. The mean value theorem then gives

$$p(\alpha, x) = \frac{c^{1/2}(0)}{c^{1/2}(x)} + \frac{x}{c^{1/2}(x)}F'(\alpha)\frac{c'(\lambda_M)}{c^{5/2}(\lambda_M)}$$

for some value of λ_M between $\lambda(0)$ and $\lambda(x)$. Then by (8.2) $c(\lambda_M) = 1 + N\lambda_M + O(\lambda_M^2)\ldots$, with $\lambda_M \ll 1$, so that

$$c'(\lambda_M) = N + O(\lambda_M), \;\; c^n = 1 + O(\lambda) + \cdots \text{ any } n,$$

so

$$\frac{1}{c^{1/2}(x)}\frac{c'(\lambda_M)}{c^{5/2}(\lambda_M)} \backsim N + O(\lambda_M) \text{ and } \frac{c^{1/2}(0)}{c^{1/2}(x)} \backsim 1 + O(\lambda).$$

Hence, a uniformly valid first approximation to $p(\alpha, x)$ is

$$p(\alpha, x) = 1 + F'(\alpha)Nx \qquad (8.15)$$

which integrates to

$$t = \alpha + x + NF(\alpha)x, \qquad (8.16)$$

defining the arbitrary function of α by selecting $t = x + \alpha$ at some wavelet where $u = a = 0$. Equation (8.16) is a uniformly valid first approximation to $\alpha(x, t)$ and indicates that, like a simple wave, the α wave, traveling to the right, depends only on the signal $F(\alpha)$ that it carries. Similarly for the left traveling wave, a uniformly valid implicit equation for the first approximation to $\beta(x, t)$ is given by

$$t = \beta - (x - 1) + NG(\beta)(x - 1). \qquad (8.17)$$

Equations (8.4), (8.16) and (8.17) define a *finite rate* approximation, since there is no restriction on F' and G'. Thus, by (8.4), (8.16) and (8.17) a uniform approximation describing the motion in the finite elastic medium is the superposition of small amplitude noninteracting simple waves.

8.1.1. *Fixed and Stress-free Boundaries*

The solution given by (8.4), (8.16) and (8.17) is a uniformly valid first approximation to describe nonlinear waves in bounded media. This representation, which is suitable for use in the examples considered in Chapter 2, is described as the superposition of noninteracting simple waves. The two unknown functions $F(\alpha)$ and $G(\beta)$ are determined from the boundary conditions.

First, we consider the simple case of an unforced homogeneous elastic panel or rod with **fixed ends**, described by (8.1)–(8.4) with the boundary conditions $u = 0$ at both $x = 0$ and $x = 1$ and

$$u = G(\beta) + F(\alpha) = 0, \ t > 0, \qquad (8.18)$$

where α and β are given implicitly by (8.16) and (8.17). Typical initial conditions at $t = 0$ are

$$F = \phi(x) \text{ and } G = \psi(x), \ 0 \le x \le 1.$$

For the linear problem, when $N = 0$ in (8.16) and (8.17), setting $x = 0$ implies

$$t = \beta + 1 = \alpha,$$

then (8.18) implies that

$$F(R) + G(R - 1) = 0. \tag{8.19}$$

Similarly setting $x = 1$ yields

$$t = \alpha + 1 = \beta$$

and (8.18) gives

$$F(S - 1) + G(S) = 0, \tag{8.20}$$

so that, on eliminating F and G from (8.19) and (8.20),

$$F(\theta + 2) = F(\theta), \quad G(\theta + 2) = G(\theta) \tag{8.21}$$

and the solution is periodic with period 2.

In the *finite rate theory*, when $N \neq 0$ in (8.16) and (8.17), for an unforced homogeneous elastic panel or rod with fixed ends setting $x = 0$ implies

$$t = \alpha = \beta + 1 - NG(\beta), \quad G(\beta) = -F(\alpha),$$

while setting $x = 1$ gives

$$t = \beta = \alpha + 1 + NF(\alpha), \quad F(\alpha) = -G(\beta).$$

Then

$$F(R) + G(R - 1 - NF(R)) = 0 \tag{8.22}$$

and

$$F(S - 1 + NG(S)) + G(S) = 0. \tag{8.23}$$

Eliminating $G(S)$ from (8.22) and (8.23) yields

$$F(\alpha + P(\alpha)) = F(\alpha), \quad P(\alpha) = 2[1 + NF(\alpha)]. \tag{8.24}$$

Eliminating $F(R)$ yields

$$G(\beta + Q(\beta)) = G(\beta), \ Q(\beta) = 2[1 - NG(\beta)]. \qquad (8.25)$$

We recall that $\alpha = t$ on $x = 0$ and $\beta = t$ on $x = 1$, so that (8.24) and (8.25) are the approximate difference equations on $x = 0$ and $x = 1$ respectively. Equations (8.24) and (8.25) are *nonlinear differ- ence equations* since the difference depends on the solution.

We have shown that in a finite elastic panel u and a are given by the solution (8.4) and the first approximation to the linear charac- teristics are given by (8.16) and (8.17). The nonlinear travel time of a wave from $x = 1$ to $x = 0$ and back is $P(\alpha) = 2[1 + NF(\alpha)]$, given by (8.24), while the equivalent linear travel time is $P_L(\alpha) = 2$ in the linear difference equation (8.21).

If we extend Whitham's nonlinearization by replacing the linear travel time by the nonlinear travel time in the linear difference equa- tions (8.21) then we have exactly the nonlinear difference equations (8.24) and (8.25). Thus, the extension of Whitham's nonlinearization technique to waves in a bounded medium is equivalent to saying that the motion is the superposition of noninteracting simple waves trav- elling in opposite directions. However, this statement is conditional on the boundary conditions, as the next example shows.

By contrast to (8.21), if one end of the bar is **stress-free** instead of fixed, so that say $a(0, t) = 0$, then (8.4) implies (8.19) becomes

$$F(R) - G(R - 1) = 0 \qquad (8.26)$$

which in turn implies that

$$F(\theta + 2) = F(\theta - 2), \ G(\theta + 2) = G(\theta - 2) \qquad (8.27)$$

and the linear solution is periodic with period 4.

In the nonlinear case, when $N \neq 0$ in (8.16) and (8.17), an unforced homogeneous elastic panel or rod stress-free at $x = 0$ implies

$$t = \alpha = \beta + 1 - NG(\beta), \ G(\beta) = F(\alpha), \qquad (8.28)$$

while at $x = 1$

$$t = \beta = \alpha + 1 + NF(\alpha), \ F(\alpha) = -G(\beta). \qquad (8.29)$$

Motivated by the linear result, we use (8.28) and (8.29) to follow reflections through two cycles of a wave initially leaving $x = 1$ at time t_0. It is reflected at $x = 0$ at times t_1, t_3 and at $x = 1$ at times t_2, t_4. Then, using just the first correction to the linear characteristics, the times and amplitudes are related through:

$$t_1 = t_0 + 1 - NG(t_0), \quad F(t_1) = G(t_0)$$
$$t_2 = t_1 + 1 + NF(t_1), \quad G(t_2) = -F(t_1),$$
$$t_3 = t_2 + 1 - NG(t_2), \quad G(t_2) = F(t_3),$$
$$t_4 = t_3 + 1 + NF(t_3), \quad G(t_4) = -F(t_3)$$

and hence

$$G(t_4) = G(t_0), \quad t_4 = t_0 + 4. \tag{8.30}$$

The conclusion is that for a stress-free boundary at $x = 0$ and a fixed boundary at $x = 1$, the first correction to the characteristics over two cycles vanishes. Hence, the correction must be at a higher order and so will involve wave interactions. We give fuller consideration to this case in Chapter 14 in the context of a resonance tube open at one end.

8.1.2. *Radiation Boundary Condition*

We now consider the case when energy is allowed to radiate out through the surface at $x = 0$ while the end $x = 1$ is kept fixed. The interface conditions at $x = 0$ are continuity of stress and velocity, written in dimensional variables as:

$$\overline{\sigma}(0^-, t) = \overline{\sigma}(0^+, t) \quad \text{and} \quad \overline{u}(0^-, t) = \overline{u}(0^+, t), \tag{8.31}$$

hence, the impedance boundary condition at $x = 0$ is, see (3.20),

$$\overline{\sigma}(0, t) = i_1 \overline{u}(0, t), \tag{8.32}$$

where $i_1 = \sqrt{E_1 \rho_1} > 0$ is the impedance coefficient for the material in $x < 0$. For the dimensionless stress and velocity in $x > 0$, nondimensionalized with respect to E_2 and $\sqrt{E_2/\rho_2}$ respectively, the condition

(8.32) becomes

$$\sigma(0, t) = iu(0, t), \tag{8.33}$$

where $i = i_1/i_2$ is the ratio of impedances.

Since $a = \sigma + O(\lambda^2)$, the first-order solution (8.4) is

$$u = G(\beta) + F(\alpha), \quad \sigma = G(\beta) - F(\alpha),$$

so that at $x = 0$,

$$F(\alpha) = -kG(\beta), \quad k = \frac{i-1}{i+1}, \quad 0 \le i < \infty, \tag{8.34}$$

where k $(-1 \le k \le 1)$ is the reflection coefficient. For a fixed end $u(0, t) = 0$, $i = \infty$ $(k = 1)$ and for a stress-free end $\sigma(0, t) = 0$, $i = 0$ $(k = -1)$.

The boundary condition (8.33) leaves (8.25) unchanged, but (8.24) becomes

$$F(R) + kG\left(R - 1 - \frac{1}{k}NF(R)\right) = 0.$$

Then the nonlinear difference equations (8.24) and (8.25) become

$$F(\alpha + P(\alpha)) = kF(\alpha), \quad P(\alpha) = 2 + \left(1 + \frac{1}{k}\right)NF(\alpha) \tag{8.35}$$

and

$$G(\beta + Q(\beta)) = kG(\beta), \quad Q(\beta) = 2 - (k+1)NG(\beta). \tag{8.36}$$

To solve (8.36) for G, we define the initial value of G as

$$G(t_0) = M\psi(t_0), \quad 0 \le t < 2, \tag{8.37}$$

where $M \ll 1$ is the Mach number. Then (8.36) give

$$G(t_{n+1}) = kG(t_n) = k^{n+1}M\psi(t_0), \tag{8.38}$$

where

$$t_{n+1} = t_n + 2 - (k+1)NG(t_n)$$
$$= t_n + 2 - (k+1)k^n NM\psi(t_0).$$

Also

$$t_n = t_{n-1} + 2 - (k+1)k^{n-1}NM\psi(t_0),$$

then

$$t_{n+1} = t_0 + 2n - (k+1)NM\psi(t_0)\sum_{i=0}^{n} k^i$$

$$= t_0 + 2n - \frac{k+1}{1-k}[1 - k^{n+1}]NM\psi(t_0). \qquad (8.39)$$

Thus, (8.38) and (8.39) give the solution to (8.36) subject to the initial condition (8.37) prior to the formation of a shock; see Sec. 8.14 for the solution containing shocks.

A shock forms when $\frac{dG(t_{n+1})}{dt_{n+1}} \to \infty$. But (8.38) implies that

$$\frac{dG(t_{n+1})}{dt_{n+1}} = k^{n+1}M\psi'(t_0)\frac{dt_0}{dt_{n+1}}$$

so that a shock will form when $\frac{dt_{n+1}}{dt_0} = 0$. Now by (8.39)

$$\frac{dt_{n+1}}{dt_0} = 1 - \frac{k+1}{1-k}[1 - k^{n+1}]NM\psi'(t_0). \qquad (8.40)$$

For a shock to form, when $|k| < 1$, we require

$$\left|NM\psi'(t_0)\right| > \left|\frac{1-k}{1+k}\right| = i \qquad (8.41)$$

for some t_0 with $0 < t_0 < 2$. This says that there is a critical initial acceleration level, and a shock will only form for an initial acceleration above this level. Otherwise the signal is damped out before a shock can form. When $|k| > 1$ the signal is amplified and a shock always forms.

We note that $u = 0$ on $x = 0$ implies $i = \infty$ or $k = 1$, so we will now consider the case when there is a small amount of energy leaking from the end:

$$k = 1 - 2M\delta, \; M \ll 1. \qquad (8.42)$$

Then there is a leakage of energy at $x = 0$ when $\delta > 0$ or an input of energy when $\delta < 0$. For the latter case, when $|k| > 1$, see Sec. 8.1.4.

Now (8.41) implies that a shock will form only if

$$\left| N\psi'(t_0) \right| > \delta. \tag{8.43}$$

Equations (8.36), with k given by (8.42), imply

$$G(\eta) = (1 - 2M\delta)G(s), \tag{8.44}$$

where

$$\eta = s + 2 - 2NG(s) + 2M\delta NG(s). \tag{8.45}$$

An alternative way of deriving (8.43) from (8.44) and (8.45) is by means of a multiscale expansion. Let,

$$G(s) = MG_0(s,\tau) + M^2 G_1(s,\tau) + \cdots, \tag{8.46}$$

where we introduce the second scale $\tau = Ms$. The time τ is associated with the assumed shock formation time $t = O(M^{-1})$. Then

$$G(s+2) = MG_0(s+2,\tau) + M^2 \left[2\frac{\partial G_0}{\partial \tau}(s+2,\tau) + G_1(s+2,\tau) \right] + \cdots$$

and at $O(M)$ (8.44) and (8.45) give

$$G_0(s+2,\tau) = G_0(s,\tau). \tag{8.47}$$

At $O(M^2)$ we get

$$G_1(s+2,\tau) - G_1(s,\tau)$$
$$= -2\frac{\partial G_0}{\partial \tau}(s+2,\tau) + 2NG_0(s,\tau)\frac{\partial G_0}{\partial s}(s+2,\tau) - 2\delta G_0(s,\tau). \tag{8.48}$$

We take

$$G_1(s+2,\tau) = G_1(s,\tau) \tag{8.49}$$

and then

$$0 = \frac{\partial G_0}{\partial \tau}(s,\tau) - NG_0(s,\tau)\frac{\partial G_0}{\partial s}(s,\tau) + \delta G_0(s,\tau). \tag{8.50}$$

This is Eq. (6.24) with c replaced by $-NG_0$. When $\delta > 0$ the signal is damped and a shock will form only if $|N\psi'| > \delta$, which agrees with (8.43).

Note that in deriving (8.48) we have assumed that $G'(s) = O(M)$. Then the resulting p.d.e. (8.50) is a small amplitude *small rate* approximation as distinct from the small amplitude *finite rate* approximation given by the difference equation (8.44) and (8.45). By setting $\delta = 0$ we get the small rate approximation to the nonlinear difference equation (8.25), and in this case a shock will always form.

8.1.3. *Periodicity and Shocks*

For a time-periodic continuous oscillation of period T_P, integration of equations (8.1) over one period immediately implies that the mean velocity and stress (or pressure for a gas) are independent of x. However, Eq. (8.1) are nonlinear and discontinuous solutions may develop. If in one period the oscillation contains K shocks and $[g]_i$ denotes the jump in g across the ith shock, which travels with speed U_i, integration of (8.1) yields

$$\frac{d}{dx}\int_0^{T_P} \sigma(x,t)dt = -\sum_1^K ([u]_i + U_i[\sigma]_i,$$

$$\frac{d}{dx}\int_0^{T_P} u(x,t)dt = -\sum_1^K ([\lambda]_i + U_i[u]_i$$

and then the shock relations (6.49) imply that the mean stress and velocity are constants. For boundary conditions like (8.31) and (8.32) this implies that the system (8.1) can only sustain time-periodic oscillations that have zero mean when measured from the initial equilibrium state. An important implication of this result for later problems are the restrictions periodicity imposes on the functions $F(\alpha)$ and $G(\beta)$ in the general solution (8.4). Since the mean stress and velocity are constants, then

$$\int_0^{T_P} F(s)ds = \int_0^{T_P} G(s)ds = 0. \tag{8.51}$$

When there is no energy loss at the boundaries $x = 0$ and $x = 1$, up to the time shocks form the total energy in the panel, which is a measure of the amplitude of the disturbance, does not change. To see this we define the strain energy as $W(\lambda) = \int_0^\lambda \sigma(s)ds$, then the rate of change of total energy in the panel is equal to the net mechanical power that is fed across the boundaries:

$$E'(t) = \frac{d}{dt} \int_0^1 \left[W(\lambda) + \frac{1}{2}u^2 \right] dx = u\sigma|_{x=0}^{x=1}. \qquad (8.52)$$

Hence, for stationary boundaries $E(t) = $ constant, and the amplitude does not attenuate before shocks form. However, once shocks form (8.52) must be modified and, in the small strain limit, (8.52) becomes

$$E'(t) = u\sigma|_{x=0}^{x=1} - \Phi(t), \qquad (8.53)$$

where for K shocks

$$\Phi(t) = -\frac{1}{6}N \left(\sum_1^K [\lambda]_i^3 \right) \geq 0. \qquad (8.54)$$

$\Phi(t)$ is nonnegative because in a medium which softens in extension, such as a gas, N is negative $(N = -\frac{\gamma+1}{2})$ and shocks are compressive $([\lambda]_i > 0)$, while in a medium which hardens in extension, like many solids, N is positive and shocks are expansive, $([\lambda]_i < 0)$. Hence, the rate of change of total energy in the panel is a balance between the net energy flowing across the boundaries and the shock dissipation, $\Phi(t)$.

8.1.4. *A Self-sustained Oscillation*

For the case $|k| > 1$ in Sec. 8.1.2 the signal is amplified geometrically. We see this by considering linear theory, when $N = 0$ in Eq. (8.35). This implies that F satisfies

$$F(r + 2) = kF(r),$$

so that the amplitude of F grows like k^n, where n is the number of cycles. This is an example of a system that is linearly unstable to perturbations about the initial state. However, within nonlinear theory, when $N \neq 0$, a shock always forms. Since shocks dissipate energy, this raises the possibility of a balance between the energy flowing in across the boundary and the shock dissipation. As observed by Chu [1963] and Chu and Ying [1963], one such example is the oscillation of a gas in a straight tube into which there is an input of energy due to a pressure-sensitive heat source at the end $x = 0$. The characteristic of the heater defines a feedback parameter, $j > 0$, related to the impedance i, as in (8.32), but in a way to **add** energy to the system. The equivalent boundary condition for the system in Sec. 8.1.2 is

$$u(0, t) = -j\sigma(0, t), \tag{8.55}$$

where in (8.35) and (8.36),

$$k = \frac{1 + j}{1 - j} > 1, \ 0 < j < 1.$$

If $\epsilon = \max |\sigma|$, the feedback energy is $O(j\epsilon^2)$ and for a stable oscillation this has to balance shock dissipation at $O(\epsilon^3)$. Hence, when $j = O(\epsilon)$ and so $k \backsim 1 + 2j > 1$, a periodic solution is possible. This is an example of a system that is linearly unstable to perturbations about the initial state, but the nonlinear system can asymptote to a stable time-periodic oscillation. The periodic state is a resonant phenomenon in the sense that a velocity on the boundary of $O(\epsilon^2)$ maintains a motion of $O(\epsilon)$.

As in the previous section, for a small amplitude disturbance in $0 \le x \le 1$, the full solution is determined by (8.4), (8.16) and (8.17), where $F(\alpha)$ and $G(\beta)$ are given by (8.35) and (8.36). However, conditions (8.51) imply that for the oscillation to asymptote to a stable time-periodic state, the initial state must also satisfy (8.51). Then, as shown in Mortell and Seymour [1973], by considering the long time behavior of the shock equations, the amplitude of a shock asymptotes to a constant nonzero value, which is the amplitude of the periodic motion. The periodic solution depends on the location of the zeros of the initial perturbation.

Here, we first give a simple way to determine a possible periodic solution, then follow the evolution to the periodic state for a particular initial state.

For solutions that are periodic of period 2, (8.35) can be written as

$$F(t) = kF(r), \ t = r + (1 + k^{-1})NF(r), \tag{8.56}$$

where $(1 + \frac{1}{k}) \backsim 2$ for $j \ll 1$. In Sec. 10.2.1 we show that the concept of critical points of a nonlinear difference equation can be used to determine the form of resonant periodic motions when the difference equation contains a periodic forcing function. For the simpler nonlinear difference equation (8.56) that contains no forcing function, critical points help to construct an exact periodic solution. We define the critical points of (8.56) as: points on the solution curve where $t = r$ and $F(t) = F(r)$. For Eq. (8.56) these are the points $t = t_c$ such that $F(t_c) = 0$, of which there must be at least one such point since the initial state has zero mean value. If (8.56) is differentiated with respect to r and evaluated at $t = t_c$ we obtain a quadratic equation for $\lambda = F'(t_c)$:

$$\lambda\left(\lambda - \frac{k(k-1)}{(k+1)N}\right) = 0. \tag{8.57}$$

In general, the nature of the roots of this equation will classify each critical point in a manner similar to that of a nonlinear ordinary differential equation. In this case the roots of (8.57) are simply $\lambda = 0$ and $\lambda = \frac{k(k-1)}{(k+1)N} > 0$ for $N > 0$. The root $\lambda = 0$ yields the trivial solution $F(t) = 0$, while the nonzero root suggests, on noting (6.27),

$$F(t) = \frac{k(k-1)}{(k+1)N}(t - t_c), \quad F(t + 2) = F(t). \tag{8.58}$$

To check that this is an exact periodic solution, consider $F(t)$ given by (8.58), using the second of (8.56):

$$F(t) = \frac{k(k-1)}{(k+1)N}[r + (1 + k^{-1})NF(r) - t_c]$$

$$= \frac{k(k-1)}{(k+1)N}\left[r + (1 + k^{-1})N\frac{k(k-1)}{(k+1)N}(r - t_c) - t_c\right]$$

$$= \frac{k^2(k-1)}{(k+1)N}(r - r_c) \text{ as } t_c = r_c$$

$$= kF(r), \text{ as required.}$$

Thus, substitution of (8.58) into (8.56) shows that this is an exact solution of the difference equation. The final periodic state is therefore piecewise linear, passing through alternate zeros of the original signal function and joined by shocks which are of constant strength. The possible steady states are composed of N-waves whose number and amplitudes depend on the zeros of the initial disturbance.

We now follow the evolution to such a steady state for a particular initial state. We first need initial values of $F(r) = F_0(r)$ for $0 \leq r \leq 2$. These can be found either from conditions along $0 \leq x \leq 1$ at $t = 0$, or conditions on a boundary over a time interval of length 2. Note that $F_0(r)$ must have zero mean value over a period. The signal $F(t)$ on $x = 0$ at any subsequent time is found in terms of $F_0(r)$ from the difference equation (8.56). From Sec. 8.1.2, the solution after n cycles is

$$F(t) = k^n F_0(r), \quad 2(n-1) \leq t \leq 2n, \quad n = 0, 1, 2, \ldots, \qquad (8.59)$$

with

$$t = r + 2n + (1 + k)NH(n)F_0(r), \quad H(n) = \frac{1 - k^n}{k(1 - k)}. \qquad (8.60)$$

We now choose an initial condition that is piecewise linear over $0 \leq r \leq 2$ for which we can explicitly calculate the evolution of the signal and the shock strength cycle by cycle:

$$F_0(r) = \begin{cases} 2r & \text{if } 0 \leq r \leq 1/2, \\ 2(1 - r) & \text{if } 1/2 \leq r \leq 3/2, \\ 2(2 - r) & \text{if } 3/2 \leq r \leq 2. \end{cases} \qquad (8.61)$$

Due to the antisymmetric form about $r = 1$ of the initial shape (8.61), with $N > 0$ a shock forms, and is stationary, at $r = 1$, and the points $F(0) = 0$, $F(2) = 0$ are fixed. The shock is joined by a straight line from $F(0) = 0$ to the top of the shock and by a second straight line from the bottom of the shock to $F(2) = 0$.

We now follow the evolution of the shock height, $h_s(n)$, at $r = 1$ in the nth cycle. By symmetry this is half of the shock strength and the steady periodic response follows by letting $n \to \infty$. Firstly, the term $2n$ in (8.60) can be neglected as it is simply the translation at the linear sound speed unity and does not contribute to the distortion of the signal. To find the shock height at $r = 1$ we follow the vertex $F_0(1/2) = 1$ as n increases. After n cycles the amplitude has increased to k^n and it is located at $V_+(n) = 1/2 + (1+k)NH(n)$. By similar triangles

$$h_s(n) = \frac{k^n}{V_+(n)}. \tag{8.62}$$

Similarly, the negative part of $F_0(r)$, for $1 \le r \le 2$, has moved in the negative direction with the vertex originally at $r = 3/2$ now located at $V_-(n) = 3/2 - (1+k)NH(n)$. Note that for all n, $V_+(n) + V_-(n) = 2$.

The evolving signal shape is an N-wave given by

$$F_n(t) = \begin{cases} h_s(n)t & \text{if } 0 \le t \le 1, \\ h_s(n)(t-2) & \text{if } 1 \le t \le 2 \end{cases}$$

since $2 - V_-(n) = V_+(n)$. As $n \to \infty$, $h_s(n) \to h_s(\infty) = \frac{k(k-1)}{N(1+k)}$, the full shock strength is $2h_s(\infty)$, and

$$F_\infty(t) = \begin{cases} h_s(\infty)t & \text{if } 0 \le t \le 1, \\ h_s(\infty)(t-2) & \text{if } 1 \le t \le 2, \end{cases} \qquad F_\infty(t+2) = F_\infty(t). \tag{8.63}$$

By direct comparison with the exact solution (8.58) it is easy to see that the periodic signal given by (8.63) is a solution of the original nonlinear difference equation (8.56).

8.1.5. *Standing Wave with Cross Section:* $s(x) = (1 + ax)^{-2}$

The examples in Sec. 8.1 are concerned with homogeneous materials so the original equations all have constant coefficients. However, using the same stratification and exact linear solution as in Sec. 5.6.3.3 it is possible to find a nonlinear standing wave for a system with variable coefficients.

The dimensionless equations governing the motion of a polytropic gas contained in a closed tube with cross-sectional area $s(x) = (1 + ax)^{-2}$ (where a is a constant) are (2.17) with $c_0 = 1$. The equations are nondimensionalized in the manner described in Sec. 4.1 for a material of length l, with $L = l$ and $T = lc_0^{-1}$ is the travel time across $[0, 1]$. The corresponding linear equations are (2.18):

$$se_t + (su)_x = 0, \ u_t + e_x = 0, \quad \text{for } 0 \le x \le 1, t > 0, \qquad (8.64)$$

where u is particle velocity and e is the condensation. The boundary conditions are

$$u(0, t) = u(1, t) = 0. \qquad (8.65)$$

The general solution to the linear problem was derived in Sec. 5.6.4, see (5.118) and (5.119):

$$u = (1 + ax)[F(\alpha) + G(\beta)], \qquad (8.66)$$

$$e = (1 + ax)[F(\alpha) - G(\beta)] + a\left[\int^\alpha F(s)\, ds + \int^\beta G(s)\, ds\right], \qquad (8.67)$$

where $\alpha = t - x$, and $\beta = t + x - 1$ by picking $\alpha = t$ on $x = 0$, and $\beta = t$ on $x = 1$.

The exact linear solutions (8.66) and (8.67) are also, in this special case, the two-term linear geometrical acoustics solution, see Sec. 5.4.

To construct the nonlinear solution, we need the first nonlinear correction to the linear characteristics so that we can use Whitham's nonlinearization technique. Following the method given in Sec. 7.2.6, for the α-wave traveling to the right

$$t_x = 1 - u - (\gamma - 1)e/2 \qquad (8.68)$$

and for the β-wave traveling to the left

$$t_x = -[1 + u - (\gamma - 1)e/2], \qquad (8.69)$$

where u and e are given by (8.66) and (8.67) respectively and γ is the polytropic gas constant. On integrating (8.68) and (8.69), for the

α-wave (with $N = -\frac{\gamma+1}{2}$)

$$t = \alpha + x + NA_1(x)F(\alpha) + (N+1)ax \int_0^\alpha F(s)\,ds \qquad (8.70)$$

with $\alpha = t$ on $x = 0$, and for the β-wave

$$t = \beta - (x - 1) + NA_2(x)G(\beta) - (N+1)a(x-1)\int_0^\beta G(s)\,ds$$

$$(8.71)$$

with $\beta = t$ on $x = 1$. In (8.70) and (8.71)

$$A_1(x) = (2a)^{-1}[(1 + ax)^2 - 1],$$
$$A_2(x) = (2a)^{-1}[(1 + ax)^2 - (1 + a)^2]. \qquad (8.72)$$

The linear solution from (8.66) with the boundary conditions (8.65) gives

$$G(t) = -F(t + 1) \quad \text{and} \quad F(t + 2) = F(t). \qquad (8.73)$$

We let

$$F_1(\alpha) = \int_1^\alpha F(s)\,ds \quad \text{and} \quad G_1(\beta) = \int_0^\beta G(s)\,ds \qquad (8.74)$$

and note that $u(\alpha, \beta, x)$ given by (8.66) with (8.70) and (8.71) is the superposition of modulated simple waves, i.e., u changes shape with x but the α-wave does not interact with the oppositely traveling β-wave. We now invoke a generalization of Whitham's nonlinearization technique: the motion is a superposition of oppositely traveling modulated simples waves.

If we consider the β-wave that leaves $x = 1$ at time $t = s$, then the arrival time on $x = 0$ is by (8.71)

$$t = s + 1 + NA_2(0)G(s) + (N+1)aG_1(s). \qquad (8.75)$$

The α-wave leaves $x = 0$ at the time given by (8.75), i.e., $t = \alpha$ in (8.75). This wave arrives at $x = 1$ at time $t = \eta$ where, by (8.70),

$$\eta = \alpha + 1 + NA_1(1)F(\alpha) + (N+1)aF_1(\alpha). \qquad (8.76)$$

Substituting for $\alpha = t$ from (8.75) into (8.76), we find the round trip travel time for the wave is

$$\eta = s + 2 + N\left[A_2(0)G(s) + A_1(1)F(\alpha)\right] + (N+1)a\left[G_1(s) + F_1(\alpha)\right].$$
$$(8.77)$$

Using the first of (8.73),

$$G_1(s) + F_1(\alpha) = \int_0^s G(t)\,dt + \int_1^\alpha F(t)\,dt$$

$$= -\int_1^{s+1} F(t)\,dt + \int_1^{s+1} F(t)\,dt = 0$$

and then

$$\eta = s + 2 + N\left[A_2(0) - A_1(1)\right]G(s). \qquad (8.78)$$

The nonlinear difference equation is

$$G(\eta) = G(s), \qquad (8.79)$$

where $\eta - s$ is given by (8.78). We can write (8.78) and (8.79) as

$$G(s + Q(s)) = G(s) \qquad (8.80)$$

with

$$Q(s) = 2[1 - (1 + a/2)\,NG(s)]. \qquad (8.81)$$

The only difference made by the tube shape $(1 + ax)^{-2}$ compared with the constant area $s(x) \equiv 1$ is the introduction of the factor $a/2$ in $Q(s)$. The variable area changes the round trip travel time of a wavelet, but shocks can still form. However, we must be careful not to use this example to imply anything about shock formation in standing waves in general for tubes with variable cross sections. The exact linear solution in Sec. 5.6.3.3 with $s(x) = (1 + ax)^{-2}$ is not typical of problems with other cross sections as the eigenvalues of the linear problem are $\omega_n = n\pi$ for integer n and the eigenvalues are commensurate, even though there is stratification. In general the eigenvalues are incommensurate.

8.1.6. *Standing Wave in an Inhomogeneous Elastic Panel*

We consider an inhomogeneous elastic panel of length l with fixed ends when the inhomogeneity scales both density and Young's modulus in the same way, so that $\rho(x) = E(x)$ with $E(0) = 1$, as in Sec. 7.2.1. The objective here is to calculate the effect of the slow inhomogeneity on a standing wave.

Length is nondimensionalized w.r.t. l, so that $E = E(\varepsilon x)$ where $\varepsilon = l/D$ and now $0 < x < 1$. By adding the corresponding β wave to (5.35) and (7.22), the first-order solution is

$$u = E^{-1/2}(\varepsilon x)(F(\alpha) + E_1^{1/2}G(\beta)),$$

$$\lambda = E^{-1/2}(\varepsilon x)(-F(\alpha) + E_1^{1/2}G(\beta)),$$

$$\alpha = t - x - NF(\alpha)\int_0^x E^{-1/2}(\varepsilon y)dy,$$

$$\beta = t + x - 1 - NG(\beta)E_1^{1/2}\int_1^x E^{-1/2}(\varepsilon y)dy,$$

where $E_1 = E(1)$.

To find the difference equation, consider a wavelet leaving $x = 1$ at time s, reflecting from $x = 0$ at time y, then reflecting from $x = 1$ at time η. The amplitude satisfies

$$F(y) + E_1^{1/2}G(s) = 0 \quad \text{at } x = 0$$
$$F(y) + E_1^{1/2}G(\eta) = 0 \quad \text{at } x = 1,$$

where

$$y = \eta - 1 - NF(y)\int_0^1 E^{-1/2}(\varepsilon y)dy,$$

$$s = y - 1 - NG(s)E_1^{1/2}\int_1^0 E^{-1/2}(\varepsilon y)dy.$$

Hence,

$$G(\eta) - G(s) = 0, \quad \eta = s + 2 - 2NG(s)E_1^{1/2}\int_0^1 E^{-1/2}(\varepsilon y)dy$$

or

$$G(s + Q(s)) - G(s) = 0, \ Q(s) = 2(1 - NG(s)E_1^{1/2} \int_0^1 E^{-1/2}(\varepsilon y)dy).$$

$$(8.82)$$

It is noted from (8.82) that there is no attenuation of the signal G over one cycle, so that a small inhomogeneity does not affect the signal amplitude and hence a shock will form in a finite time.

8.1.7. *Standing Wave in a Maxwell Solid*

We consider an homogeneous viscoelastic Maxwell panel, $0 < \bar{x} < l$, with fixed ends as in Sec. 7.2.2. The objective here is to calculate the effect of the rate dependence on a standing wave.

This is an interesting and more complicated example as there are three length scales in the problem: the panel length, l, the relaxation length, $c_0\tau_r$, and the input wavelength, $c_0\tau_p$. We prefer to use $L = l$ so that $0 < x < 1$, but the geometric acoustics solution uses $L = 2c_0\tau_r$ and $T = 2\tau_r$, so we need to build up the solution in stages.

The linear geometric acoustics solution with the restriction $\tau_p \ll \tau_r$ on a semi-infinite region with length nondimensionalized w.r.t. $L = 2c_0\tau_r$ and $T = 2\tau_r$ is

$$u = e^{-x}F(\alpha), \ \lambda = -e^{-x}F(\alpha),$$
$$\alpha = \omega(t - x), \ \omega = 2\tau_r/\tau_p,$$

where $\omega = 2\tau_r/\tau_p \gg 1$ and $u(0, t) = F(\omega t)$. The factor of 2 in L and T is included to make the exponent $-x$. In dimensional variables this solution is

$$u = e^{-\bar{x}\delta/l}F(\alpha), \ \lambda = -e^{-\bar{x}\delta/l}F(\alpha),$$
$$\alpha = \bar{t}/\tau_p - \bar{x}/(c_0\tau_p),$$

where $\delta = l/(2c_0\tau_r)$. If we pick $L = l$ when we nondimensionalize, so that now $0 < x < 1$, then the solution becomes

$$u = e^{-\delta x}F(\alpha), \ \lambda = -e^{-\delta x}F(\alpha),$$
$$\alpha = \omega_l(t - x),$$

where $\omega_l = l/(c_0\tau_p)$.

Hence, from (7.7) and (7.8) the first order nonlinear solution for waves in the finite panel is

$$u = e^{-x\delta}F(\alpha) + e^{\delta(x-1)}G(\beta), \quad \lambda = -e^{-\delta x}F(\alpha) + e^{\delta(x-1)}G(\beta),$$
$$\alpha = t - x - NF(\alpha)(1 - e^{-x\delta})\delta^{-1},$$
$$\beta = t + x - 1 + NG(\beta)(1 - e^{\delta(x-1)})\delta^{-1}.$$

To find the difference equation, consider a wavelet leaving $x = 1$ at time s, reflecting from $x = 0$ at time y, then reflecting from $x = 1$ at time η. The amplitude satisfies

$$F(y) + e^{-\delta}G(s) = 0 \text{ at } x = 0$$
$$F(y) + e^{\delta}G(\eta) = 0 \text{ at } x = 1,$$

where

$$y = \eta - 1 - \delta^{-1}NF(y)(1 - e^{-\delta}),$$
$$s = y - 1 - \delta^{-1}NG(s)(e^{-\delta} - 1).$$

Hence,

$$G(\eta) - e^{-2\delta}G(s) = 0, \quad \eta = s + 2 - \delta^{-1}NG(s)(1 - e^{-2\delta}),$$

or

$$G(s+Q(s)) - e^{-2\delta}G(s) = 0, \quad Q(s) = 2 - \delta^{-1}NG(s)(1 - e^{-2\delta}). \quad (8.83)$$

The signal G is attenuated by a factor $e^{-2\delta}$ over one cycle.

To solve the initial value problem for G, we define the initial value of as

$$G(t_0) = M\psi(\omega_l t_0), \qquad 0 \le t < 2, \qquad (8.84)$$

where $M \ll 1$ is the Mach number. Then, with $k = e^{-2\delta}$ (8.83) give

$$G(t_{n+1}) = kG(t_n) = k^{n+1}M\psi(\omega_l t_0), \qquad (8.85)$$

$$t_{n+1} = t_n + 2 - (1 - k)\delta^{-1}NG(\omega_l t_n)$$
$$= t_n + 2 - (1 - k)k^n\delta^{-1}NM\psi(\omega_l t_0).$$

Then

$$t_{n+1} = t_0 + 2n - (1-k)\delta^{-1} NM\psi(\omega_l t_0) \sum_{i=0}^{n} k^i$$

$$t_{n+1} = t_0 + 2n - [1 - k^{n+1}]\delta^{-1} NM\psi(\omega_l t_0). \tag{8.86}$$

A shock will form when $\frac{dt_{n+1}}{dt_0} = 0$. Now

$$\frac{dt_{n+1}}{dt_0} = 1 - [1 - k^{n+1}]\delta^{-1}\omega_l NM\psi'(\omega_l t_0). \tag{8.87}$$

Hence, for a shock to form we require

$$\left| NM\omega_l \psi'(\omega_l t_0) \right| > \delta, \text{ or} \tag{8.88}$$

$$\left| NM\omega\psi'(\omega_l t_0) \right| > 1. \tag{8.89}$$

The inclusion of viscoelasticity results in an attenuation of the signal and the possible prevention of shock formation. In the limit as $\delta \to 0$ we recover the elastic result that a shock always forms.

8.2. Multiple Scale Examples

In Sec. 8.1 we showed how a uniform approximation to the small amplitude motion in a finite homogeneous elastic panel was represented by the superposition of two noninteracting simple waves. This led to a nonlinear difference equation on the boundary for the Riemann Invariants $F(\alpha)$ and $G(\beta)$. We also showed how, in the small rate limit, we can derive, using a multiple scale expansion, the nonlinear p.d.e. whose solutions agreed with those of the difference equation in the appropriate limit.

Here we show how to use a multiple scale expansion directly on the governing equations to produce an approximating p.d.e. and give a number of examples. Because we introduce two time scales, e.g., the travel time and the shock formation time, which are of different magnitudes, we are automatically dealing with a small amplitude, small rate theory. Under this restriction a shock cannot form in one travel time.

The nonlinear difference equation was derived from an hyperbolic system and this technique is not applicable to elliptic equations, like Laplace's equation. The sloshing of shallow water in a tank is such a situation, but the multiple scales technique allows us to set up an approximating p.d.e., in this case a KdV equation. In this context the multiple scales expansion is more general.

In a sense we follow the approach used in Sec. 8.1 in that we set up an equation that determines the evolution of a signal, e.g., the Riemann "invariant", but now *on a boundary*. We do not get involved in tracking waves in the interior of the medium. Once we know the solution on the boundary we can use the characteristics to determine the solution between the boundaries.

8.2.1. *Homogeneous Elastic Panel*

We consider one-dimensional longitudinal disturbances in an elastic panel of finite length L. The dimensionless equations in Lagrangian coordinates relating stress, strain and velocity are (see (2.3))

$$u_t = \sigma_x, \quad \text{and} \quad \lambda_t = u_x, \quad 0 \le x \le 1, \quad t > 0 \tag{8.90}$$

with stress related to strain by

$$\sigma = \lambda + N\lambda^2 + O(\lambda^3), \tag{8.91}$$

where N is an $O(1)$ dimensionless material constant and $\lambda \ll 1$.

In (8.90) and (8.91) we have chosen L as the unit of length and L/c_0 as the unit of time, where $c_0 = (\frac{E_0}{\rho_0})^{1/2}$ is the sound speed in the reference state. E_0 is Young's modulus, which is taken as the unit of stress, and ρ_0 is the constant density.

To complete the problem, the ends $x = 0$ and $x = 1$ are rigidly bonded so that

$$u(0, t) = u(1, t) = 0, \quad t \ge 2 \tag{8.92}$$

$$u(0, t) = 0, \quad\quad\quad 0 \le t \le 2. \tag{8.93}$$

The strain at $x = 1$ is specified by

$$\lambda(1, t) = -M\psi(t), \quad\quad 0 \le t \le 2, \tag{8.94}$$

where the Mach number is $M \ll 1$. The problem now is to describe the evolution of the standing wave $\lambda(x, t)$ in the elastic panel for $t \geq 2$.

According to linear theory,

$$-\sigma = -\lambda = M[f(t - x) + f(t + x)], \quad t > 2, \qquad (8.95)$$

where $f = \psi$, and $f(t + 2) = f(t)$.

Thus, the motion of any particle in the material is periodic with period 2. But for the nonlinear system (8.90) and (8.91) Lax [1957] proved that nontrivial periodic solutions cannot exist, and Ludford [1952] used the hodograph transformation to calculate the time at which shocks form. Mortell [1977] showed how to calculate the evolution using multiple scales and Lardner [1975] used a Krylov–Bogoliubov technique to examine the standing wave. Thus we have an example in which linear theory gives results which, while self-consistent, are at variance with those of the nonlinear theory that it approximates. It will emerge that linear theory must inevitably fail in this case when the time becomes large.

For the small amplitude problem (8.90)–(8.94), the results of Lax [1957] show that a singularity, i.e., a shock, develops in the medium. Then there are two natural time scales — the time to travel the width of the medium and the time for the shock formation. In order to apply a multi-scale perturbation multi-scale perturbation technique the time for shock formation must be much greater than the travel time in the medium. An equivalent statement is that the initial acceleration must satisfy $\left|\frac{\partial u}{\partial t}\right| \ll c_0^2/L$, where here u, t are dimensional variables. Linear theory is valid when $|u/c_0| \ll 1$ and $|u_t| L/c_0^2 \ll 1$. The second of these is violated when a shock forms, and thus linear theory is valid on a time scale that is small compared with the shock formation time.

We now introduce a slow time scale $\tau = Mt$, so that a shock forms when $\tau = O(1)$. Then we assume a perturbation expansion

$$u(t, x, M) = M u_1(t, x, \tau) + M^2 u_2(t, x, \tau) + \cdots, \qquad (8.96)$$

$$\lambda(t, x, M) = M \lambda_1(t, x, \tau) + M^2 \lambda_2(t, x, \tau) + \cdots \qquad (8.97)$$

and note

$$\left.\frac{\partial}{\partial t}\right|_x = \left.\frac{\partial}{\partial t}\right|_{x,\tau} + M\left.\frac{\partial}{\partial \tau}\right|_{x,t}.$$

Equations (8.90)–(8.92) imply, at $O(M)$, that for $t > 0$

$$\frac{\partial \lambda_1}{\partial x} - \frac{\partial u_1}{\partial t} = 0, \qquad \frac{\partial \lambda_1}{\partial t} - \frac{\partial u_1}{\partial x} = 0 \tag{8.98}$$

with

$$u_1(0, t, \tau) = u_1(1, t, \tau) = 0. \tag{8.99}$$

At $O(M^2)$ the equations are

$$\frac{\partial \lambda_2}{\partial x} - \frac{\partial u_2}{\partial t} = -\frac{\partial}{\partial x}(N\lambda_1^2) + \frac{\partial u_1}{\partial \tau}, \qquad \frac{\partial \lambda_2}{\partial t} - \frac{\partial u_2}{\partial x} = -\frac{\partial \lambda_1}{\partial \tau} \tag{8.100}$$

with the boundary conditions

$$u_2(0, t, \tau) = u_2(1, t, \tau) = 0. \tag{8.101}$$

The solution to (8.98) and (8.99) is

$$u_1 = f(\xi, \tau) - f(\eta, \tau), \qquad -\lambda_1 = f(\xi, \tau) + f(\eta, \tau), \tag{8.102}$$

where $\xi = t - x$, $\eta = t + x$ and f is arbitrary except that is must satisfy the periodicity condition in the fast variable t,

$$f(t + 2, \tau) = f(t, \tau). \tag{8.103}$$

Then (8.102) says that u_1 and λ_1 travel as *linear* waves in any cycle. To solve (8.100), λ_2 is eliminated to give

$$\frac{\partial^2 u_2}{\partial x^2} - \frac{\partial^2 u_2}{\partial t^2} = -\frac{\partial^2}{\partial x \partial t}(N\lambda_1^2) + \frac{\partial^2 u_1}{\partial t \partial \tau} + \frac{\partial^2 \lambda_1}{\partial x \partial \tau} \tag{8.104}$$

and this becomes

$$-4\frac{\partial^2 u_2}{\partial\xi\partial\eta} = \left(\frac{\partial^2}{\partial\xi^2} - \frac{\partial^2}{\partial\eta^2}\right) N\lambda_1^2 + \left(\frac{\partial}{\partial\xi} + \frac{\partial}{\partial\eta}\right)\frac{\partial u_1}{\partial\tau}$$
$$+ \left(\frac{\partial}{\partial\eta} - \frac{\partial}{\partial\xi}\right)\frac{\partial\lambda_1}{\partial\tau}.$$

On substituting for λ_1 using (8.102) we get

$$-4\frac{\partial^2 u_2}{\partial\xi\partial\eta} = -N\left[\frac{\partial^2}{\partial\eta^2}f^2(\eta,\tau) + 2f(\xi,\tau)\frac{\partial^2}{\partial\eta^2}f(\eta,\tau)\right]$$
$$+ N\left[\frac{\partial^2}{\partial\xi^2}f^2(\xi,\tau) + 2f(\eta,\tau)\frac{\partial^2}{\partial\xi^2}f(\xi,\tau)\right]$$
$$+ 2\left[\frac{\partial^2 f}{\partial\xi\partial\tau}(\xi,\tau) - \frac{\partial^2 f}{\partial\eta\partial\tau}(\eta,\tau)\right].$$

Integration then gives

$$2u_2(\xi,\eta,\tau) = \xi\frac{\partial f}{\partial\tau}(\eta,\tau) - \eta\frac{\partial f}{\partial\tau}(\xi,\tau)$$
$$- N\left[\eta f(\xi,\tau)\frac{\partial f}{\partial\xi} - \xi f(\eta,\tau)\frac{\partial f}{\partial\eta}\right.$$
$$\left. + \frac{\partial f}{\partial\xi}\int^\eta f(s,\tau)\,ds - \frac{\partial f}{\partial\eta}\int^\xi f(s,\tau)\,ds\right]$$
$$+ 2f_2(\xi,\tau) + 2g_2(\eta,\tau), \tag{8.105}$$

where f_2, g_2 are arbitrary solutions of the homogeneous equation.

Since $\xi = \eta = t$ at $x = 0$, the boundary condition $u_2(0,t,\tau) = 0$ is automatically satisfied provided

$$f_2(t,\tau) + g_2(t,\tau) = 0.$$

We insist that $f_2(t+2,\tau) = f_2(t,\tau)$, i.e., f_2 is one component of the second-order standing wave, and then the boundary condition $u_2(1,t,\tau) = 0$ implies that $f(t,\tau)$ must satisfy

$$\frac{\partial f}{\partial\tau} + Nf\frac{\partial f}{\partial t} + \frac{1}{2}N\int_{t-1}^{t+1} f(s,\tau)\,ds \cdot \frac{\partial f}{\partial t} = 0. \tag{8.106}$$

Due to (8.103), the integral term in (8.106) is independent of t. Integration of (8.106) over a time interval of length 2 implies that

$\int_{t-1}^{t+1} f(s,\tau)\,ds$ is also independent of τ, since

$$\frac{\partial}{\partial\tau}\int_{t-1}^{t+1} f(s,\tau)\,ds + \frac{1}{2}N\left[f^2(t+1) - f^2(t-1)\right]$$
$$+ \frac{1}{2}N\int_{t-1}^{t+1} f(s,\tau)\,ds \cdot [f(t+1) - f(t-1)] = 0$$

and the result $\frac{\partial}{\partial\tau}\int_{t-1}^{t+1} f(s,\tau)\,ds = 0$ follows from (8.103).

The specification of the initial input, $F(t)$, implies $f(t,0) = F(t)$, $0 \leq t \leq 2$. Then, at least up until the time a shock forms, (8.106) may be replaced by

$$\frac{\partial f}{\partial\tau} + Nf\frac{\partial f}{\partial t} = 0, \tag{8.107}$$

where f is the amplitude of the signal as measured from the mean of $F(t)$. Now the integral or interaction term is gone and the result (8.107) is just a simple wave in the variables t and τ. This tells us how the signal f (or the strain λ) evolves as a simple wave on the fixed boundary $x = 1$, where $\lambda_1(t,\tau) = -2f(t-1,\tau)$. The signal f is determined by the nonlinear equation (8.107) on the fixed boundary $x = 1$, while it propagates as a linear wave in the medium, see (8.102).

It should be noted that the physical solution associated with (8.107) requires that $f(t,\tau)$ be evaluated along the lines $\tau = Mt$, $t \geq 0$. The function $f(t,\tau)$ need only be evaluated for $0 \leq t \leq 2$ due to the periodicity condition (8.103). We then evaluate $f(t,\tau)$ along the lines $\tau = Mt + 2(n-1)M$, $n = 1,2,3,\ldots(0 \leq t \leq 2)$, where n is the number of round trips of the linear wave after startup.

A shock wave forms when

$$1 + N\tau\psi'(t) = 0 \tag{8.108}$$

which agrees with the prediction of Lax [1957] about the presence of a singularity in the flow and linear theory is valid only when $\tau = Mt \ll 1$. Linear theory becomes invalid when the acceleration of the motion, $\frac{\partial f}{\partial t} = \frac{\psi'(\alpha)}{1+N\tau\psi'(\alpha)}$ becomes unbounded i.e., when $\tau = O(1)$ or equivalently when $t = O(M^{-1})$.

Since $\xi = t - x$, $\eta = t + x$, the form of $u_2(\xi, \eta, \tau)$ given by (8.105) would seem to have a secular or growth term in t. However, the boundary condition is that $u_2 = 0$ at $x = 0$ and at $x = 1$, so the growth in t occurs only in one travel time in the panel. In other words the clock starts again for t each time the wave hits a boundary and this prevents secular growth.

The form of the particular integral u_2 given by (8.105) is not unique. With some trial and error we can produce

$$u_2 = -Nx[f(t - x, \tau)f'(t - x, \tau) + f(t + x, \tau)f'(t + x, \tau)]$$
$$-\frac{1}{2}N\left[f'(t - x, \tau)\int^{t+x} f(s, \tau)\,ds - f'(t + x, \tau)\int^{t-x} f(s, \tau)\,ds\right]$$
$$-x[f_\tau(t + x, \tau) + f_\tau(t - x, \tau)] \tag{8.109}$$

that is also a solution of (8.104) and this form of u_2 also leads to Eq. (8.107). Since $0 \leq x \leq 1$, there is no growth term in u_2 given by (8.109). In (8.109) the prime means differentiation w.r.t. the first variable shown. The form (8.109) allows us to calculate the uniform correction to u_2 without a secular term.

The formula (8.109) for the particular integral for (8.104) follows from (8.105) by replacing ξ by $-2x$ and η by $2x$, where ξ, η appear as multiplying factors, e.g., $\xi\frac{\partial f}{\partial \tau}(\eta, \tau)$ becomes $-2x\frac{\partial f}{\partial \tau}(t + x, \tau)$ and $\eta\frac{\partial f}{\partial \tau}(\xi, \tau)$ becomes $2x\frac{\partial f}{\partial \tau}(t - x, \tau)$. The basis for this observation is as follows. Consider the equation

$$\left(\frac{\partial}{\partial t} + \frac{\partial}{\partial x}\right)U = F(t - x) \tag{8.110}$$

and note that the right-hand side is a solution of the homogeneous equation. With $\xi = t - x$ and $\eta = t + x$, (8.110) becomes

$$2U_\eta = F(\xi)$$

which integrates to

$$U = \frac{1}{2}\eta F(\xi). \tag{8.111}$$

Now observe that for *any* function $F(t - x)$

$$\left(\frac{\partial}{\partial t} + \frac{\partial}{\partial x}\right)[xF(t - x)] = F(t - x)$$

and then a particular integral of (8.110) is $U = xF(t - x)$, which is exactly (8.111) when η is replaced by $2x$.

A similar results follows, replacing ξ by $-2x$, for

$$\left(\frac{\partial}{\partial t} - \frac{\partial}{\partial x}\right)U = G(t + x), \tag{8.112}$$

on noting that $U = xG(t + x)$ is a solution of (8.112), and that $U = \frac{1}{2}\xi G(\eta)$.

When we know the signal f, then u_2 at a point (t, x) within the medium is given by (8.109) and u_1 by (8.102).

8.2.2. *Finite Length Maxwell Solid*

The governing equations for the one-dimensional motion of a viscoelastic panel of length l were derived in 7.1.2; u, σ and λ are particle velocity, stress and strain. The dimensionless variables are selected using $T = l/c_0$, $L = l$, where $c_0 = \sqrt{E_0/\rho}$, then the basic equations are

$$u_t - \sigma_x = 0, \ \ \lambda_t = u_x, \ \ \sigma_t = \Gamma'(\lambda)\lambda_t - \varepsilon\mu\sigma, \ \ 0 \leq x \leq 1, \ t > 0, \tag{8.113}$$

where $\Gamma(\lambda) = \lambda + N\lambda^2 + O(\lambda^2)$, $\varepsilon\mu = l/(c_0\tau_r) = 2\delta$, and τ_r is the relaxation time. The small parameter $\varepsilon \ll 1$ is the ratio of the travel time in the material to the relaxation time. We also assume that the signal amplitude (or Mach number, M) is $O(\varepsilon)$.

The boundary condition for fixed ends is

$$u(0, t) = u(1, t) = 0 \tag{8.114}$$

We expect the relaxation and/or shock formation to take effect over the time scale $\tau = \varepsilon t = O(1)$. So we assume an expansion of the form

$$u(t, x; \varepsilon) = \varepsilon u_1(t, x, \tau) + \varepsilon^2 u_2(t, x, \tau) + \cdots \tag{8.115}$$

$$\lambda(t, x; \varepsilon) = \varepsilon \lambda_1(t, x, \tau) + \varepsilon^2 \lambda_2(t, x, \tau) + \cdots \tag{8.116}$$

$$\sigma(t, x; \varepsilon) = \varepsilon \sigma_1(t, x, \tau) + \varepsilon^2 \sigma_2(t, x, \tau) + \cdots \tag{8.117}$$

Then substituting (8.115)–(8.117) into (8.113), we get at $O(\varepsilon)$

$$\frac{\partial \sigma_1}{\partial x} = \frac{\partial u_1}{\partial t}, \ \frac{\partial u_1}{\partial x} = \frac{\partial \lambda_1}{\partial t}, \ \frac{\partial \sigma_1}{\partial x} = \frac{\partial \lambda_1}{\partial t} \tag{8.118}$$

and

$$u_1(0, t, \tau) = u_1(1, t, \tau) = 0. \tag{8.119}$$

Then,

$$u_1(x, t, \tau) = f(\xi, \tau) - f(\eta, \tau), \tag{8.120}$$

where $\xi = t - x$, and $\eta = t + x$ and $f(t + 2, \tau) = f(t, \tau)$. So u_1 is a linear wave modulated on a time scale τ.

At $O(\varepsilon^2)$ we get

$$\frac{\partial \sigma_2}{\partial x} = \frac{\partial u_2}{\partial t} + \frac{\partial u_1}{\partial \tau}, \ \frac{\lambda_2}{\partial t} + \frac{\partial \lambda_1}{\partial \tau} = \frac{\partial u_2}{\partial x}, \tag{8.121}$$

where

$$\frac{\partial \sigma_1}{\partial \tau} + \frac{\partial \sigma_2}{\partial t} = \frac{\partial \lambda_2}{\partial t} + \frac{\partial \lambda_1}{\partial \tau} + 2N\lambda_1 \frac{\partial \lambda_1}{\partial t} - \mu\sigma_1. \tag{8.122}$$

Eliminating σ_2 and λ_2 in (8.121) and (8.122), the result is

$$\frac{\partial^2 u_2}{\partial t^2} - \frac{\partial^2 u_2}{\partial x^2} = -2\frac{\partial^2 u_1}{\partial t \partial \tau} + 2N\frac{\partial}{\partial x}\left(\lambda_1 \frac{\partial \lambda_1}{\partial t}\right) - \mu\frac{\partial u_1}{\partial t} \tag{8.123}$$

and this becomes

$$4\frac{\partial^2 u_2}{\partial \xi \partial \eta} = -2\left[\frac{\partial^2 f}{\partial \xi \partial \tau}(\xi, \tau) - \frac{\partial^2 f}{\partial \eta \partial \tau}(\eta, \tau)\right]$$

$$+ 2N\left[\left(\frac{\partial f}{\partial \eta}\right)^2 + f(\eta, \tau)\frac{\partial^2 f}{\partial \eta^2} - \left(\frac{\partial f}{\partial \xi}\right)^2 - f(\xi, \tau)\frac{\partial^2 f}{\partial \xi^2}\right.$$

$$\left. + f(\xi, \tau)\frac{\partial^2 f}{\partial \eta^2} - f(\eta, \tau)\frac{\partial^2 f}{\partial \xi^2}\right] - \mu\left[\frac{\partial f}{\partial \xi}(\xi, \tau) - \frac{\partial f}{\partial \eta}(\eta, \tau)\right].$$

Then the latter equation integrates to

$$4u_2 = -2\left[\eta\frac{\partial f}{\partial \tau}(\xi, \tau) - \xi\frac{\partial f}{\partial \tau}(\eta, \tau)\right]$$

$$+ 2N\left[\xi f\frac{\partial f}{\partial \eta}(\eta,\tau) - \eta f\frac{\partial f}{\partial \xi}(\xi,\tau) + \frac{\partial f}{\partial \eta}\int^\xi f\,ds - \frac{\partial f}{\partial \xi}\int^\eta f\,ds\right]$$
$$- \mu[\eta f(\xi,\tau) - \xi f(\eta,\tau)].$$

The condition that $u_2 = 0$ on $x = 0$ is automatically satisfied, while the condition that $u_2 = 0$ on $x = 1$ implies

$$\frac{\partial f}{\partial \tau}(t,\tau) + Nf\frac{\partial f}{\partial t} + \frac{1}{2}\mu f = 0, \tag{8.124}$$

on using the periodicity of f and measuring f from its mean so that $\int_{t-1}^{t+1} f\,ds = 0$. This is the same equation as (8.50), where there is damping by radiation, with $f \equiv -G_0$, since $\delta = \varepsilon\mu/2$.

Thus, in both cases there is a *cumulative* effect of damping which can prevent a shock forming — see (8.41). We can compare this with the effect of a variable cross section in Sec. 8.1.5, or an inhomogeneous elastic panel in 8.1.6 where the result is a simple wave and there is no cumulative effect of the variable cross section or inhomogeneity in preventing a shock.

We can also use a nonlinear difference equation to solve (7.4)–(7.5), and this is sketched here. The equations we need are (8.83) with $\delta \ll 1$, then

$$g(y) - (1 - 2\delta)g(s) = 0, \quad y = s + 2 - \delta^{-1}Ng(s)(1 - [1 - 2\delta + 2\delta^2])$$

or

$$g(y) - (1 - 2\delta)g(s) = 0, \quad y = s + 2 - 2Ng(s)(1 - \delta). \tag{8.125}$$

Then, with $F = -g$ and $2\delta = \varepsilon\mu$, the multiple scale expansion

$$F(t) = \varepsilon f(t,\tau) + \varepsilon^2 f_1(t,\tau) + \cdots, \quad \tau = \varepsilon t,$$

yields the result at $O(\varepsilon^2)$

$$\frac{\partial f}{\partial \tau} + Nf\frac{\partial f}{\partial t} + \frac{\mu}{2}f = 0. \tag{8.126}$$

Now solve (8.126) with $f(t,0) = \psi(\omega_l t)$: i.e.,

$$f(t,\tau) = \psi(\alpha)e^{-\frac{\mu}{2}\tau},$$
$$t = \alpha\omega_l^{-1} + N\psi(\alpha)(1 - e^{-\frac{\mu}{2}\tau})\frac{2}{\mu}.$$

Hence, a shock will form if

$$\left|N\psi'(\omega_l t)\right| > \frac{\mu}{2}\omega_l^{-1}, \quad \text{or}$$

$$\left|\varepsilon N\psi'(\omega_l t)\right| > \frac{\varepsilon\mu}{2}\omega_l^{-1} = \delta\omega_l^{-1} = \tau_p/2\tau_r = 1/\omega$$

since $\delta = l/(2c_0\tau_r)$ and $\omega_l = l/(c_0\tau_p)$. This is the same as (8.89) with $M = \varepsilon$.

8.2.3. *Sloshing in a Shallow Tank*

We are concerned here with one-dimensional long waves on shallow water in a tank of finite length l. From the background given in 7.2.3, by "shallow" we mean that the wavelength is long compared with the water depth. Hence, if the wavelength is of the same order as the tank length, we can define the dimensionless shallow water parameter as $\delta = H_0/l \ll 1$, where H_0 is the undisturbed fluid depth.

If waves are initiated at $x = 0$ and propagate to infinity the disturbance in the water is described by the Korteweg–de Vries (KdV) equation in the variables x and t, see Whitham [1974]. When the waves are confined to a tank of finite length it is shown here that the evolution of a component of the motion on a boundary again satisfies a KdV equation, but in the variables t and $\tau = \varepsilon t$.

The equations for the velocity potential $\Phi(x, y, t)$ and the surface displacement $\eta(x, t)$ in the two-dimensional irrotational motion of an incompressible inviscid liquid are

$$\delta^2\frac{\partial^2\Phi}{\partial x^2} + \frac{\partial^2\Phi}{\partial y^2} = 0, \quad 0 < x < 1, \quad -1 < y < \eta(x, t), \qquad (8.127)$$

$$\frac{\partial\Phi}{\partial y} = 0 \quad \text{on } y = -1, \qquad (8.128)$$

$$\frac{\partial\Phi}{\partial t} + \frac{1}{2}\left[\left(\frac{\partial\Phi}{\partial x}\right)^2 + \delta^{-2}\left(\frac{\partial\Phi}{\partial y}\right)^2\right] + \eta = 0 \quad \text{on } y = \eta(x, t),$$

$$(8.129)$$

$$\frac{\partial\Phi}{\partial y} - \delta^2\left[\frac{\partial\eta}{\partial t} + \frac{\partial\Phi}{\partial x}\frac{\partial\eta}{\partial x}\right] = 0 \quad \text{on } y = \eta(x, t). \qquad (8.130)$$

Equations (8.129) and (8.130) are the Bernoulli free surface condition and the kinematic free surface condition, respectively. In (8.127)–(8.130) the x is normalized by the tank length l, and y and the surface displacement η by H_0. The time t is measured in units of l/c_0, where $c_0 = \sqrt{gH_0}$ is the linear surface wave speed, g is the acceleration due to gravity, and the potential Φ has been normalized by $c_0 l$.

The second dimensionless parameter is $\varepsilon = a/H_0 \ll 1$, where a is the maximum amplitude of the surface elevation. In this example we assume that δ and ε are related by

$$\delta^2 = \kappa\varepsilon, \tag{8.131}$$

where $\kappa = O(1)$. The details of the normalization may be found in Cole [1968] or Whitham [1974]. The relation (8.131) between the dimensionless parameters ensures that the frequency dispersion induced by the effect of the tank bottom and the nonlinearity take effect and balance over the same time scale *viz.*, $\varepsilon t = O(1)$.

We now introduce the slow time scale $\tau = \varepsilon t$ and assume a small amplitude expansion of the form

$$\Phi(x, y, t, \varepsilon) = \varepsilon\Phi_0(x, y, t, \tau) + \varepsilon^2\Phi_1(x, y, t, \tau) + \cdots$$
$$\eta(x, t, \varepsilon) = \varepsilon\eta_0(x, t, \tau) + \varepsilon^2\eta_1(x, y, t, \tau) + \cdots .$$

On substituting the expansion for Φ into (8.127) and (8.128), using (8.131), and integrating gives

$$\Phi_0 = \varphi_0(x, t, \tau), \tag{8.132}$$

$$\Phi_1 = \varphi_1(x, t, \tau) - \frac{1}{2}\kappa(y + 1)^2\frac{\partial^2\varphi_0}{\partial x^2}, \tag{8.133}$$

$$\Phi_2 = \varphi_2(x, t, \tau) - \frac{1}{2}\kappa(y + 1)^2\frac{\partial^2\varphi_1}{\partial x^2} + \frac{1}{24}\kappa^2(y + 1)^2\frac{\partial^4\varphi_1}{\partial x^4}. \tag{8.134}$$

The Bernoulli free surface condition (8.129) at $O(\varepsilon)$ now yields the following equation, on noting $\frac{\partial}{\partial t} \to \frac{\partial}{\partial \tau} + \varepsilon\frac{\partial}{\partial \tau}$,

$$\frac{\partial\Phi_0}{\partial t} + \eta_0 = 0 \quad \text{on } y = \varepsilon\eta_0, \tag{8.135}$$

since $\frac{\partial \Phi_0}{\partial y} = 0$ from (8.132). Then

$$\eta_0 = -\frac{\partial \varphi_0}{\partial t}. \tag{8.136}$$

At the the next order we get

$$\frac{\partial \Phi_1}{\partial t} + \frac{\partial \Phi_0}{\partial \tau} + \frac{1}{2}\left(\frac{\partial \Phi_0}{\partial x}\right)^2 + \eta_1 = 0$$

on $y = \varepsilon \eta_0$, and using (8.132) and (8.133) this is

$$\frac{\partial \varphi_1}{\partial t} - \frac{1}{2}\kappa \frac{\partial^3 \varphi_0}{\partial x^2 \partial t} + \frac{\partial \varphi_0}{\partial \tau} + \frac{1}{2}\left(\frac{\partial \varphi_0}{\partial x}\right)^2 + \eta_1 = 0. \tag{8.137}$$

The kinematic free surface condition (8.130) gives at $O(\varepsilon)$:

$$\frac{\partial \Phi_0}{\partial y} = 0$$

and at $O(\varepsilon^2)$:

$$\frac{\partial \Phi_1}{\partial y} - \kappa \frac{\partial \eta_0}{\partial t} = 0 \quad \text{on } y = \varepsilon \eta_0$$

or using (8.133)

$$\frac{\partial \eta_0}{\partial t} = -\frac{\partial^2 \varphi_0}{\partial x^2}, \tag{8.138}$$

Now (8.136) and (8.138) give the linear wave equation

$$\frac{\partial^2 \varphi_0}{\partial t^2} - \frac{\partial^2 \varphi_0}{\partial x^2} = 0. \tag{8.139}$$

At $O(\varepsilon^2)$ we get

$$\frac{\partial \Phi_2}{\partial y} - \kappa \left[\frac{\partial \eta_1}{\partial t} + \frac{\partial \eta_0}{\partial t} + \frac{\partial \Phi_0}{\partial x}\frac{\partial \eta_0}{\partial x}\right] - \kappa \eta_0 \frac{\partial^2 \varphi_0}{\partial x^2} = 0.$$

Note that the last term here comes from the ε^3 term in $\varepsilon^2 \frac{\partial \Phi_1}{\partial y} = -\varepsilon^2 \kappa (1 + \varepsilon \eta_0)\frac{\partial^2 \varphi_0}{\partial x^2}$. On using (8.132), (8.134) and (8.136) we get

$$\frac{\partial \eta_0}{\partial \tau} + \frac{\partial \eta_1}{\partial t} = -\frac{\partial^2 \varphi_1}{\partial x^2} + \frac{\kappa}{6}\frac{\partial^4 \varphi_0}{\partial x^4} + \frac{\partial \varphi_0}{\partial x}\frac{\partial^2 \varphi_0}{\partial x \partial t} + \frac{\partial \varphi_0}{\partial t}\frac{\partial^2 \varphi_0}{\partial x^2}. \tag{8.140}$$

We use (8.137) to eliminate η_1 and (8.136) to eliminate η_0 and the result is

$$\frac{\partial^2 \varphi_1}{\partial x^2} - \frac{\partial^2 \varphi_1}{\partial t^2} = \kappa \left(\frac{1}{6} \frac{\partial^4 \varphi_0}{\partial x^4} - \frac{1}{2} \frac{\partial^4 \varphi_0}{\partial x^2 \partial t^2} \right)$$
$$+ 2\frac{\partial \varphi_0}{\partial x} \frac{\partial^2 \varphi_0}{\partial x \partial t} + \frac{\partial \varphi_0}{\partial t} \frac{\partial^2 \varphi_0}{\partial x^2} + 2\frac{\partial^2 \varphi_0}{\partial t \partial \tau}. \qquad (8.141)$$

We now apply the fixed end boundary conditions, i.e.,

$$\frac{\partial \varphi_0}{\partial x} = \frac{\partial \varphi_1}{\partial x} = 0 \quad \text{on } x = 0 \text{ and } x = 1. \qquad (8.142)$$

Equation (8.139) with $\frac{\partial \varphi_0}{\partial x} = 0$ on $x = 0$, $x = 1$ implies

$$\varphi_0(t, x, \tau) = f(\xi, \tau) + g(\eta, \tau), \qquad (8.143)$$

where $\xi = t - x$ and $\eta = t + x$, with

$$\frac{\partial f}{\partial t}(t, \tau) = \frac{\partial g}{\partial t}(t, \tau) \quad \text{and} \quad \frac{\partial f}{\partial t}(t + 2, \tau) = \frac{\partial f}{\partial t}(t, \tau). \qquad (8.144)$$

Thus the signal propagates as a linear wave.

In terms of ξ, η the left-hand side of (8.141) is $-4\frac{\partial^2 \varphi_1}{\partial \xi \partial \eta}$ and the first term on the right-hand side of (8.141) becomes

$$\kappa \left[\frac{1}{6} \left(\frac{\partial^4 f}{\partial \xi^4} + \frac{\partial^4 f}{\partial \xi^4} \right) - \frac{1}{2} \left(\frac{\partial^4 f}{\partial \xi^4} + \frac{\partial^4 g}{\partial \eta^4} \right) \right] = -\frac{\kappa}{3} \left(\frac{\partial^4 f}{\partial \xi^4} + \frac{\partial^4 g}{\partial \eta^4} \right).$$

Integrating this part of (8.141) with respect to ξ and η implies

$$-4\varphi_1 = -\frac{\kappa}{3} \left(\eta \frac{\partial^3 f}{\partial \xi^3} + \xi \frac{\partial^3 g}{\partial \eta^3} \right)$$

and then the first term on the right-hand side of (8.141) gives

$$-4\frac{\partial \varphi_1}{\partial x} = -\frac{\kappa}{3} \left[-\eta \frac{\partial^4 f}{\partial \xi^4} - \frac{\partial^3 g}{\partial \eta^3} + \frac{\partial^3 f}{\partial \xi^3} + \xi \frac{\partial^4 g}{\partial \eta^4} \right]. \qquad (8.145)$$

Treating the remaining terms on the right-hand side of (8.141) in a similar way, we arrive at the result

$$
\begin{aligned}
-4\frac{\partial \varphi_1}{\partial x} = {} & -\frac{\kappa}{3}\left[-\eta\frac{\partial^4 f}{\partial \xi^4} - \frac{\partial^3 g}{\partial \eta^3} + \frac{\partial^3 f}{\partial \xi^3} + \xi\frac{\partial^4 g}{\partial \eta^4}\right] \\
& + \frac{3}{2}\left[-2\eta\frac{\partial f}{\partial \xi}\frac{\partial^2 f}{\partial \xi^2} + 2\xi\frac{\partial g}{\partial \eta}\frac{\partial^2 g}{\partial \eta^2} + \left(\frac{\partial f}{\partial \xi}\right)^2 - \left(\frac{\partial g}{\partial \eta}\right)^2\right] \\
& - f\frac{\partial^2 g}{\partial \eta^2} + g\frac{\partial^2 f}{\partial \xi^2} \\
& + 2\left[-\eta\frac{\partial^2 f}{\partial \xi \partial \tau} - \frac{\partial g}{\partial \tau} + \frac{\partial f}{\partial \tau} + \xi\frac{\partial^2 g}{\partial \eta \partial \tau}\right].
\end{aligned}
\tag{8.146}
$$

The condition $\frac{\partial \varphi_1}{\partial x} = 0$ on $x = 0$ is automatically satisfied by (8.146) on noting that $\xi = \eta = t$ and the first of (8.144).

On $x = 1$, $\xi = t - 1$ and $\eta = t + 1$ and we use both of (8.144) to get

$$
\frac{2}{3}\kappa\frac{\partial^4 f}{\partial t^4}(t - 1, \tau) - 6\frac{\partial f}{\partial t}\frac{\partial^2 f}{\partial^2 t}(t - 1, \tau) - 4\frac{\partial^2 f}{\partial t \partial \tau}(t - 1, \tau) = 0,
\tag{8.147}
$$

where we have also used $g(t, \tau) = f(t, \tau)$ and $f(t + 2, \tau) = f(t, \tau)$ which are compatible with (8.144).

Now let $\frac{\partial f}{\partial t}(t, \tau) = w(t, \tau)$ and (8.147) becomes

$$
\frac{\partial w}{\partial \tau} + \frac{3}{2}w\frac{\partial w}{\partial t} - \frac{\kappa}{6}\frac{\partial^3 w}{\partial t^3} = 0, \qquad t > 0, \quad \tau > 0.
\tag{8.148}
$$

This is the KdV equation for the variable $w(t, \tau)$. So the KdV equation that governs the flow in a semi-infinite tank appears again as governing the evolution of one component of the signal on the boundary $x = 1$, while the wave in the medium propagates as a linear wave.

One component of the wave motion evolves on the boundary $x = 1$ according to the KdV equation. The motion is periodic in the fast time t by (8.144), but is modulated on the slow time τ. We note that (8.148) does not contain an integral term and thus the oppositely traveling waves do not interact, in contrast to Example 1.

Hence, the motion in the tank consists of oppositely traveling waves that do not interact and evolve according to the KdV equation.

An alternative form of $\frac{\partial \varphi_1}{\partial x}$ to that given by (8.146) is, on using the first of (8.144) to eliminate g,

$$
\begin{aligned}
-4\frac{\partial \varphi_1}{\partial x} =\ & \frac{\kappa}{3}[2x\{f''''(t-x,\tau) + f''''(t+x,\tau)\} \\
& + f'''(t+x,\tau) - f'''(t-x,\tau)] \\
& -\frac{3}{2}[4x\{f'(t-x,\tau)f''(t-x,\tau) + f'(t+x,\tau)f''(t+x,\tau)\} \\
& + f'^2(t+x,\tau) - f'^2(t-x,\tau)] \\
& -f(t+x,\tau)f''(t-x,\tau) + f(t-x,\tau)f''(t+x,\tau) \\
& +2[2x\{f'_\tau(t-x,\tau) + f'_\tau(t+x,\tau)\} \\
& +f_\tau(t+x,\tau) - f_\tau(t-x,\tau)],
\end{aligned}
\tag{8.149}
$$

where prime means differentiation w.r.t. the first argument, see the comment after (8.109). Then (8.149) with the boundary conditions (8.144) again give the result (8.148).

8.2.4. *Alternative Derivation for a Nonlinear Standing Surface Wave in a Shallow Tank*

We know that the relevant Eq. (8.148) is a balance between nonlinearity and dispersion. Now consider the nonlinear differential–difference equation

$$
w(t + Q(t)) - w(t) - \frac{\kappa}{3}\frac{\partial^3 w}{\partial t^3} = 0,
\tag{8.150}
$$

where

$$
Q(t) = 2\left[1 + \frac{3}{2}w(t)\right].
\tag{8.151}
$$

Note that (8.150) and (8.151) are a particular form of (8.25) (that describes the nonlinearity) with the addition of a dispersive term. On expanding $w(t + Q(t))$ for small $|w|$, the equation for the motion of period 2 is

$$\frac{3}{2}w\frac{\partial w}{\partial t} - \frac{\kappa}{6}\frac{\partial^3 w}{\partial t^3} = 0 \qquad (8.152)$$

which is the steady state KdV equation. If we now assume that

$$w(t;\varepsilon) = \varepsilon w_0(t,\tau) + \varepsilon^2 w_1(t,\tau) + \cdots, \qquad \tau = \varepsilon t,$$

a multiscale expansion of the nonlinear term in (8.150) yields $w_0\,(t+2,\tau) = w_0(t,\tau)$, on noting $\delta^2 = \kappa\varepsilon$, with

$$\frac{\partial w_0}{\partial \tau} + \frac{3}{2}w_0\frac{\partial w_0}{\partial t} - \frac{\kappa}{6}\frac{\partial^3 w_0}{\partial t^3} = 0 \qquad (8.153)$$

and the KdV equation (8.148) is recovered.

8.2.5. *Nonlinear Hydraulic or Long Wave Sloshing in a Tank*

When nonlinearity dominates frequency dispersion the relevant equations are the *hydraulic equations*, the dimensionless versions of (2.27):

$$\eta_t + [(1+\eta)u]_x = 0, \quad u_t + uu_x + \eta_x = 0. \qquad (8.154)$$

These are nondimensionalized in the same way as the previous section, so $H_0 = 1$ and $L = l$.

As before, let $\tau = \varepsilon t$, where $\varepsilon = a/H_0 \ll 1$, and assume the expansion

$$u(t,x;\varepsilon) = \varepsilon u_1(t,x,\tau) + \varepsilon^2 u_2(t,x,\tau) + \cdots \qquad (8.155)$$

$$\eta(t,x;\varepsilon) = \varepsilon\eta_1(t,x,\tau) + \varepsilon^2\eta_2(t,x,\tau) + \cdots . \qquad (8.156)$$

The boundary conditions for a tank closed at both ends are $u(0,t) = u(1,t) = 0$.

Since $\frac{\partial}{\partial t} \to \frac{\partial}{\partial \tau} + \varepsilon \frac{\partial}{\partial \tau}$ we obtain at $O(\varepsilon)$:

$$\frac{\partial \eta_1}{\partial t} + \frac{\partial u_1}{\partial x} = 0, \quad \frac{\partial u_1}{\partial t} + \frac{\partial \eta_1}{\partial x} = 0,$$

where $u_1(0, t, \tau) = u_1(1, t, \tau) = 0$. Then

$$\frac{\partial^2 \eta_1}{\partial t^2} - \frac{\partial^2 \eta_1}{\partial x^2} = 0,$$

so that

$$\eta_1 = f(\alpha, \tau) - g(\beta, \tau), \; u_1 = f(\alpha, \tau) + g(\beta, \tau),$$

where $\alpha = t - x$, and $\beta = t + x$.

Now $u_1(0, t, \tau) = 0$ implies $f(t, \tau) + g(t, \tau) = 0$, i.e., $g = -f$. Further, $u_1(1, t, \tau) = 0$ implies $f(t - 1, \tau) = f(t + 1, \tau)$. Then

$$u_1 = f(\alpha, \tau) - f(\beta, \tau), \; \eta_1 = f(\alpha, \tau) + f(\beta, \tau). \tag{8.157}$$

At $O(\varepsilon^2)$ we find that

$$\frac{\partial \eta_2}{\partial t} + \frac{\partial u_2}{\partial x} = -\frac{\partial \eta_1}{\partial \tau} - \frac{\partial}{\partial x}(\eta_1 u_1) \tag{8.158}$$

and

$$\frac{\partial u_2}{\partial t} + \frac{\partial \eta_2}{\partial x} = -\frac{\partial u_1}{\partial \tau} - u_1 \frac{\partial u_1}{\partial x}. \tag{8.159}$$

The boundary conditions are $u_2(0, t) = u_2(1, t) = 0$.

Eliminating η_2 from (8.158) and (8.159) gives

$$\frac{\partial^2 u_2}{\partial t^2} - \frac{\partial^2 u_2}{\partial x^2} = -\frac{\partial^2 \eta_1}{\partial x \partial \tau} - \frac{\partial^2}{\partial x^2}(\eta_1 u_1) + \frac{\partial^2 u_1}{\partial x \partial \tau} + \frac{\partial}{\partial t}\left(u_1 \frac{\partial u_1}{\partial x} \right),$$

or switching to α, β variables,

$$-4 \frac{\partial^2 u_2}{\partial \alpha \partial \beta} = -\frac{\partial}{\partial \tau}\left(-\frac{\partial}{\partial \alpha} + \frac{\partial}{\partial \beta} \right) \eta_1 - \left(-\frac{\partial}{\partial \alpha} + \frac{\partial}{\partial \beta} \right)^2 (\eta_1 u_1)$$

$$+ \frac{\partial}{\partial \tau}\left(\frac{\partial}{\partial \alpha} + \frac{\partial}{\partial \beta} \right) u_1 + \left(\frac{\partial}{\partial \alpha} + \frac{\partial}{\partial \beta} \right) (u_1(u_1)_x)$$

$$= 2f_{\alpha\tau} - 2f_{\beta\tau} - \left[2\frac{\partial}{\partial\alpha}(f(\alpha)f_\alpha) - 2\frac{\partial}{\partial\beta}(f(\beta)f_\beta)\right]$$

$$- \frac{\partial}{\partial\alpha}(f(\alpha)f'(\alpha)) + \frac{\partial}{\partial\alpha}f'(\alpha)f(\beta) - \frac{\partial}{\partial\alpha}f(\alpha)f'(\beta)$$

$$+ \frac{\partial}{\partial\beta}(f(\beta)f'(\beta)) + \frac{\partial}{\partial\beta}f(\beta)f'(\alpha) - \frac{\partial}{\partial\beta}f'(\beta)f(\alpha).$$

On integrating we get

$$-4u_2 = 2\beta f_\tau(\alpha) - 2\alpha f_\tau(\beta) - 2\left[\beta f f_\alpha - \alpha f f_\beta\right]$$

$$+ \alpha f f_\beta - f'(\beta)\int^\alpha f d\alpha - \beta f f_\alpha + f'(\alpha)\int^\beta f d\beta. \quad (8.160)$$

On $x = 0$, $\alpha = \beta = t$ and hence $u_2 = 0$ on $x = 0$ is automatically satisfied.

On $x = 1$, $\alpha = t - 1$, $\beta = t + 1$, $f(t - 1, \tau) = f(t + 1, \tau)$ and $u_2 = 0$. Then

$$0 = 2(t + 1)f_\tau(t - 1) - 2(t - 1)f_\tau(t + 1)$$

$$- 2\left[(t + 1)f f_t(t + 1) - (t - 1)f f_t(t + 1)\right]$$

$$+ (t - 1)f f_t(t + 1) - f'(t + 1)\int^{t-1} f ds$$

$$- (t + 1)f f_t(t - 1) + f'(t - 1)\int^{t+1} f ds,$$

where the third-to-last and last terms can be simplified to $f'(t - 1)\int_{t-1}^{t+1} f ds = 0$. Further, the t terms cancel, so that

$$f_\tau - \frac{3}{2}f f_t = 0. \quad (8.161)$$

For the equivalent problem for a polytropic gas

$$f_\tau + N f f_t = 0,$$

where $N = -\frac{3}{2} = -\frac{\gamma+1}{2}$, i.e., $\gamma = 2$. This is a well known result and is called the gas dynamic analogy: see Kevorkian and Cole [2011] or Stoker [1957].

An alternative form of the particular integral (8.160) in which there are no secular terms is

$$u_2 = -x[f_\tau(t-x,\tau) + f_\tau(t+x,\tau)]$$
$$+ \frac{3}{2}x[f(t-x,\tau)f'(t-x,\tau) + f(t+x,\tau)f'(t+x,\tau)]$$
$$+ \frac{1}{4}f'(t+x)\int^{t-x} f\,ds - f'(t-x)\frac{1}{4}\int^{t+x} f\,ds.$$

This again leads to (8.161).

8.2.6. *The Boussinesq Equations*

Adding frequency dispersion to the hydraulic equations in the previous section we obtain the dimensionless form of the Boussinesq equations (see Whitham [1974])

$$\eta_t + [(1+\eta)u]_x = 0, \ \ u_t + uu_x + \eta_x + \frac{\delta^3}{3}\eta_{xtt} = 0, \qquad (8.162)$$

where the undisturbed depth is $H_0 = 1$ and $\delta^2 = \kappa\varepsilon$, giving a balance between nonlinearity and dispersion. We apply the boundary conditions $u(0,t) = u(1,t) = 0$ and again assume the expansion (8.155) and (8.156) with $\tau = \varepsilon t$.

At $O(\varepsilon)$

$$\eta_1 = f(\alpha,\tau) + f(\beta,\tau), \ \ u_1 = f(\alpha,\tau) - f(\beta,\tau),$$

where $\alpha = t - x$, and $\beta = t + x$ and $f(t+2,\tau) = f(t,\tau)$. Following the procedure of the previous section, $f(t,\tau)$ satisfies

$$\frac{\partial f}{\partial \tau} - \frac{3}{2}f\frac{\partial f}{\partial t} - \frac{\kappa}{6}\frac{\partial^3 f}{\partial t^3} = 0 \qquad (8.163)$$

which agrees with (8.153) on putting $f = -w_0$. So a component of a standing wave governed by the Boussinesq equation satisfies the KdV equation, as expected.

8.2.7. *A Generalized KdV Equation: An Anharmonic Lattice or a Nonlinear Dispersive String*

A connection between the vibrations of a nonlinear string and the KdV equation (8.148) was found by Fermi, Pasta and Ulam [1955] (FPU). Here, we show how solitary waves appear naturally when one seeks a periodic solution of the vibrating string equation subject to fixed boundaries.

Fermi, Pasta and Ulam [1955] began a numerical investigation of the vibration of a nonlinear string with the purpose of studying the approach to equipartition of energy among the various modes. The FPU investigation is related to the process of relaxation to thermal equilibrium in a nonlinear lattice. The expectation was that for some nonequilibrium distributions of modal energies there would be the eventual equipartition of energy amongst all the available modes. Initially, when all the energy was in the lowest mode it was expected to rapidly distribute over the available 32 modes of the system. Instead it was shared by only a few modes.

Subsequently Kruskal and Zabusky [1964], Lax [1957] and Keller and Ting [1966] showed the impossibility of nontrivial periodic free vibrations of such a nonlinear string. Zabusky [1967] in an attempt to find a continuum model for an anharmonic (nonlinear) lattice retained a higher derivative term in the nonlinear string equation so that the linear version of the equation is frequency dispersed (or dispersive). Zabusky's equation has the form

$$\frac{\partial^2 y}{\partial t^2} = \left(1 + \varepsilon \frac{\partial y}{\partial x}\right) \frac{\partial^2 y}{\partial x^2} + \Lambda^2 \varepsilon \frac{\partial^4 y}{\partial x^4}, \quad 0 \le x \le 1, \tag{8.164}$$

where $y(t, x; \varepsilon)$ is the displacement of a particle from its equilibrium position, $\varepsilon \ll 1$ is a measure of nonlinearity, Λ^2 is a constant, and the undisturbed sound speed is normalized to unity.

We note that for a unidirectional wave, e.g., $y = f(t - x)$, then (8.164) reduces to

$$\Lambda^2 g''' - g g' = 0, \tag{8.165}$$

where $g = f'$. This is a steady state KdV equation. If we look for a solution $y = f(t + x)$, then we get

$$\Lambda^2 g''' + gg' = 0. \tag{8.166}$$

Numerical studies in Zabusky and Kruskal [1965] showed that for the KdV equation, given initial conditions evolved to a series of solitary waves in the far field. This observation culminated in a method for finding exact solutions for the KdV equation subject to specified initial data, see Gardener, Greene, Kruskal and Miura [1967].

Our interest here is to show how solitary waves appear naturally when one seeks a solution of (8.164) subject to fixed end boundary conditions

$$\frac{\partial y}{\partial t} = 0 \quad \text{at } x = 0 \text{ and } x = 1. \tag{8.167}$$

Since we expect that the nonlinear and dispersive terms balance when $\varepsilon t = O(1)$, we introduce the slow time scale $\tau = \varepsilon t$. Then we assume

$$y(t, x; \varepsilon) = y_0(t, x, \tau) + \varepsilon y_1(t, x, \tau) + \cdots . \tag{8.168}$$

At $O(1)$, (8.164) and (8.168) imply

$$\frac{\partial^2 y_0}{\partial t^2} - \frac{\partial^2 y_0}{\partial x^2} = 0 \tag{8.169}$$

with the boundary conditions

$$\frac{\partial y_0}{\partial t} = 0 \quad \text{on } x = 0 \text{ and } x = 1. \tag{8.170}$$

Then,

$$y_0(t, x, \tau) = f(t + x, \tau) - f(t - x, \tau), \tag{8.171}$$

where f is arbitrary, subject only to the condition

$$f(t - 1, \tau) = f(t + 1, \tau). \tag{8.172}$$

So f has period 2 in the fast time scale, t, and is modulated on the slow time scale τ. At $O(\varepsilon)$, (8.164) and (8.168) imply

$$\frac{\partial^2 y_1}{\partial t^2} - \frac{\partial^2 y_1}{\partial x^2} = -2\frac{\partial^2 y_0}{\partial t \partial \tau} + \frac{\partial y_0}{\partial x}\frac{\partial^2 y}{\partial x^2} + \Lambda^2\frac{\partial^4 y_0}{\partial x^4}, \tag{8.173}$$

subject to the boundary conditions

$$\frac{\partial y_1}{\partial t} = -\frac{\partial y_0}{\partial \tau} \quad \text{on } x = 0 \text{ and } x = 1. \tag{8.174}$$

Now $\frac{\partial y_0}{\partial \tau} = 0$ on $x = 0$ and $x = 1$ by (8.171) and (8.172), and then the boundary condition is $\frac{\partial y_1}{\partial t} = 0$ on $x = 0$ and $x = 1$. We introduce $\xi = t - x$, $\eta = t + x$ and (8.173) becomes

$$\begin{aligned}
4\frac{\partial^2 y_1}{\partial \xi \partial \eta} = {}&-2\frac{\partial}{\partial \tau}\left[f'(\xi,\tau) - f'(\eta,\tau)\right] \\
&+ f'(\eta,\tau)f''(\eta,\tau) - f'(\xi,\tau)f''(\xi,\tau) \\
&+ f'(\xi,\tau)f''(\eta,\tau) - f'(\eta,\tau)f''(\xi,\tau) \\
&+ \Lambda^2[f''''(\eta,\tau) - f''''(\xi,\tau)],
\end{aligned} \tag{8.175}$$

where prime indicates differentiation w.r.t. the first variable. Then (8.175) integrates to

$$\begin{aligned}
4y_1 = {}&-2\frac{\partial}{\partial \tau}\left[\eta f(\xi,\tau) - \xi f(\eta,\tau)\right] + \frac{1}{2}\xi f'^2(\eta,\tau) - \frac{1}{2}\eta f'^2(\xi,\tau) \\
&+ f(\xi,\tau)f'(\eta,\tau) - f(\eta,\tau)f'(\xi,\tau) \\
&+ \Lambda^2\left[\xi f'''(\eta,\tau) - \eta f'''(\xi,\tau)\right]
\end{aligned} \tag{8.176}$$

and

$$\begin{aligned}
4\frac{\partial y_1}{\partial t} = {}&-2\frac{\partial}{\partial \tau}\left[f(\xi,\tau) + \eta f'(\xi,\tau) - \xi f'(\eta,\tau) - f(\eta,\tau)\right] \\
&+ \frac{1}{2}\left[2\xi f'(\eta,\tau)f''(\eta,\tau) + f'^2(\eta,\tau)\right] \\
&- \frac{1}{2}\left[2\eta f'(\xi,\tau)f''(\xi,\tau) + f'^2(\xi,\tau)\right] \\
&+ f(\xi,\tau)f''(\eta,\tau) - f(\eta,\tau)f''(\xi,\tau) \\
&+ \Lambda^2[\xi f''''(\eta,\tau) + f'''(\eta,\tau) - f'''(\xi,\tau) - \eta f''''(\xi,\tau)].
\end{aligned} \tag{8.177}$$

On $x = 0$, $\xi = \eta = t$ and (8.177) satisfies $\frac{\partial y_1}{\partial t} = 0$ on $x = 0$ automatically.

On $x = 1$, $\xi = t - 1$ and $\eta = t + 1$, then (8.177) implies, on noting (8.172),

$$2\frac{\partial g}{\partial \tau}(t, \tau) + g(t, \tau)\frac{\partial g}{\partial t}(t, \tau) + \Lambda^2 \frac{\partial^3 g}{\partial t^3} = 0, \tag{8.178}$$

on setting $g(t, \tau) = \frac{\partial f}{\partial t}(t, \tau)$. So (8.178) is a KdV equation, and shows that a standing wave evolves on the boundary $x = 1$ according to a KdV equation in the fast variable t and the slow variable τ. The basic solution (8.171) with period 2 in the fast time scale is modulated according to a KdV equation. The steady state version of (8.178) when $\frac{\partial g}{\partial \tau} = 0$ agrees with (8.166).

A particular integral of (8.175) for $\frac{\partial y_1}{\partial t}$, containing no secular terms, is

$$\begin{aligned}
4\frac{\partial y_1}{\partial t} = &-2\frac{\partial}{\partial \tau}\left[f(\xi, \tau) - f(\eta, \tau) + 2x\{f'(\xi, \tau) + f'(\eta, \tau)\}\right] \\
&- \frac{1}{2}[4x\{f'(\eta, \tau)f''(\eta, \tau) + f'(\xi, \tau)f''(\xi, \tau)\} \\
&+ f'^2(\xi, \tau) - f'^2(\eta, \tau)] \\
&+ f(\xi, \tau)f''(\eta, \tau) - f(\eta, \tau)f''(\xi, \tau) \\
&+ \Lambda^2 \left[-2x\{f''''(\eta, \tau) + f''''(\xi, \tau)\} + f'''(\eta, \tau)\} - f'''(\xi, \tau)\right]
\end{aligned}$$

and this leads to (8.178).

Chapter 9

Nonlinear Resonance: Shocked Solutions

We now consider the inclusion of nonlinear effects in the classical problem of the resonant excitation of a gas in a straight cylindrical closed tube. The governing equations in Eulerian coordinates in dimensional form from Sec. 2.2 are (2.14) where $c_0 = \sqrt{\frac{\gamma p_0}{\rho_0}}$ is the linear sound speed in the reference state. The typical configuration is a tube of length l, closed at one end, containing an homogeneous gas excited by the periodic motion of a piston with amplitude δ and frequency $\overline{\omega}$. The tube is closed at $\overline{x} = 0$, while the gas is forced by a piston with period τ_p at $\overline{x} = l$, so that the boundary conditions can be written in terms of the velocity as

$$\overline{u}(0,\overline{t}) = 0 \quad \text{and} \quad \overline{u}(l,\overline{t}) = \delta\overline{\omega}\sin(2\pi\overline{t}/\tau_p), \tag{9.1}$$

where δ and τ_p are the piston displacement and period respectively, and $\overline{\omega} = 2\pi/\tau_p$.

The problem is simplified if it is converted into dimensionless variables by taking $L = l$, and $T = lc_0^{-1}$ then (9.1) become

$$u(0,t) = 0 \quad \text{and} \quad u(1,t) = M\sin(2\pi\omega t), \tag{9.2}$$

where the dimensionless frequency is $\omega = l/c_0\tau_p$ and $M = \overline{\omega}\delta/c_0 = 2\pi\varepsilon\omega \ll 1$ is the Mach number and $\varepsilon = \delta/l$. Exactly at resonance, when τ_p equals the return travel time in the tube, $\tau_p = 2l/c_0$ so that $\omega = 1/2$. The governing equations are then (2.14) with $c_0 = 1$. The effects of viscosity and heat conduction are ignored to first order

206

and may be considered as corrections. Thus, the focus is directly on the nonlinearity.

From Sec. 4.2 the solution to the linear problem subject to the boundary conditions (9.2) written in normal mode form is (4.12)

$$u(x,t) = \frac{M}{\sin(2\pi\omega)} \sin(2\pi\omega t)\sin(2\pi\omega x), \ \omega \neq n/2, \ n = 1,2,3\ldots.$$

$$(9.3)$$

It is clear that any attempt at a solution of the nonlinear resonance problem (when $\omega \backsim 1/2$) by a perturbation based on the linear approximation (9.3) must fail. Additionally, at the next order in a regular perturbation expansion in powers of the amplitude we find $u_2(x,y) \backsim \sin^{-2}(2\pi\omega)$ and the perturbation scheme also breaks down when $\omega = \Omega_0 = 1/4$. We refer to $\Omega_0 = 1/4$ as "subharmonic" or "quadratic" resonance.

This problem of nonlinear resonance has been studied experimentally and theoretically for many years. One of the earliest experimental studies was that of Lettau [1939]. His experiments involved both closed and open tubes and he found that, even for relatively small excitation amplitudes, $\varepsilon = \delta/l \ll 1$, shock waves are a feature of the motion for a straight closed tube. Lettau also found that for higher resonant modes shocks could form in one travel time in the tube. Similarly the experiments of Saenger and Hudson [1960] showed that shocks occur in a closed tube for a piston amplitude as small as $\frac{1}{8}$ inch in a tube of length 132 inches when the piston was operating at the fundamental resonant frequency. They also reported that if the driving frequency is near a higher resonance then oppositely traveling shocks collide and pass through each other in the interior of the tube. They summarized their observations: the shock waves have constant strength in the tube, travel at a constant adiabatic sound speed, and do not interact with each other. The shocks also have a noticeably greater amplitude compared with continuous oscillations at nonresonant frequencies. These are the observations that must be predicted theoretically. Other similar experiments were subsequently reported by Gulyayev and Kusnetsov [1963], Galiev, Ilgamov and Sadykov [1970], Zaripov and Ilgamov [1976] and Sturtevant [1974]. Periodic motions containing shocks were also observed at frequencies around half the fundamental in a number of experiments e.g., Galiev, Ilgamov and Sadykov [1970], Zaripov and Ilgamov [1976] and Althaus and Thomann [1987].

The early analytic approaches were those of Betchov [1958], Gorkov [1963] and Chester [1964]. Betchov [1958] first constructed an analytic solution to the closed end problem based on the assumption that it consisted of a continuous and a discontinuous component, and was derived from the inviscid equations with no dissipation. This showed that the boundedness of the solution arose from the nonlinearity that gave rise to shocks and did not require an extra dissipative mechanism. He also gave a simple energy argument, based on shock dissipation, which gave the order of magnitude of the shocked solution. Gorkov [1963] worked with the velocity potential to set up a nonlinear wave equation that was solved by successive approximations. He found the same basic equation as Chester [1964]. In 1964, Chester gave a deductive argument where shock waves appear as a natural outcome of the solution in a well defined frequency band around each resonant frequency — called the resonant band. This is the most influential paper dealing with shocks in closed tubes, and we first describe the essentials of his procedure in some detail.

9.1. Chester's Procedure

To follow Chester's procedure we replace $e(x,t)$ in (2.14) with $c(x,t)$, the dimensionless sound speed in Eulerian coordinates, and so from (2.13), $1 + e = c^{2/(\gamma-1)}$. Then (2.14) becomes

$$\frac{\partial c}{\partial t} + u\frac{\partial c}{\partial x} + \frac{\gamma-1}{2}c\frac{\partial u}{\partial x} = 0, \qquad \frac{\partial u}{\partial t} + u\frac{\partial u}{\partial x} + \frac{2}{\gamma-1}c\frac{\partial c}{\partial x} = 0. \quad (9.4)$$

Adding and subtracting the two equations in (9.4) yields

$$\left[\frac{\partial}{\partial t} + (u+c)\frac{\partial}{\partial x}\right]\left[u + \frac{2}{\gamma-1}c\right] = 0,$$

$$\left[\frac{\partial}{\partial t} + (u-c)\frac{\partial}{\partial x}\right]\left[u - \frac{2}{\gamma-1}c\right] = 0. \quad (9.5)$$

The terms $u \pm \frac{2}{\gamma-1}c$ are the Riemann Invariants of the system.

We now write $c = 1 + c_1$, where $c_1 \ll 1$ is the nonlinear correction to the linear sound speed. To keep a consistent notation for the small amplitude motion we write $u = u_1 \ll 1$. Then (9.5) are written as

$$\left(\frac{\partial}{\partial t} + \frac{\partial}{\partial x}\right)\left(u_1 + \frac{2}{\gamma - 1}c_1\right) = -(u_1 + c_1)\frac{\partial}{\partial x}\left(u_1 + \frac{2}{\gamma - 1}c_1\right),$$

$$(9.6)$$

$$\left(\frac{\partial}{\partial t} - \frac{\partial}{\partial x}\right)\left(u_1 - \frac{2}{\gamma - 1}c_1\right) = -(u_1 - c_1)\frac{\partial}{\partial x}\left(u_1 - \frac{2}{\gamma - 1}c_1\right).$$

$$(9.7)$$

We know from Saenger and Hudson [1960] that the resonant shock waves travel at a constant sound speed and do not interact. Then the acoustic (linear) equations must be part of the solution, even though the linear solution (9.3) is unbounded at resonance. The insight that resolved the difficulty was expressed by Chester [1964]:

"Although the acoustic approximation will in general be the significant part of the disturbance, this will not be so the in neighborhood of a node of that approximation. ...if a boundary condition is to be applied in such a neighborhood, a reliable solution cannot be obtained unless the first approximation is improved, and the extra terms will be all important if the first approximation is locally zero".

We focus here on the solution exactly at the fundamental resonance to see how the procedure works. Later, we will deal with the neighborhood of resonance and the resonant band. Chester [1964] assumes that $u = u_1 + u_2$ where $u_2 \ll u_1$. Then from (9.6) and (9.7) the first approximation is

$$u_1 + \frac{2}{\gamma - 1}c_1 = 2f_1(\xi), \quad u_1 - \frac{2}{\gamma - 1}c_1 = 2f_2(\eta), \qquad (9.8)$$

where f_1, f_2 are arbitrary functions representing solutions of (9.6) and (9.7) when the terms on the right-hand side are neglected, and $\xi = t - x$, $\eta = t + x$. The motion given by (9.8) is the superposition of two noninteracting waves traveling at the linear sound speed. It remains to calculate the signals f_1, f_2 carried by these waves.

At the next order the approximation is given by

$$\left(\frac{\partial}{\partial t} + \frac{\partial}{\partial x}\right)\left(u_2 + \frac{2}{\gamma - 1}c_2\right) = 2f_1'(\xi)\left(\frac{\gamma + 1}{2}f_1(\xi) + \frac{3 - \gamma}{2}f_2(\eta)\right),$$

$$(9.9)$$

$$\left(\frac{\partial}{\partial t} - \frac{\partial}{\partial x}\right)\left(u_2 - \frac{2}{\gamma - 1}c_2\right) = -2f_2'(\eta)\left(\frac{3 - \gamma}{2}f_1(\xi) + \frac{\gamma + 1}{2}f_2(\eta)\right).$$

$$(9.10)$$

To find particular integrals for (9.9) and (9.10) we rewrite them as

$$\frac{\partial}{\partial \eta}\left(u_2 + \frac{2}{\gamma - 1}c_2\right) = f_1'(\xi)\left(\frac{\gamma + 1}{2}f_1(\xi) + \frac{3 - \gamma}{2}f_2(\eta)\right),$$

$$\frac{\partial}{\partial \xi}\left(u_2 - \frac{2}{\gamma - 1}c_2\right) = -f_2'(\eta)\left(\frac{3 - \gamma}{2}f_1(\xi) + \frac{\gamma + 1}{2}f_2(\eta)\right),$$

and these integrate to

$$u_2 + \frac{2}{\gamma - 1}c_2 = f_1'(\xi)\left(\frac{\gamma + 1}{2}\eta f_1(\xi) + \frac{3 - \gamma}{2}\int^\eta f_2(s)ds\right) \qquad (9.11)$$

and

$$u_2 - \frac{2}{\gamma - 1}c_2 = -f_2'(\eta)\left(\frac{3 - \gamma}{2}\int^\xi f_1(s)ds + \frac{\gamma + 1}{2}\xi f_2(\eta)\right).$$

$$(9.12)$$

Since the presence of η in (9.11) and ξ in (9.12) gives rise to secular terms, we can replace (9.11) and (9.12) by

$$u_2 + \frac{2}{\gamma - 1}c_2 = (\gamma + 1)xf_1(\xi)f_1'(\xi) + \frac{3 - \gamma}{2}f_1'(\xi)\int^\eta f_2(s)ds \quad (9.13)$$

and

$$u_2 - \frac{2}{\gamma - 1}c_2 = (\gamma + 1)xf_2(\eta)f_2'(\eta) - \frac{3 - \gamma}{2}f_2'(\xi)\int^\xi f_1(s)ds \quad (9.14)$$

which are found by replacing η by $2x$ and ξ by $-2x$, as was demonstrated in Chapter 7. These agree with (3.7) and (3.8) in Chester

[1964], which are obtained by noting

$$\left(\frac{\partial}{\partial t} + \frac{\partial}{\partial x}\right)(x f_1(\xi)) = f_1(\xi)$$

and

$$\left(\frac{\partial}{\partial t} + \frac{\partial}{\partial x}\right)\left(\frac{1}{2}f_1(\xi)f_2(\eta)\right) = f_1(\xi)f_2'(\eta).$$

Now $u = u_1 + u_2$ is given by

$$u = u_1 + u_2 = f_1(\xi) + f_2(\eta) + \frac{\gamma+1}{2}x\left(f_1(\xi)f_1'(\xi) + f_2(\eta)f_2'(\eta)\right)$$

$$+\frac{3-\gamma}{4}\left(f_1'(\xi)\int^\eta f_2(s)ds - f_2'(\eta)\int^\xi f_1(s)ds\right).$$

Using this equation, the boundary condition $u = 0$ on $x = 0$ implies $f_1 = -f_2 = f$, say. Then

$$u = f(\xi) - f(\eta) + \frac{\gamma+1}{2}x\left(f(\xi)f'(\xi) + f(\eta)f'(\eta)\right)$$

$$+\frac{3-\gamma}{4}\left(f'(\eta)\int^\xi f(s)ds - f'(\xi)\int^\eta f(s)ds\right). \qquad (9.15)$$

The full boundary condition on $x = 1$ is (9.2).

Chester [1964] now makes the assumption "*that the appropriate asymptotic solution for f is periodic with the same period as the piston*". In our dimensionless variables the period of the piston is 2. Hence, f has period 2 and for all t, $f(t-1) - f(t+1) = 0$. Thus, u_1 is a linear standing wave, and u_2 is an iteration about it. Then the boundary condition (9.2) on $x = 1$ and (9.15) yield

$$M\sin(\pi t) = f(t-1) - f(t+1)$$

$$+\frac{\gamma+1}{2}\left(f(t-1)f'(t-1) + f(t+1)f'(t+1)\right)$$

$$+\frac{3-\gamma}{4}\left(f'(t+1)\int^{t-1} f(s)ds - f'(t-1)\int^{t+1} f(s)ds\right).$$

Using the periodicity of f with $\omega = 1/2$, this equation implies

$$M \sin(\pi t) = (\gamma+1) f(t-1) f'(t-1) + \frac{3-\gamma}{4} \left(-f'(t-1) \int_{t-1}^{t+1} f(s) ds \right).$$
$$(9.16)$$

Now f is chosen to be measured from the mean of the periodic state, $\int_{t-1}^{t+1} f(s) ds = 0$. On writing $G(t) = f(t-1)$, the signal carried by a wave (G) satisfies the nonlinear o.d.e.

$$(\gamma + 1) G(t) G'(t) = M \sin(\pi t). \qquad (9.17)$$

Though often called "Chester's Equation" (9.17) was also derived by Gorkov [1963].

The critical points of (9.17) are at $G = 0$, $\sin(\pi t) = 0$, i.e., $G = 0$, $t = 0,\ 1,\ 2$ and the corresponding slopes are given by

$$(G')^2 = \frac{\pi M}{\gamma + 1} \cos(\pi t).$$

Then $G = 0$, $t = 0$ and 2 are saddle points and $G = 0$, $t = 1$ is a center. The integral curves look like those in Fig. 9.1.

Integrating (9.17) gives, on noting $G(0) = 0$

$$G_{\pm}(t) = \pm \left[\frac{4M}{\pi(\gamma + 1)} \right]^{\frac{1}{2}} \sin \frac{\pi}{2} t. \qquad (9.18)$$

Since $\int_0^2 G(t) dt = 0$, the solution must be

$$G(t) = \begin{cases} G_+(t), & 0 \le t < 1 \\ G_-(t), & 1 < t \le 2, \end{cases} \qquad (9.19)$$

with $G(t+2) = G(t)$. Hence, the solution contains a shock of strength $4(\frac{M}{\pi(\gamma+1)})^{\frac{1}{2}}$ at $t = 1$, where $M = \frac{\delta\pi}{l}$, see Fig. 9.1. The shock strength is $O(M)^{\frac{1}{2}} \gg O(M)$. The first order solution is the standing wave

$$u_1 = f(t - x) - f(t + x), \qquad (9.20)$$

where $f(t) = G(t+1)$ and $G(t)$ is given by (9.18) and (9.19) and has period 2, the same period as the piston. This solution agrees with the experimental observations in Saenger and Hudson [1960].

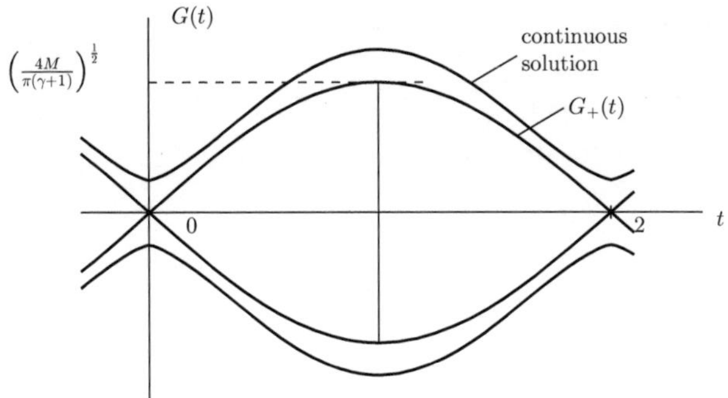

Fig. 9.1.   Integral curves (9.18).

9.2. A Perturbation Approach

We now show that the solution (9.18)–(9.20) may be found by a
different approach, a perturbation expansion. Observe that for the
solution to Duffing's equation

$$\frac{d^2x}{dt^2} + (\alpha x + \beta x^3) = M\sin(\omega t),$$

the motion takes place in the neighborhood of a linear free vibration
for $M \ll 1$. The solution is found by iterating about this, and so the
forcing term does not appear at the first order, see Stoker [1950]. In
Chester's procedure, $u_1(x,t) = f(\xi) - f(\eta)$, with $f(t+2) = f(t)$, i.e.,
u_1 is a linear standing wave. In analogy with Duffing's equation, we
develop the idea that the resonant motion in a closed tube may be
viewed as lying in the neighborhood of a linear standing wave.

The dimensionless equations are (9.5), with the boundary conditions (9.2). The fundamental resonant frequency is $\omega = 1/2$. The
first question is: what is the expansion parameter? It cannot be an
integer power of M as this would lead to a linear resonant problem
at first order. However, we see from (9.18) that the expansion parameter could be $M^{\frac{1}{2}}$, which, in turn, would lead to a standing wave at
$O(M^{\frac{1}{2}})$ since the input is $O(M)$.

Another way of arriving at the same conclusion is by considering
the energy balance in the tube. The work done by the piston is $\int pu\,dt$
over one cycle. On the piston $u = O(M)$ and otherwise u and p are

$O(M^m)$, where m is unknown. But the energy input is balanced by shock dissipation, which is $([u])^3$. Then the balance is determined by $\int p u \, dt = [u]^3$. The left-hand side is $O(M^{1+m})$ and the right-hand side is $O(M^{3m})$. This balance requires $m = \frac{1}{2}$, and the conclusion is that u and p are $O(M^{\frac{1}{2}})$.

Since the resonant frequency is $1/2$, we require that the flow has period 2, i.e.,

$$u(x, t + 2) = u(x, t). \tag{9.21}$$

Also the mean velocity and pressure are both constant and do not depend on x, so we require

$$\int_0^2 u(x, t) dt = 0,$$

since the mean velocity is zero at $x = 0$ and $x = 1$.

The assumed perturbation expansion is

$$u(x, t) = M^{\frac{1}{2}} u_1(x, t) + M u_2(x, t) + \cdots \tag{9.22}$$

$$c(x, t) = 1 + M^{\frac{1}{2}} c_1(x, t) + M c_2(x, t) + \cdots \tag{9.23}$$

with the boundary conditions (9.2).

As in Sec. 9.1 we examine only the case of exact resonance, $\omega = 1/2$. The resonant band will be examined in Sec. 9.4. Built into the perturbation scheme is that at resonance an output at $O(M^{\frac{1}{2}})$ arises from an input at $O(M)$.

The $O(M^{\frac{1}{2}})$ problem for $0 \le x \le 1, \quad t > 0$ is

$$\left(\frac{\partial}{\partial t} + \frac{\partial}{\partial x} \right) \left(u_1 + \frac{2}{\gamma - 1} c_1 \right) = 0, \quad \left(\frac{\partial}{\partial t} - \frac{\partial}{\partial x} \right) \left(u_1 - \frac{2}{\gamma - 1} c_1 \right) = 0 \tag{9.24}$$

with boundary conditions $u_1(0, t) = u_1(1, t) = 0$. The solution is

$$u_1 = f(\xi) - f(\eta), \qquad c_1 = \frac{\gamma - 1}{2} (f(\xi) + f(\eta)), \tag{9.25}$$

where f is an arbitrary function, except that $f(t + 2) = f(t)$ where $\xi = t - x$, $\eta = t + x$. So the first approximation u_1 is a linear standing wave with the signal f determined at the next order, $O(M)$, when the resonant forcing is introduced.

At $O(M)$ the problem to be solved is

$$\left(\frac{\partial}{\partial t} + \frac{\partial}{\partial x}\right)\left(u_2 + \frac{2}{\gamma - 1}c_2\right) = -(u_1 + c_1)\frac{\partial}{\partial x}\left(u_1 + \frac{2}{\gamma - 1}c_1\right)$$

$$= 2f'(\xi)\left(\frac{\gamma + 1}{2}f(\xi) - \frac{3 - \gamma}{2}f'(\xi)\right)$$

and

$$\left(\frac{\partial}{\partial t} - \frac{\partial}{\partial x}\right)\left(u_2 - \frac{2}{\gamma - 1}c_2\right) = -(u_1 - c_1)\frac{\partial}{\partial x}\left(u_1 - \frac{2}{\gamma - 1}c_1\right)$$

$$= 2f'(\eta)\left(\frac{3 - \gamma}{2}f(\xi) - \frac{\gamma + 1}{2}f(\eta)\right)$$

on using (9.25). The boundary conditions are

$$u_2(0, t) = 0, \quad u_2(1, t) = \sin(\pi t). \tag{9.26}$$

In the same way as in Sec. 9.1, integrating the $O(M)$ equations and applying the boundary conditions (9.26) gives the equation for f

$$f(t)f'(t) = -\frac{\sin(\pi t)}{\gamma + 1}. \tag{9.27}$$

This agrees with (9.17) on noting $G(t) = f(t - 1)$ and to first order the solution is the standing wave

$$u_1(x, t) = M^{\frac{1}{2}}(f(t - x) - f(t + x)),$$

where $f(t)$ satisfies (9.27), with $\int_0^2 f(t)dt = 0$ and $f(t + 2) = f(t)$.

9.3. Nonlinearization of a Difference Equation

Chester's procedure in 9.1 and the perturbation approach of 9.2 are consistent, but complicated. A simpler procedure that gives the same result is to start with the linear difference equation, but use the nonlinear travel time.

From Sec. 4.2, the linear difference equation is (4.8)

$$\bar{G}(t) - \bar{G}(t-2) = M\sin(2\pi\omega t), \qquad (9.28)$$

where $\omega = l/c_0\tau_p$ and $M = \bar{\omega}\delta/c_0 = 2\pi\varepsilon\omega \ll 1$ is the Mach number of the piston at $x = 1$ and $\varepsilon = \delta/l$. The periodic solution to (9.28) is

$$\bar{G}(t) = -\frac{M}{2\sin(2\pi\omega)}\cos(2\pi\omega t + 2\pi\omega),$$

provided $\omega \neq \omega_n = n/2$, for integer n. At the fundamental resonance the dimensionless frequency is $\omega = 1/2$ and there is no periodic solution with period 2 in t.

In dimensionless variables the tube length and reference sound speed are both unity. Waves traveling with the reference sound speed by definition carry zero amplitude measured from the reference state and have a round-trip travel time of 2, the period of the piston at the fundamental resonance. However, within nonlinear theory, a wave carrying a small, but nonzero, amplitude travels with a speed different from unity and has a round-trip travel time different from 2. Thus, at exactly the fundamental resonance, while the wave with zero-amplitude is resonating, all other amplitudes are not.

By changing the piston frequency a nonzero amplitude can be in resonance; i.e., the travel time of this amplitude matches the period of the piston. Thus, nonlinearity leads us to consider a band of resonant frequencies rather than an isolated, discrete, resonant frequency. Thus, we have a "resonant band" of frequencies.

The critical observation is that the nonlinear travel times (i.e., the travel times of the amplitudes other than zero) are all important in constructing a bounded solution. This is not unlike the observation in nonlinear oscillations that the frequency depends on the amplitude. It is also reminiscent of Whitham's nonlinearization technique, since we get the corrected (nonlinear) travel time that depends on the corrected (nonlinear) characteristics. Now the focus shifts to getting a better (nonlinear) approximation to the travel time.

From Sec. 8.1.1, the Lagrangian nonlinear travel time $Q(t) = 2[1 - N\bar{G}(t)]$ for an unforced oscillation replaces 2 in the linear difference equation (9.28) that becomes

$$\bar{G}(t) - \bar{G}(t - Q(t)) = M \sin(2\pi\omega t), \tag{9.29}$$

where $N = -\frac{\gamma+1}{2}$.

Since $|N\bar{G}(t)| \ll 1$, for $\omega = 1/2$, (9.29) is approximated by

$$\bar{G}(t)\bar{G}'(t) = \frac{M}{\gamma+1}\sin(\pi t) + O(M^2) \tag{9.30}$$

on noting $\bar{G}(t+2) = \bar{G}(t)$ and assuming the small rate condition $\bar{G}'(t)| \ll 1$, so that (9.30) agrees with previous calculations (9.17) and (9.27) at resonance.

The critical insight here is to use the nonlinear travel time $Q(t)$ in the linear difference equation (9.28), and then the equation for $\bar{G}$ follows immediately.

9.4. The Resonant Band

We now examine the case of an oscillation in a tube when the driver operates in a band of frequencies about the resonant frequency for a general forcing function $\bar{H}(\omega t)$. Experiments by Saenger and Hudson [1960] show that shocks occur in such a band of frequencies, and the strength of a shock is attenuated as the frequency moves away from the resonant frequency until eventually the shock disappears and the motion is continuous.

For the next sections, it is convenient to rescale the independent variable with ω so that the oscillation has unit period. This is done by redefining the β-characteristic so that $\beta = \omega t$ on $x = 1$.

The difference equation is derived by following a wavelet leaving $x = 1$ at time r, carrying the amplitude $\bar{G}(\omega r)$, through a full cycle, arriving back at $x = 1$ at time t. The equation is now (9.29) in the form

$$\bar{G}(\omega t) = \bar{G}(\omega r) + M\sin(2\pi\omega t), \ t = r + Q(\omega r). \tag{9.31}$$

We can generalize the piston motion to any continuous function with unit period, $M\bar{H}(\omega t)$, where $\bar{H}(y+1) = \bar{H}(y)$ and we note the restriction that $\int_0^1 \bar{H}(y)dy = 0$. Writing $\omega t = y$, $\omega r = s$ and assuming the period of $\bar{G}(y)$ is the same as that of the driver $\bar{H}(y)$ i.e., $\bar{G}(y + 1) = \bar{G}(y)$, then (9.31) becomes

$$\bar{G}(y) = \bar{G}(s) + M\bar{H}(y), \ \ y = s + 2\omega[1 - N\bar{G}(s)] \tag{9.32}$$

and for simplicity we confine our attention to the neighborhood of the fundamental resonance $\omega = \omega_1 = \frac{1}{2}$. The result for higher harmonics follows in a similar way.

To account for the off-resonant frequencies, let

$$\omega = \frac{1}{2}(1 + \Delta) \tag{9.33}$$

and so that $\frac{1}{2}\Delta$ is the detuning from resonance. Defining $H = -2\omega N M\bar{H}$ and $G(y)$ by

$$G(y) = \Delta - 2\omega N\bar{G}(y), \tag{9.34}$$

Eq. (9.32) become

$$G(y) - G(s) = H(y), \ \ y = s + G(s). \tag{9.35}$$

Equation (9.35) is called the *Standard Mapping of chaos theory.*

Assuming $|G| \ll 1$ and $|G'| \ll 1$, (9.35) reduces to the o.d.e.

$$G(y)G'(y) = H(y) \tag{9.36}$$

with

$$\int_0^1 G(y)dy = \Delta, \tag{9.37}$$

since $\int_0^1 \bar{G}(y)dy = 0$.

To construct shocked solutions, we make a distinction between the integral curves of (9.36) designated as $Z(y)$ and the signal $G(y)$ that must additionally satisfy the mean condition (9.37), thus

$$Z(y)Z'(y) = H(y). \tag{9.38}$$

We can choose the origin so that

$$H(0) = H(y_1) = H(1) = 0,$$

where $0 < y_1 < 1$, with $H'(0) = H'(1) > 0$ and $H'(y_1) < 0$. Then $(0,0)$, $(1,0)$ and $(y_1,0)$ are the singular points in the $y - Z$ plane.

By differentiating (9.38), the slope at a singular point is given by $Z'^2 = H'$. Then $(0,0)$ and $(0,1)$ are saddle points and $(y_1, 0)$ is a center. Integrating (9.38) and noting that $Z(0) = 0$, yields

$$Z^{\pm}(y) = \pm 2^{\frac{1}{2}} (h(y) - h(0))^{\frac{1}{2}}, \qquad (9.39)$$

where $H(y) = h'(y)$. These are the separatrices that connect the saddle points $(0,0)$ and $(1,0)$. Since $Z^+(y) > 0$ for $0 < y < 1$, with $Z(0) = 0$, any solution $G(y)$ with $G(0) > 0$ is periodic in y and $G(y) > Z^+(y)$. Then G satisfies

$$\int_0^1 G(s)ds > \int_0^1 Z^+(s)ds.$$

For an applied frequency $\omega = \frac{1}{2}(1 + \Delta)$ such that

$$\Delta > \int_0^1 Z^+(s)ds > 0$$

there is a unique, *continuous*, periodic solution $G(y)$. For this continuous solution

$$G(y) = 2^{\frac{1}{2}} \left[h(y) - h(0) + C(\Delta)\right]^{\frac{1}{2}}, \ \ C(\Delta) > 0,$$

and $C(\Delta)$ is determined by the mean condition (9.37). Similarly there are *continuous*, periodic solutions $G(y)$ and $G(0) < 0$ with $G(y) < Z^-(y)$ when

$$\Delta < \int_0^1 Z^-(s)ds < 0.$$

As a consequence no single integral curve $G(y)$ will satisfy the mean condition, and the signal function will be *discontinuous*, when

$$\int_0^1 Z^-(s)ds < \Delta < \int_0^1 Z^+(s)ds. \qquad (9.40)$$

The range of frequencies, defined by (9.40), for which the signal is *discontinuous*, is called the *resonant band*. If the resonant band is given by $\Delta^- < \Delta < \Delta^+$, then from (9.39) and (9.40)

$$\Delta^\pm = \pm(2)^{\frac{1}{2}}\bar{h}(1), \tag{9.41}$$

where $\bar{h}(y) = \int_0^y [h(s) - h(0)]^{\frac{1}{2}} ds$. Then the signal $G(y)$ is given by

$$G(y) = \begin{cases} Z^+(y), & 0 \le y < y_s \\ Z^-(y), & y_s < y \le 1 \end{cases} \tag{9.42}$$

and the position of the shock at $y = y_s$ is chosen to satisfy the mean condition (9.37):

$$\int_0^{y_s} Z^+(t)dt + \int_{y_s}^1 Z^-(t)dt = \Delta. \tag{9.43}$$

For $\Delta^- < \Delta < \Delta^+$ the shock strength is

$$Z^+(y_s) - Z^-(y_s) = 2\,(2)^{\frac{1}{2}}\,[h(y_s) - h(0)]^{\frac{1}{2}} \tag{9.44}$$

and the shock position is given by

$$\bar{h}(y_s) = \frac{1}{2}\bar{h}(1) + (2)^{-\frac{3}{2}}\,\Delta \tag{9.45}$$

on using (9.39) and (9.43).

For the special forcing function $h(y) = \frac{\omega MN}{\pi}\cos(2\pi y)$, i.e., $\overline{H}(y) = \sin(2\pi y)$, $\bar{h}(y) = (-\frac{2\omega MN}{\pi^3})^{\frac{1}{2}}(1 - \cos\pi y)$, the above results are particularly simple. The edges of the resonant band are, by (9.41)

$$\Delta^\pm = \pm 4\left(-\frac{\omega MN}{\pi^3}\right)^{\frac{1}{2}} \backsim \pm\frac{2}{\pi}(\varepsilon(\gamma+1))^{\frac{1}{2}} \tag{9.46}$$

with $M = 2\pi\varepsilon\omega$, $\omega = 1/2$ and $\varepsilon = \delta/l$. Equation (9.46) can also be written as

$$\Delta^\pm = \pm\left(\frac{8A}{\pi^3}\right)^{\frac{1}{2}} \quad \text{or} \quad A = \frac{\pi^3}{8}\Delta^2. \tag{9.47}$$

Using (9.45) the shock is located at y_s where

$$\cos(\pi y_s) = -\Delta\frac{1}{4}\left(-\frac{\pi^3}{\omega MN}\right)^{\frac{1}{2}} \backsim -\frac{\Delta\pi}{2}\left(\frac{1}{\varepsilon(\gamma+1)}\right)^{\frac{1}{2}}, \tag{9.48}$$

and the shock strength is, from (9.44),

$$4\left(\frac{-\omega MN}{\pi}\right)^{\frac{1}{2}}\sin(\pi y_s) \frown 2\left(\varepsilon(\gamma+1)\right)^{\frac{1}{2}}\sin(\pi y_s). \qquad (9.49)$$

9.4.1. *Damped Oscillation*

For an undamped standing wave the nonlinear difference equations are (8.24) and (8.25). We now examine the case of damped oscillations in a tube when the driver operates in a band of frequencies about the resonant frequency. We consider the case when energy is allowed to radiate out through the surface at $x = 0$, while the periodic piston operates at the end $x = 1$. This situation was considered in Seymour and Mortell [1973a].

Following the notation in Sec. 9.4, the equations used are (8.36) in the form

$$\bar{G}(y) = k\bar{G}(s) + M\bar{H}(y), \ y = s + \omega[2 - (1+k)N\bar{G}(s)], \qquad (9.50)$$

where k $(-1 \leq k \leq 1)$ is the reflection coefficient at $x = 0$ and $M\bar{H}(\omega t)$ is the velocity of the piston. In (9.50) the β-characteristic has been normalized so that $\beta = \omega t$ on $x = 1$. Also, $\bar{H}(y)$ is normalized so that $\bar{H}(y+1) = \bar{H}(y)$, and we note that $\int_0^1 \bar{H}(y)dy = 0$.

The linear resonant frequencies are $\omega = \omega_n = \frac{1}{2}n$, $n = 1, 2, 3, \ldots$. and we confine our attention to the fundamental resonance $\omega_1 = \frac{1}{2}$ and set $\omega = \frac{1}{2}(1 + \Delta)$ so that $\frac{1}{2}\Delta$ is the detuning from resonance. Then (9.50) become

$$\bar{G}(y) = k\bar{G}(s) + M\bar{H}(y), \ y = s + \Delta - \omega(1+k)N\bar{G}(s)], \qquad (9.51)$$

on assuming the period of $\bar{G}(y)$ is the same as that of the driver $\bar{H}(y)$ i.e., $\bar{G}(y+1) = \bar{G}(y)$.

Defining

$$G = \Delta - (1+k)\omega N\bar{G}, \ H = \mu\Delta - (1+k)\omega N M\bar{H}, \qquad (9.52)$$

where $\mu = 1 - k$, Eqs.(9.51) become

$$G(y) - kG(s) = H(y), \ y = s + G(s). \qquad (9.53)$$

When $\Delta = 0$ Eq. (9.53) is the *Dissipative Standard Mapping of chaos theory*, see Lichtenberg and Lieberman [1992].

Assuming $|G| \ll 1$ and $|G'| \ll 1$, (9.53) reduce to the o.d.e.

$$G(y)G'(y) + \mu G(y) = H(y) \tag{9.54}$$

with

$$\int_0^1 G(y)dy = \Delta. \tag{9.55}$$

Equation (9.54) is well known: it is a generalization of the equation that describes the motion of a viscously damped pendulum under a constant external moment and also occurs in the study of the pull-out torque of a synchronous motor, see Stoker [1950] or Minorsky [1962].

Case $\mu_c < \mu = 1 - k < 1$. The equation for the integral curves of (9.54) is

$$Z(y)Z'(y) + \mu Z(y) = H(y). \tag{9.56}$$

The term containing μ in (9.56) introduces damping into the equation, and we wish to see what the effect of this damping is. We show that for a given forcing term $H(y)$ there is a critical amount of damping, $\mu = \mu_C$, such that for $\mu > \mu_C$ the solution of (9.56) is continuous for all frequencies. Thus, the damping here kills the shock that is present for $k = 1$ ($\mu = 0$).

We again assume that the zeros of $H(y)$ are $y = y_0, y_1, y_2$, where $y_0 < y_1 < y_2$, $H'(y_0) = H'(y_2) > 0$ and $H'(y_1) < 0$. The singular points are labeled $A_0(\mu, \Delta)$, $A_2(\mu, \Delta)$ and $B_1(\mu, \Delta)$, where $A_0(\mu, \Delta) = (0, 0)$ by definition, $A_2(\mu, 0) = (1, 0)$ by periodicity and $B_1(\mu, 0) = (t_1, 0)$. Then, as before, A_0 and A_2 are saddle points and the separatrices through A_i, $i = 0, 2$, (denoted by $Z_i^{\pm}(y)$) have slopes

$$\lambda^{\pm}(y_i) = \frac{1}{2}(-\mu \pm \sqrt{\mu^2 + 4H'(y_i)}), \tag{9.57}$$

where $\lambda^+(y_i) > 0 > \lambda^-(y_i)$ since $H'(y_i) > 0$ for $i = 0, 2$. The slopes at B_1 are also given by (9.57), where $H'(y_i) < 0$. If

$$I(y_i) = \mu^2 + 4H'(y_i) > 0, \tag{9.58}$$

then B_1 is a node, while if $I(y_1)$ is negative, then B_1 is a local spiral point.

Thus, for a given forcing function $H(t)$, B_1 is a node if there is sufficient damping. This gives the possibility of constructing a continuous solution through A_0, B_1 and A_2 for any value of the detuning parameter Δ. The nodal condition is clearly *necessary* for the existence of such a solution, and it turns out that a *sufficient* condition is

$$\mu^2 > \mu_c^2 = \max_y[-4H'(y)] > 0.$$

This is proved in Seymour and Mortell [1973a]. Thus for $\mu > \mu_C$, $G(y) = Z^+(t)$ is continuous and satisfies the mean condition (9.55) as can be seen by integrating (9.54) over $[0,1]$. In this case linear theory gives a good approximation. This solution is confirmed by experiment in Sturtevant [1974].

Case $0 < \mu < \mu_c$. In this case there is not enough damping to produce a shockless solution for all frequencies. For example, referring to Fig. 9.2, when $\Delta = 0$, $Z_0^+(y_1) > 0$, $Z_2^-(y_1) < 0$ and $Z_0^+(y_0) = Z_2^-(y_2) = 0$, where $0 < y_2 < y_1 < y_0 < 1$, so that the separatrix Z_0^+ through A_0 does not reach A_2, but gets as far as $y_0 < 1$, and, similarly Z_2^- through A_2 gets as far as $y_2 > 0$. Hence, the separatrices do not connect the saddle points. As Δ is increased through the resonant band, so that $\int_0^1 \bar{H}(t)dt > 0$, there is a unique frequency, $\Delta = \Delta^+$, so that $Z_0^+(t) = Z_2^+(t)$, $0 \le t \le 1$. That is, for $\Delta = \Delta^+ > 0$ the positive separatrix connects the saddle points A_0 and A_2, see Fig. 9.3. Then for $\Delta > \Delta^+ > 0$, there is a unique continuous periodic solution $Z = Z_\Delta(y) > 0$, $0 \le y \le 1$. Similarly, there is a $\Delta = \Delta^- < 0$ for which $Z_0^-(y) = Z_1^-(y)$, $0 \le y \le 1$; that is the negative separatrix connects the saddle points A_0 and A_2. For $\Delta < \Delta^-$, there is a unique, continuous periodic solution $Z = Z_\Delta(y) < 0$, $0 \le y \le 1$. These results can be inferred from theorems in Sansone and Conti [1964]. In Figs. 9.2 and 9.3 y in the text here replaces η in the original figures in Seymour and Mortell [1973a].

When $\Delta^- < \Delta < \Delta^+$ there is no continuous periodic solution of (9.36). The signal function $G(y)$ is constructed from a composition

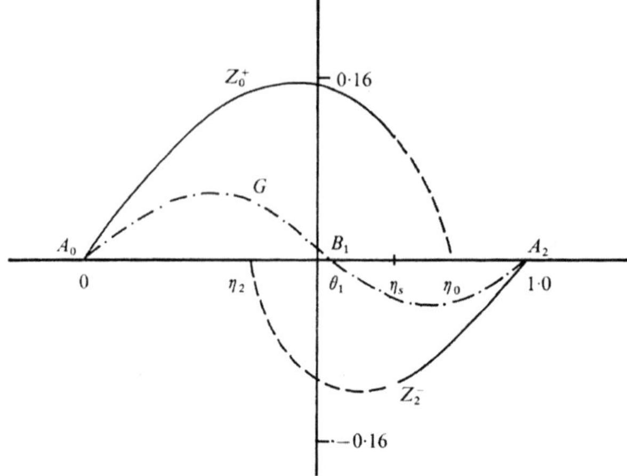

Fig. 9.2.   Separatrices for $\Delta = 0.07$.

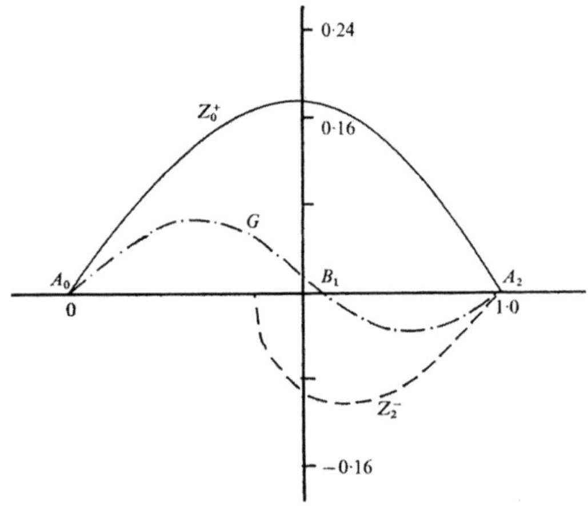

Fig. 9.3.   Separatrices for $\Delta = \Delta^+$.

of the integral curves $Z_0^+(y)$ and $Z_2^-(y)$. Then

$$G(y) = \begin{cases} Z_0^+(y), & y_0 \leq y < y_s \\ Z_0^-(y), & y_s < y \leq y_2, \end{cases} \tag{9.59}$$

where $y = y_s$ is the shock position. Now it is possible to chose y_s so that the mean condition (9.55) is satisfied. All that remains is to show that if the mean condition is satisfied the shock condition is automatically satisfied.

If $S(x)$ is the arrival time at x of a weak shock traveling in the negative x direction and $\beta^{\pm}(x)$ are the wavelets immediately ahead of and behind the shock, then the weak-shock relations imply that

$$\frac{dS}{dx} = -1 + \frac{1}{2}N[\bar{G}(\beta^+) + \bar{G}(\beta^-)]. \tag{9.60}$$

Similarly, for shocks moving to the right:

$$\frac{dS}{dx} = 1 + \frac{1}{2}N[\overline{F}(\alpha^+) + \overline{F}(\alpha^-)]. \tag{9.61}$$

In the periodic state there is negligible distortion of the signal. Therefore $\beta^{\pm}$ are independent of x. Equations (9.60) and (9.61) can each be integrated to give the travel time of a shock from $x = 1$ to $x = 0$. Then, for a shock moving left with $S(1) = 0$,

$$S(x) = -x + 1 + \frac{1}{2}N(x - 1)[\bar{G}(\beta^+) + \bar{G}(\beta^-)], \quad \text{and}$$

$$S(0) = 1 - \frac{1}{2}N[\bar{G}(\beta^+) + \bar{G}(\beta^-)].$$

For a shock moving right with $S(0) = 0$,

$$S(x) = x + \frac{x}{2}N[\overline{F}(\alpha^+) + \overline{F}(\alpha^-)], \quad \text{and} \quad S(1) = 1 + \frac{1}{2}N[\overline{F}(\alpha^+) + \overline{F}(\alpha^-)].$$

The impedance boundary condition at $x = 0$ is $\overline{F} = -k\bar{G}$, $k = \frac{i-1}{i+1}$, and the travel times from $x = 0$ to $x = 1$ then give the total round trip travel time as

$$S(0) + S(1) = T_t = 2 - \frac{1}{2}N(1 + k)[\bar{G}(\beta^+) + \bar{G}(\beta^-)]. \tag{9.62}$$

The periodicity condition is $T_t = \frac{1}{\omega} = \frac{2}{1+\Delta}$, hence, (9.62) becomes

$$4\left(1 - \frac{1}{1 + \Delta}\right) = N(1 + k)[\bar{G}(\beta^+) + \bar{G}(\beta^-)]$$

and the definition (9.52) of G implies that at a shock.

$$G(\beta^+) + G(\beta^-) = 0. \tag{9.63}$$

Integration of (9.54) over $[y_0, y_0 + 1]$ implies

$$\frac{1}{2}[G^2(y_s^+) - G^2(y_s^-)] + \mu \int_{y_0}^{y_0+1} G(t)dt = \int_{y_0}^{y_0+1} H(t)dt. \qquad (9.64)$$

Now the mean condition (9.55) and the definition H in (9.52) imply that the shock condition (9.63) is satisfied by (9.64). Thus, a *shock* is fitted into the solution by satisfying the mean condition. For $\Delta = \Delta^+$, the solution curve $Z_0^+(y, \Delta^+)$, $0 \leq y \leq 1$ is a separatrix, and similarly for $\Delta = \Delta^-$, the solution curve $Z_2^-(y, \Delta^-)$, $0 \leq y \leq 1$, is a separatrix. Then the implicit conditions for the edge of the resonant band are

$$\int_0^1 Z_0^+(y, \Delta^+)dy = \Delta^+ \quad \text{and} \quad \int_0^1 Z_2^-(t, \Delta^-)dt = \Delta^-.$$

For Δ outside $[\Delta^-, \Delta^+]$ the integral curves and the solutions are continuous; for Δ inside $[\Delta^-, \Delta^+]$ the solution contains a shock. For $\Delta = \Delta^\pm$ the solution is continuous with a cusp at $t = 0$ and $t = 1$.

9.4.2. *A Bouncing Ball Problem and Chaos*

A simple physical set-up that gives rise to the dissipative standard mapping concerns the dynamics of a bouncing ball, see Holmes [1982]. A ball bounces on a massive vibrating table, where the mass of the table greatly exceeds that of the ball. The impact relationship is, see Meriam and Kraige [2012],

$$V(t_j) - W(t_j) = -\alpha[U(t_j) - W(t_j)], \qquad (9.65)$$

where U, V, W are the absolute velocities of the approaching ball, the receding ball and the table, α ($0 \leq \alpha \leq 1$) is the coefficient of restitution and $t = t_j$ is the time of the jth impact. We assume that the distance the ball travels under the influence of gravity, g, is large compared with the displacement of the table. Then the time interval between impacts can be approximated by

$$t_{j+1} - t_j = 2V(t_j)/g \qquad (9.66)$$

and by definition

$$U(t_{j+1}) = -V(t_j). \tag{9.67}$$

We assume the displacement of the table is given by $-\frac{A}{2\pi}\cos(2\pi\omega t)$, so that $W_j = W(t_j) = A\omega\sin(2\pi\omega t_j)$. Then (9.66) implies, with $\phi = \omega t$,

$$\phi_{j+1} - \phi_j = v_j, \quad v_j = 2\omega V_j/g. \tag{9.68}$$

Now (9.65) implies, using (9.67),

$$V_{j+1} = \alpha V_j + (1 + \alpha)W_{j+1}$$

or

$$v_{j+1} - \alpha v_j = \gamma\sin(2\pi[\phi_j + v_j]) \tag{9.69}$$

on using (9.68), where $\gamma = 2(1 + \alpha)A\omega^2/g$. Equations (9.68) and (9.69) constitute the dissipative standard mapping, see Guckenheimer and Holmes [1983], where γ is the amplitude of the input velocity and $\alpha < 1$ is the dissipation due to the coefficient of restitution. The continuous form of (9.68) and (9.69) can be written as

$$F(y) - \alpha F(x) = \gamma\sin(2\pi y), \quad y = x + F(x)$$

which is the dissipative standard mapping, see (9.53).

The determinant of the Jacobian for the mapping (9.68) and (9.69) is

$$\begin{vmatrix} 1 & 1 \\ 2\pi\gamma\cos(2\pi[\phi_j + v_j]) & \alpha + 2\pi\gamma\cos(2\pi[\phi_j + v_j]) \end{vmatrix} = \alpha.$$

Thus, there is dissipation for $\alpha < 1$, i.e., the mapping contracts area uniformly, while for $\alpha = 1$ the standard mapping preserves area.

We can deduce from Sec. 9.4 that for $\alpha_c < \alpha \leq 1$ there is a range of ω (the resonant band) about the fundamental resonant frequency for which shocks appear and this, in the case of (9.68) and (9.69), implies the motion of the bouncing ball is chaotic. This is confirmed by the experiments of Wood and Byrne [1981]. Outside the resonant band there are periodic motions of the ball. For $\alpha < \alpha_c$ the motion is also periodic. A detailed analysis of the dissipative standard mapping is given in Guckenheimer and Holmes [1983].

9.4.3. *Resonance with Varying Cross Section* $s(x) = (1 + ax)^{-2}$

We now consider a special case based on the analysis in Sec. 8.1.5: the resonant oscillations in a tube with cross-sectional areas $s(x) = (1 + ax)^{-2}$. From (8.64), the linear problem in dimensionless variables, is

$$se_t + (su)_x = 0, \quad u_t + e_x = 0, \quad 0 \le x \le 1, \quad t > 0$$

with $u(0,t) = 0$, $u(1,t) = MH(\omega t)$. M is the Mach number and $H(y+1) = H(y)$. Putting $v(x,t) = \frac{u}{1+ax}$, v satisfies

$$v_{tt} - v_{xx} = 0$$

with $v(0,t) = 0$, $v(1,t) = \frac{M}{1+a}H(\omega t)$, which leads to the linear difference equation (4.8) in the form

$$G(y) - G(s) = \frac{M}{1+a}H(y), \quad s = y - 2\omega, \tag{9.70}$$

where $y = \omega t$. There is no periodic solution G with the same period as the driver when $\omega = \frac{1}{2}n$, $n = 1, 2, 3 \dots$. These are the commensurate resonant frequencies. The case $s(x) = (1 + ax)^{-2}$ is exceptional as usually a variable cross section leads to incommensurate resonant frequencies.

To nonlinearize (9.70), we use (8.81) where the nonlinear travel time in the tube is given by

$$Q(s) = 2\left[1 - \left(1 + \frac{a}{2}\right)NG(s)\right]$$

and then the nonlinear difference equation is

$$G(y) - G(y - Q(t)) = \frac{M}{1+a}H(y), \quad Q(t) = 2\omega - 2\omega\left(1 + \frac{a}{2}\right)NG(y).$$

This can be generalized to include radiation damping at $x = 0$ to give

$$G(\bar{t}) - kG(\bar{t} - Q(\bar{t})) = \frac{M}{1+a}H(\bar{t}) \tag{9.71}$$

with

$$Q(\bar{t}) = 2\omega - \omega(1+k)\left(1 + \frac{a}{2}\right)NG(\bar{t}),$$

where the reflection coefficient at $x = 0$ is $k = \frac{1-i}{1+i}$. The analysis and results for (9.71) are exactly as in Sec. 9.4.1 with M replaced by $\frac{M}{1+a}$, and N replaced by $(1 + \frac{a}{2})N$. There are shocks in the flow when the damping is less than a critical number, and a resonant band in which shocks form. An alternative approach for the undamped case $k = 1$ is given in Keller [1977].

9.5. Small Rate Subharmonic Oscillations

We have seen earlier in this chapter that periodic solutions that satisfy the boundary conditions

$$u(0,1) = 0 \quad \text{and} \quad u(1,t) = 2\pi\varepsilon\omega\sin(2\pi y), \tag{9.72}$$

where $\varepsilon = \delta/l$ (δ is the maximum piston displacement) and $y = \omega t$, become unbounded within linear theory when $\omega = \omega_n = \frac{1}{2}n, \quad n = 1, 2, 3, \ldots$ At the next order in a perturbation expansion in powers of ε, we find

$$u_2(x,y) = \frac{M(\pi\omega)^3}{\sin^2(2\pi\omega)}\big([\cos(4\pi\alpha) - \cos(4\pi\beta)]\cot(4\pi\omega)$$

$$- x[\sin(4\pi\alpha) + \sin(4\pi\beta)]\big) \tag{9.73}$$

with $\alpha = y - \omega x$, $\beta = y + \omega x$, provided

$$\omega \neq \Omega_n = \frac{1}{4}(2n+1), \quad n = 0, 1, 2, \ldots. \tag{9.74}$$

Hence, the perturbation scheme also breaks down when $\omega = \Omega_0 = 1/4$. We refer to $\Omega_0 = 1/4$ as a "subharmonic" or "quadratic" resonance. Periodic motions containing shocks have been observed in a closed tube at frequencies around half the fundamental in a number of experiments e.g., Galiev, Ilgamov and Sadykov [1970], Zaripov and Ilgamov [1976] and Althaus and Thomann [1987]. Yet linear theory predicts a continuous solution.

To describe a consistent nonlinear theory of subharmonic resonance, we begin with the basic difference equation derived in Sec. 9.4, for a general forcing function $h(y)$ with unit period and $\int_0^1 h(y)dy = 0$:

$$g(y) = g(s) + Mh(y), \quad y = s + 2\omega[1 - Ng(s)]. \tag{9.75}$$

We define

$$G(y) = -2\omega N g(y) + \Delta, \quad H(y) = -2\omega N M h(y), \tag{9.76}$$

where G has unit period and satisfies the mean condition $\int_0^1 G(\eta)\, d\eta = \Delta$ and $\Delta = 2(\omega - \frac{1}{4})$, so that the detuning is $\Delta/2$. Then Equation (9.75) becomes

$$G(y) = G(s) + H(y), \quad y = s + G(s). \tag{9.77}$$

Hence, the quadratic resonance region for small A is $|\frac{\Delta}{2}| = \frac{A}{2\pi}$, or $|\Delta| = \frac{A}{\pi}$. For the sinusoidal piston velocity $Mh(y) = 2\pi\varepsilon\omega\sin(2\pi y)$, i.e., $M = 2\pi\varepsilon\omega$, normally used in experiments,

$$H(y) = A\sin(2\pi y), \tag{9.78}$$

where $A = -4\pi N\varepsilon\omega^2$ is a similarity parameter.

For linear resonance, when ω lies within the resonant band about $\omega = 1/2$, there is always an amplitude of the propagating signal that completes *one* cycle in the tube during a period of the piston. This property characterizes linear resonance. For subharmonic resonance, when ω lies within the resonant band about $\omega = \frac{1}{4}$, there is always an amplitude of the propagating signal that completes *two* cycles in the tube during one period of the piston. This latter observation is used to set up the appropriate nonlinear difference equation.

Since the interest is in two cycles, but over a single period, Eqs.(9.77) are written

$$G(y) = G(r) + H(y), \quad y = r + \frac{1}{2} + G(r), \tag{9.79}$$

$$G(r) = G(s) + H(r), \quad r = s + \frac{1}{2} + G(s). \tag{9.80}$$

So r is the intermediate "time" between the initial and final "times" s and y. The equivalent linear equation for $\omega = \Omega_0$ is $G(y) - G(y - \frac{1}{2}) = H(y)$ that has the solution

$$G(y) = \frac{1}{2} H(y), \tag{9.81}$$

when, like $\sin(2\pi y)$, $H(y + \frac{1}{2}) = -H(y)$, and $G(y)$ has unit period. Now define the new dependent variable $Z(y)$

$$Z(y) = G(y) - \frac{1}{2} H(y), \tag{9.82}$$

so that Z is the difference between the nonlinear and the linear solutions. We note that in this case the linear solution exists, in contrast to the case of linear resonance.

Substituting (9.82) into (9.79) yields

$$Z(y) = Z(r) + \frac{1}{2} H(r) + \frac{1}{2} H(y) \tag{9.83}$$

with a similar result for $Z(r)$ in terms of $Z(s)$ from (9.80). Then on eliminating $Z(r)$, we get

$$Z(y) = Z(s) + \frac{1}{2} H(s) + \frac{1}{2} H(y) + H(r). \tag{9.84}$$

Similarly eliminating $Z(r)$ from the second of (9.79) and (9.80) gives

$$y = s + 1 + 2Z(s) + H(s) + H(r), \quad r = s + \frac{1}{2} + Z(s) + \frac{1}{2} H(s). \tag{9.85}$$

Equations (9.84) and (9.85) are now used to derive the small rate equation for $Z(y)$ when $|Z|, |Z'| \ll 1$ and $H(y + \frac{1}{2}) = -H(y)$. From (9.84) and (9.85) we get

$$s = y - 1 - 2Z(y - 1) \quad \text{and} \quad r = y - \frac{1}{2} - Z(y - 1) + \frac{1}{2} H(y - 1). \tag{9.86}$$

Then (9.84) gives

$$Z(s) = Z(y - 1) - 2Z(y - 1)Z'(y - 1) + \cdots$$

and

$$\frac{1}{2}H(s) + \frac{1}{2}H(y) + H(r) = -\frac{1}{2}H(y-1)H'(y-1) + \cdots$$

on noting $H(y - \frac{1}{2}) + H(y-1) = 0$. Then (9.84) becomes

$$Z(y)Z'(y) = -\frac{1}{4}H(y)H'(y), \tag{9.87}$$

since $Z(y+1) = Z(y)$.

With $H(y) = A\sin(2\pi y)$, the equation for Z is

$$Z(y)Z'(y) = \frac{1}{2}\frac{d}{dy}(Z^2) = -\frac{\pi}{4}A^2\sin(4\pi y) \tag{9.88}$$

and this integrates to give the separatrices

$$Z^{\pm}(y) = \pm\frac{A}{2}\left|\cos(2\pi y)\right|, \quad 0 \le y \le 1,$$

see Fig. 10.16 for $A = 0.008$. There are two shocks of equal strength in the solution $Z(y)$ that are half a period apart. A possible continuous solution

$$Z(y) = \begin{cases} Z^+(y), \ 1/4 \le y < 3/4 \\ Z^-(y), \ 3/4 < y \le 5/4 \end{cases}$$

is excluded by experiments showing shocks, e.g., Althaus and Thomann [1987], and by evolution, see Cox and Mortell [1992] and Fig. 11.5.

The mean condition on G, $\int_0^1 G(y)dy = \Delta$, implies, on using (9.82), that

$$|\Delta| < \int_0^1 Z^+(y)dy = \frac{A}{\pi}. \tag{9.89}$$

Hence, for small A, if $|\Delta| > \frac{A}{\pi}$, then the solution $Z(y)$ is continuous and periodic. If $|\Delta| < \frac{A}{\pi}$, discontinuities connecting the separatrices must be inserted. Thus, for small A, $\Delta_q(A) = \frac{1}{2} - \frac{A}{\pi} + O(A^2)$ is the edge of the resonant band where shocks occur, and marks out the boundary between continuous and discontinuous solutions, see Fig. 9.4. We note that the nonlinear "correction", Z, has the same order of magnitude as the linear solution (9.81).

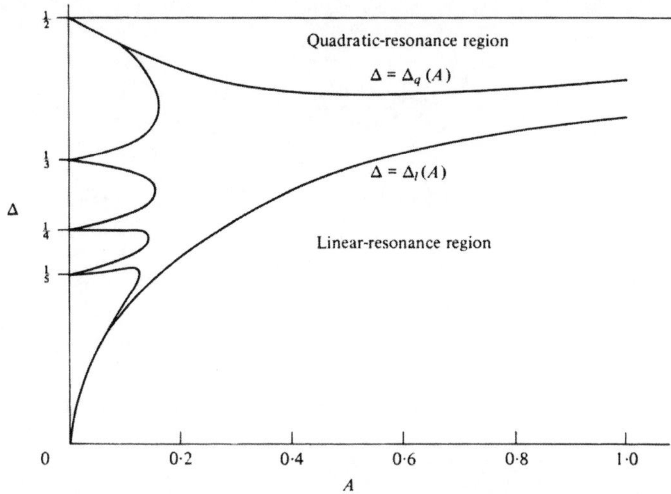

Fig. 9.4.   Quadratic resonance region $\Delta = \Delta_q(A)$.

9.5.1. *Nonlinearization*

From Eq. (9.84), the linear difference equation is

$$Z(y) = Z(s) + \frac{1}{2}H(s) + \frac{1}{2}H(y) + H(r) \qquad (9.90)$$

but with

$$y = s + 1 \quad \text{and} \quad r = s + \frac{1}{2}. \qquad (9.91)$$

Since $H(y)$ has unit period and $H(y+\frac{1}{2}) = -H(y)$, Eq. (9.90) implies that $Z(y)$ has unit period, i.e., the solution of the linear difference equation (9.90) is an arbitrary function of Z that has unit period, a standing wave. Now (9.90) is nonlinearized by using the nonlinear travel times for s and r in terms of y given by (9.85). Then the result (9.87) follows, which yields a shocked solution within the resonant band $|\Delta| < \frac{A}{\pi}$.

Chapter 10

Finite Rate Oscillations

In the previous chapter we used the small amplitude equations

$$G(y) - G(s) = H(y), \ y = s + G(s). \tag{10.1}$$

where $H(y+1) = H(y)$, to describe resonant oscillations of a gas in a closed tube. Equation (10.1) is a nonlinear difference or functional equation.

When $H(y) = A\sin(2\pi y)$, Eq. (10.1) has been given the name the *standard mapping* or the *Chirikov-Taylor mapping*. This mapping, which exhibits chaotic solutions, has been studied intensively, see for example, Chirikov [1979], Guckenheimer and Holmes [1983] and Lichtenberg and Liebermann [1992]. In Chapter 9, in the small rate case, where $|G'| \ll 1$, we have constructed continuous and shocked solutions of (10.1) for resonance at the fundamental frequency and at half the fundamental.

The shocked solutions for resonance in a tube correspond to stochastic (chaotic) solutions of (10.1). The parameter A in the periodic forcing $H(y)$ is called the *stochastic parameter*. The analysis of damped resonance of a gas in a closed tube leads to the *dissipative standard map*, see (9.53) and Seymour and Mortell [1973a].

Equation (10.1) describes nonlinear waves traveling over and back in a closed tube driven by the periodic forcing $H(y)$. The fundamental difficulty, which has been evident since the pioneering work of Riemann, is the integration of the characteristic equations corresponding to waves propagating in opposite directions where normally there is an interaction of the waves. Contrast this with linear

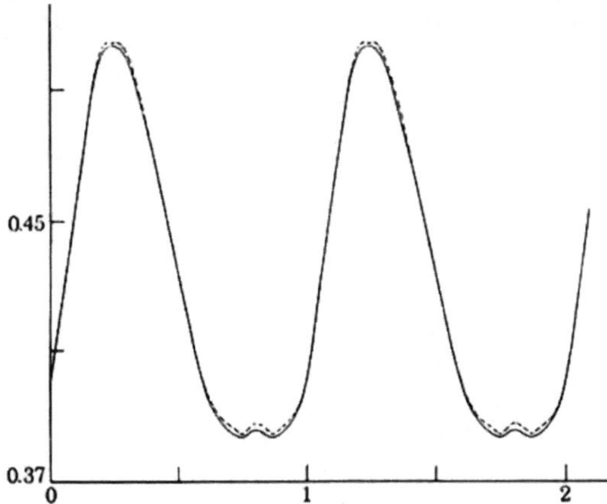

Fig. 10.1. $A = 0.12$, $\omega = 0.715$.

waves that pass through each other and either wave is unaware of
the presence of the other. In our case, the characteristic equation
is the second equation in (10.1), and clearly there is no interaction.
Equations (10.1) are a set of nonlinear coupled equations in which
there is no interaction of the waves. This question of how the interaction disappears is addressed in Mortell and Seymour [1979] where
it is shown that for small amplitude motions of the gas, the effect
of the interaction term on continuous solutions of (10.1) is negligible, see Fig. 10.1, which shows the effect of the interaction term
on continuous solutions for $A = 0.12$ and $\omega = 0.715$; solid curve
with interactions, dashed without. The equations we now examine
are (10.1) with $H(y) = A\sin(2\pi y)$ under the constraint (or mean
condition)

$$\int_0^1 G(y)dy = \Delta. \tag{10.2}$$

The detuning parameter $\Delta/2$ is defined for each mode number by

$$\Delta = 2(\omega - \omega_n), \ \ |\Delta| \le 1/2 \tag{10.3}$$

and $\omega_n = n/2$, $n = 1, 2, 3, \ldots$. Also $A = 2\pi\varepsilon(\gamma+1)\omega^2$ for a gas, where
ε is the ratio of the piston amplitude to the length of the tube. The
small amplitude Mach number of the input corresponds to $\varepsilon\omega \ll 1$.

Small rate corresponds to $|A| \ll 1$, which, it is important to note, is not implied by small amplitude.

In physical terms small rate means that a wave makes many traversals of the tube before a shock forms. For finite rate a shock can form in one travel time. In Chapter 9, when faced with solving (10.1) we invoked the small amplitude, small rate assumptions, $|G| \ll 1$ and $|G'| \ll 1$ to yield the approximating nonlinear o.d.e.

$$G(y)G'(y) = H(y).$$

We must now address Eq. (10.1) directly.

10.1. Continuous Solutions

We have seen that a regular perturbation on the governing equations breaks down at various orders when

$$\sin(n\pi\omega) = 0, \ \ n = 1, 2, 3, \ldots \tag{10.4}$$

see (4.11) or Zarembo [1967]. When seeking continuous solutions of the standard mapping, we should again find the exceptions given by (10.4). The intention is to have a consistent, small-amplitude, nonlinear approximation to the equations governing the resonant oscillations of a gas in a closed tube which is valid for all frequencies i.e., it should yield both the continuous periodic motions, and the discontinuous solutions at resonance. The requisite approximation results in a representation for u, p, e in terms of the solutions of a nonlinear difference equation with an associated mean condition. The *standard mapping* is such an approximation.

The two parameters involved in (10.1)–(10.3) are the stochastic parameter A, $0 \le A < \infty$, and the detuning parameter Δ, $-1/2 \le \Delta \le 1/2$. When the piston motion has the symmetry $H(y) = -H(1-y)$, like $A\sin(2\pi y)$, then it is easy to check that $-G(1-y)$ is a solution of (10.1) when $G(y)$ is. Then, on noting (10.2) and (10.3), all solutions can be written in terms of those for which $0 \le \Delta \le 1/2$. Thus, we need only examine the strip in the $A - \Delta$ plane: $0 \le \Delta \le 1/2$, $0 \le A < \infty$.

The essence of the technique used to generate periodic solutions with unit period is the observation that if $(x, G(x))$ lies on a periodic solution curve then $(x - m, G(x))$, for integer m, lies on the same curve. The technique overcomes the inherent difficulty of Eqs. (10.1): that to calculate the solution through a particular point the solution G must initially be known over some non-zero interval. Equations (10.1) are written in the discrete form

$$x_{n+1} = x_n + G_n, \ \ G_{n+1} = G_n + H_{n+1}, \ \ n = 0, 1, 2, \ldots, N, \quad (10.5)$$

where $G_n = G(x_n)$ and $H_n = H(x_n)$. By taking $x_0 = 0$ and specifying a value of $G_0 \neq 0$, Eqs. (10.5) generate a sequence of discrete points (x_n, G_n), $0 \leq x_n < \infty$, on the solution curve $G(x)$, $G(0) = G_0$. It is important to note that (x_n, G_n) are points on the exact invariant curve through $(0, G_0)$ — no approximation is made in going from Eqs. (10.1) to Eqs. (10.5). However, the distance between consecutive points is not at our disposal, but is determined by Eqs. (10.1), i.e., the amplitudes of the solution and the step size are coupled. A new sequence of points (y_n, G_n) is defined by

$$y_n = x_n - [x_n] + m, \ \ G_{n+1} = G_n + H_{n+1}, \ \ n = 0, 1, 2, \ldots, N$$

where $[x_n] = $ integer part of x_n and m is randomly chosen to be 0, 1 or 2 (m is random to avoid any coincidental spacing.) Then $0 \leq y_n \leq 3$. If $G(x)$, $0 \leq x \leq x_N$, is a single-valued periodic solution of Eqs. (10.1) with unit period, then (y_n, G_n) lies on $G(y)$, $0 \leq y \leq 3$. On increasing N, sufficient points (y_n, G_n) are generated to either produce a smooth periodic function $G_s(y)$, $0 \leq y \leq 3$, or establish that $G(y)$ is not periodic. If $G_s(y) = G_s(y + m)$, $0 \leq y \leq 1$, and $m = 1, 2$, we say that the solution $G(y)$ is periodic and is approximated by $G_s(y)$. Two criteria are used to check the periodicity of $G_s(y)$: (i) comparison of $G_s(y)$ and $G_s(y + 1)$ for various y and (ii) comparison of $\int_0^1 G(y)dy$ and $\int_1^2 G(y)dy$. These integrals are computed by a cubic spline and also by Simpson's rule from the exact points (y_n, G_n). The results of the integration are used to approximate $\Delta = \int_0^1 G_s(y)dy$, the mean of the periodic solution corresponding to a piston frequency $\omega = \omega_n + \Delta/2$. An example of the scatter of the points (y_n, G_n) when $G(y)$ is not

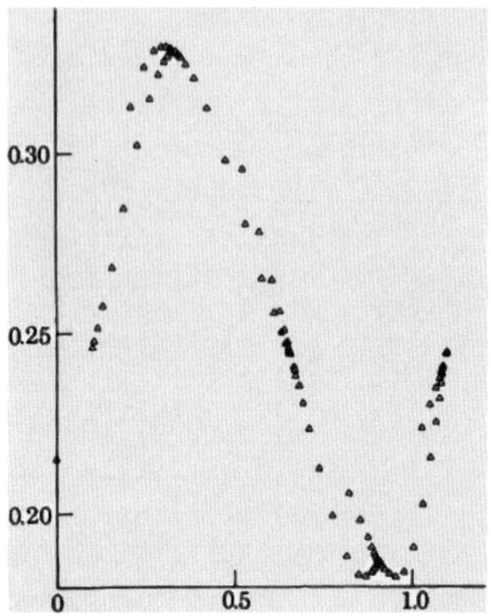

Fig. 10.2. $A = 0.12$, $G_0 = 0.215$.

periodic is given in Fig. 10.2 for $A = 0.1$ and $G(0) = 0.215$. On the basis of the scheme outlined above for the standard map (10.1), the $A - \Delta$ strip for $H(y) = A\sin(2\pi y)$ can be divided into a region in which continuous periodic solutions exist (the shaded region in Fig. 10.3) and a region in which they do not. This latter region corresponds to shocked solutions for a gas and is known as the stochastic region for the standard map. We refer to the curve separating the two regions as a (stochastic) transition curve, denoted by $A = \mathcal{A}(\Delta)$, $0 \leq \Delta \leq \frac{1}{2}$. The function $\mathcal{A}(\Delta)$ has unit period and is even about $\Delta = 0$ and $\Delta = \frac{1}{2}$. This follows from the fact that Δ depends only on the combination $\omega - \omega_n$ and the symmetry of $H(y) = A\sin(2\pi y)$. When $A > \mathcal{A}(\Delta)$ there are no continuous periodic solutions of the standard map (10.1). Then shocks are characteristic of the physical system.

Region I in Fig. 10.3 around $\Delta = 0$ is the resonant band about $\omega = \omega_n$, $n = 1, 2, 3, \ldots$, i.e., the fundamental and higher modes. The extensive experimental evidence for disturbances containing shocks when the parameters are in Region I has been given in Chapter 9.

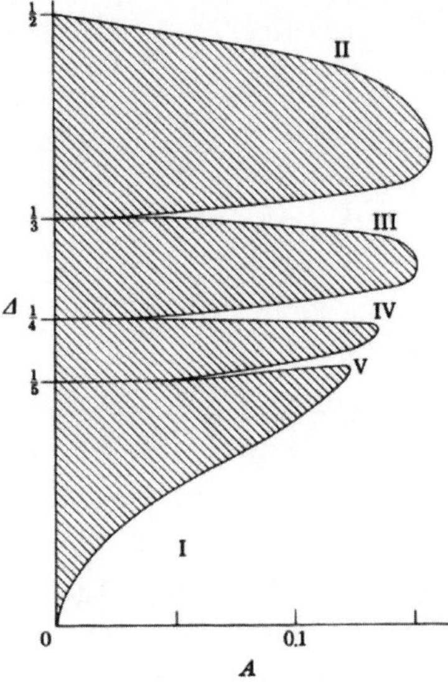

Fig. 10.3. $A = \mathcal{A}(\Delta)$.

Region II in Fig. 10.3, around the line $\Delta = \frac{1}{2}$, is the resonant band about the subharmonic frequencies $\omega = \Omega_n$, and shocks have been observed by Galiev, Ilgamov and Sadykov [1970], Zaripov and Ilgamov [1976] and Althaus and Thomann [1987]. Regions III, IV and V are resonant bands about the subharmonic frequencies given by $2(\omega - \omega_n) = m^{-1}$, $m = 3, 4, 5$. These correspond to $\sin(2m\pi\omega) = 0$, $m = 3, 4, 5$, where the regular perturbation breaks down. Again, shocks are expected but we are not aware of experimental evidence.

The analytical structure of the transition curve for small A near $\Delta = 0, \frac{1}{2}, \frac{1}{3}, \ldots$, is found by approximating (10.1) by a differential equation appropriate to each region. This is a small rate calculation and has been given for $\Delta = 0$ and $\Delta = \frac{1}{2}$, see (9.47) and (9.89), respectively. The results for $\Delta = 0, \frac{1}{2}, \frac{1}{3}, \frac{1}{4}$ in Regions I, II, III, IV are, see Mortell and Seymour [1988].

$$A = \frac{\pi^3}{8}\Delta^2 + o(\Delta)^2,$$

$$A = \pi \left|\Delta - \frac{1}{2}\right| + o\left(\left|\Delta - \frac{1}{2}\right|\right),$$

$$A = \left(\frac{81}{4}\pi\right)^{\frac{1}{3}}\left(\Delta - \frac{1}{3}\right)^{\frac{2}{3}} + o\left(\left(\Delta - \frac{1}{3}\right)^{\frac{2}{3}}\right),$$

$$A = \left(\frac{4}{\pi - 2\left(\frac{5}{3}\right)^{\frac{1}{2}}}\right)^{\frac{1}{2}}\left(\Delta - \frac{1}{4}\right)^{\frac{1}{2}} + o\left(\Delta - \frac{1}{4}\right)^{\frac{1}{2}}.$$

These boundaries correspond to the transition between continuous and stochastic solutions of the standard map. The comparison with Fig. 10.3 is excellent for $A < 0.1$, but deteriorates for larger A. It is observed from Fig. 10.3 that the maximum value of A for which shockless (i.e., continuous) motion is possible is $A \approx 0.1575$. This is the approximate value of A that marks the transition from local to global stochasticity for the standard mapping. In terms of the parameter $k = 2\pi A$ used in Lichtenberg and Liebermann [1992], this value of A yields $k = 0.989$ and compares well with the best estimate of 0.9716, with less than 2% difference.

A rough estimate of the critical value of A may be found by calculating how the "breaking length" of a wave depends on A. On using the second equation in (10.1). The condition that the applied signal $H(y)$ breaks in one cycle is

$$\max_{y}[-H'(y)] \geq 1.$$

For $H(y) = A\sin(2\pi y)$, this condition gives $A \geq (2\pi)^{-1} = 0.1592$ or $k = 2\pi A \geq 1$. The structure of the stochastic boundaries is clearly

$$A = K_m\left|\Delta - \frac{1}{m}\right|^{\frac{2}{m}} + o\left\{\left|\Delta - \frac{1}{m}\right|^{\frac{2}{m}}\right\},$$

$m = 2, 3, 4$, where K_m is a constant for each m.

In order to compare the linear and nonlinear results, we recap some previous results. The linear Eq. (4.8) is

$$\bar{G}_L(y) = \bar{G}_L(y - 2\omega) + M\sin(2\pi y),$$

so that the periodic solution is

$$\bar{G}_L = -\frac{M}{2\sin(2\pi\omega)}\cos(2\pi y + 2\pi\omega), \quad \omega \neq \frac{1}{2}n.$$

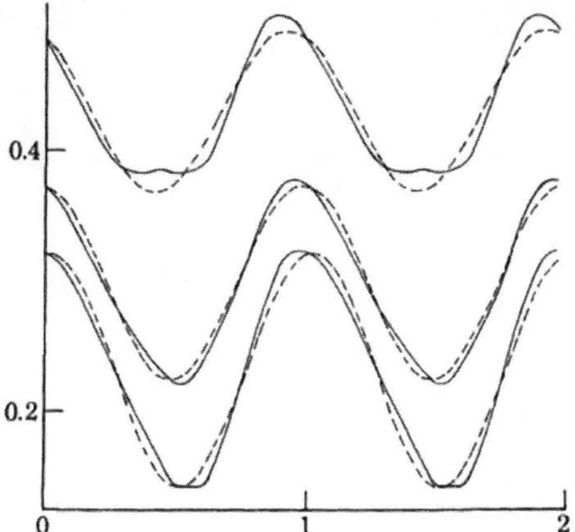

Fig. 10.4. $G(y)$, $A = 0.12$, $\Delta = 0.233, 0.298, 0.430$.

The dimensionless piston displacement is $d = -\varepsilon \cos(2\pi y)$, where $\varepsilon = \frac{\delta}{l}$, $y = \omega t$. Then the piston velocity is $u = 2\pi\varepsilon\omega \sin(2\pi y)$, and thus, the Mach number is $M = 2\pi\varepsilon\omega$. With $\omega = \frac{1}{2}(1 + \Delta)$, by (9.34) $G = \Delta + \omega(\gamma + 1)\bar{G}$, and also $H = \omega(\gamma + 1)M\bar{H} = A\sin(2\pi y)$, with $A = 2\pi(\gamma + 1)\varepsilon\omega^2$, since $\bar{H} = \sin(2\pi y)$. Then $G_L(y) = \Delta + \omega(\gamma + 1)\bar{G}_L(y) = \Delta - \frac{A}{2\sin(\pi\Delta)}[\cos(2\pi y) + \pi\Delta]$, with $y = \omega t$, and we can compare $G_L(y)$ with $G(y)$.

Figures 10.4, 10.5 and 10.6 are comparisons of the linear solution $G_L(y)$ (dashed lines) with the nonlinear solution $G(y)$. Figure 10.4, with $\Delta = 0.233, 0.298, 0.430$, indicates that the comparison is acceptable except near the quadratic resonance ($\Delta = 0.5$) in the upper graph; the vertical axis is Δ. Figure 10.5 gives a comparison between the linear and nonlinear continuous solutions for $A = 0.08$, $\Delta = 0.465$, just outside the resonant band given by $\Delta_q(0.08) = 0.475$, and Figure 10.6 shows linear and nonlinear curves at both sides of the subharmonic resonance $\Delta = \frac{1}{3}$, for $A = 0.11$. Here $\Delta = 0.325$, $0.333, 0.35$. The effect of higher harmonics is clearly discernible. Note that the values of A in these figures are well within the stochastic transition.

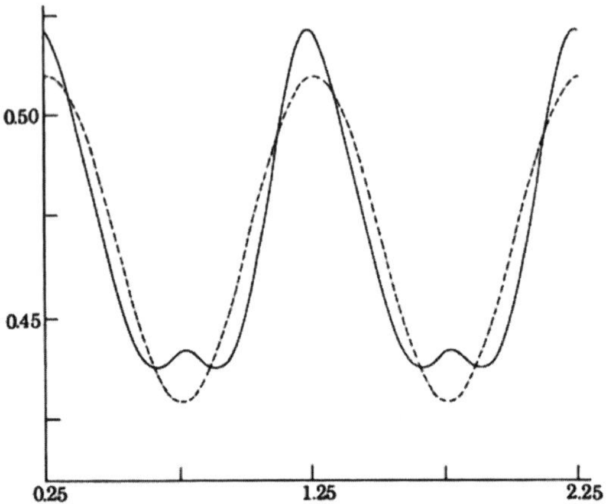

Fig. 10.5. $A = 0.08$, $\Delta = 0.465$.

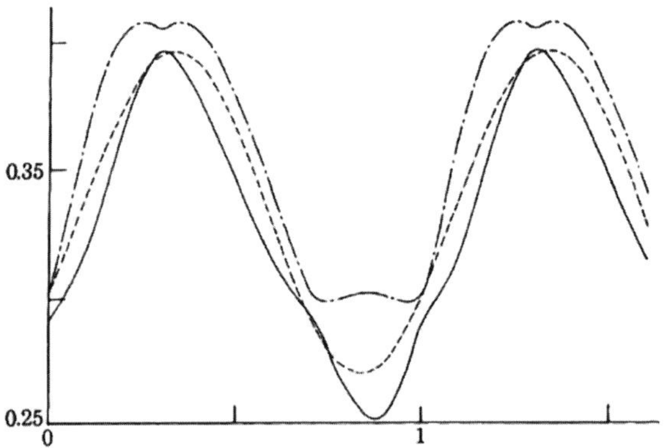

Fig. 10.6. $A = 0.11$, $\Delta \backsim 0.33$, G_L dashed lines.

10.2. Functional Equation: Discontinuous Solution

The resonant motions in a closed tube that are dealt with here are
those for which the propagating waves are intrinsically nonlinear,
i.e., there is amplitude distortion of the wave as it travels along the
tube. We refer to such oscillations as having finite rate, which can be
increased by increasing the amplitude or frequency of the piston. The
acceleration rate in an experiment is characterized by the parameter

$A = 2\pi(\gamma+1)\varepsilon\omega^2$ where ε is the ratio of the piston amplitude to the tube length, ω is the dimensionless applied frequency and γ is the gas constant.

When $A \ll 1$, the standard mapping can be approximated by a nonlinear o.d.e., see for example (9.35) and (9.36), the characteristics in the periodic state are linear and the motion is said to be small rate. Thus, the periodic signal does not distort as it travels in the tube. The experiments of Saenger and Hudson [1960] are good examples of a small rate resonant motions. Their experiments were conducted at the fundamental frequency with $\varepsilon = 0.0019$ and the conclusion that the periodic shock waves have constant strength and travel at constant speed is consistent with their small rate value of $A = 0.0075$. Sturtevant [1974] conducted experiments with CO_2 for which $\varepsilon = 0.0268$, $A = 0.096$ and with Freon 11 for which $\varepsilon = 0.0457$, $A = 0.160$. Both of these are in the finite rate range. In the experiments of Zaripov and Ilgamov [1976] it is noted that for amplitudes of 0.4 bar and higher changes of waveform, i.e., distortion, take place. Lettau [1939] showed experiments in both small and finite rate ranges, $0.01 < A < 0.17$.

The solution given by Gorkov [1963] and Chester [1964] postulates that the disturbances in the gas is an acoustic oscillation, so that the particle velocity has the form

$$u(x, t) = f(t - x) - f(t + x).$$

The implication of this form is that the waves do not distort as they travel, and the theory is thus small rate.

10.2.1. *Mapping Details*

The equations for finite rate oscillations were derived in Sec. 9.4. They are

$$G(y) = G(s) + H(y), \quad y = s + G(s), \tag{10.6}$$

where

$$G = \Delta - 2\omega N\bar{G}, \quad H = -2\omega NM\bar{H}, \quad \omega = \frac{1}{2}(1 + \Delta) \tag{10.7}$$

with $N = -(1 + \gamma)/2$ and $\int_0^1 G(y)dy = \Delta$ since $\int_0^1 \bar{H}(y)dy = 0$ and $\int_0^1 \bar{G}(y)dy = 0$.

The basic nonlinear difference equation, or functional equation, (10.6) can be regarded as the product of two mappings:

$$T_1 : (s, G(s)) \rightarrow (y, \widehat{G}(y)), \tag{10.8}$$

where $\widehat{G}(y) = G(s(y))$, $y = s + G(s)$ and

$$T_2 : (y, \widehat{G}(y)) \rightarrow (y, G(y)), \tag{10.9}$$

where $G(y) = \widehat{G}(y) + H(y)$.

T_1 can be regarded as a 'simple-wave mapping'. The function $\widehat{G}(y)$ represents the distorted signal returning to the piston after reflection from $x = 0$, but before it has been reinforced by the piston motion. T_2 then represents the action of the piston on $\widehat{G}(y)$. A given function, defined over a unit interval, is a solution of Eqs. (10.6) if it maps onto itself under the product T_2T_1. To be an acceptable physical solution it must also satisfy the periodicity condition and the mean condition.

The algorithm given in Sec. 10.1 for constructing continuous periodic solutions of Eqs.(10.6) fails to yield solutions in the linear resonance region bounded by the transition curve $A(\Delta) = \frac{\pi^3}{8}\Delta^2 + o(\Delta^2)$. The key to constructing exact solutions in this region is in recognizing that the role of the fixed points of the invariant curves is analogous to that of the critical points of the corresponding small-rate differential equation. Thus, the local structure of the invariant curves can be classified in the neighborhood of the fixed points. Two invariant curves emanating from each saddle point are found for $0 \leq y \leq 1$. When $|H| \ll 1$ their structure is similar to the separatrices of the small rate Eq. (8.42). However, as $|H|$ increases, the invariant curves become multivalued.

It is convenient to distinguish between the invariant curves of (10.6), which we denote by $Z(y)$, and the physical solution, $G(y)$, which must be single-valued and also satisfy the mean condition $\int_0^1 G(y)dy = \Delta$. A fixed point (or critical point) of (10.6), is a point

(y_c, Z_c) defined by

$$y = s = y_c \text{ when } Z(y) = Z(s) = Z_c. \tag{10.10}$$

For Eq. (10.6) they are the points where

$$Z_c = H(y_c) = 0 \tag{10.11}$$

and are thus located at the zeros of the forcing function. Any physical solution containing a critical point must include a resonating wavelet in the propagating signal with the value $Z = 0$, we expand for $|Z| \ll 1$ and obtain corresponding to the amplitude

$$\bar{G} = \frac{\Delta}{2\omega N}, \tag{10.12}$$

by the definition of G in (10.7). The frequency range for Δ about the linear resonant frequencies for which there is a resonating wavelet in $\bar{G}$, as in (10.12), is the linear resonance region (resonant band) where shocks are a feature of the motion.

In order to classify the critical points and thereby study the local structure of solutions, (10.6) is linearized about a critical point. Then (10.6) is written as $Z(s+Z(s)) = Z(s)+H(s+Z(s))$. Since at a critical point $Z=0$, we expand for $|Z| \ll 1$ and obtain, noting $H(s_c) = 0$,

$$\begin{aligned}
Z(s)Z'(s) &= H(s) + H'(s)Z(s) + \cdots \\
&= H(s_c) + (s - s_c)H'(s_c) + H'(s_c)Z(s) + \cdots \\
&= H'(s_c)(Z(s) + s - s_c) + \cdots .
\end{aligned}$$

Then the structure near a critical point is given by

$$\frac{dZ}{dx} = \mu \frac{(Z + x)}{Z}, \tag{10.13}$$

where $x = s - s_c$ and $\mu = H'(s_c)$. Differentiating (10.13) gives $\lambda^\pm = Z'(s_c)$ as

$$\lambda^\pm = \frac{1}{2}[\mu \pm (\mu^2 + 4\mu)^{1/2}]. \tag{10.14}$$

For $\mu > 0$ the critical point is a saddle point at $y = y_c$, and for $\mu < 0$ it is either a stable spiral or a node. Without loss of generality we take $y_c = 0$ with $H(y_c) = 0$ and, since H has zero mean, there is at least one saddle point in each period.

By taking the second derivative, with respect to s, of the first of (10.6) and eliminating dy/ds we obtain $Z''(y_c)$ as follows:

$$Z'(y)\frac{dy}{ds} - Z'(s) = H'(y)\frac{dy}{ds}, \quad \frac{dy}{ds} = 1 + Z'(s). \qquad (10.15)$$

Differentiating again we get, with $\frac{d^2y}{ds^2} = Z''(s)$,

$$Z''(s) = [1 + Z'(s)]^3[Z''(y) - H''(y)], \qquad (10.16)$$

on using the first of (10.15) in the form $[Z'(y) - H'(y)][1+ Z'(s)] = Z'(s)$.

On setting $y = s = y_c$, we get the second derivative at the saddle point as

$$Z''(y_c) = \left[\frac{(1 + \lambda^\pm)^3}{(1 + \lambda^\pm)^3 - 1}\right] H''(y_c). \qquad (10.17)$$

By continuing this process, we can calculate as many derivatives at a saddle point as there are derivatives of H there. Hence, the Taylor approximation to an invariant curve can be computed to any accuracy in the neighborhood of a saddle point. This then overcomes the inherent difficulty of equations of the form (10.6), *viz.*, that the solution must be known over a nonzero interval before it can be extended. Knowing the Taylor approximation in an interval about the saddle point, using the exact Eq. (10.6) allows us to march forward from the saddle point without any further errors. Two solutions emanate from any saddle point in the direction of increasing y, one with positive and one with negative slope, see Fig. 10.7. If $Z_0^+(s)$ is a known positive solution of Eqs.(10.6) for $0 = y_c \leq s \leq s_1$ then Eqs.(10.6) imply that $y > s$ and that $Z_0^+(s)$ is known for $0 \leq y \leq y_1$ where $y_1 = s_1 + Z_0^+(s_1) > s_1$. By repeating this procedure the solution can be further extended to the interval $0 \leq y \leq y_2$, where $y_2 = y_1 + Z_0^+(y_1) > y_1$. Hence, by repeated use of the mapping (10.6), any nonzero segment of solution curve containing the saddle point at $y = 0$ may be extended to the interval $0 \leq y \leq 1$. The initial segment containing the saddle point is approximated by the Taylor series. We note that the subsequent mappings are exact. The only error introduced is in truncating the Taylor series.

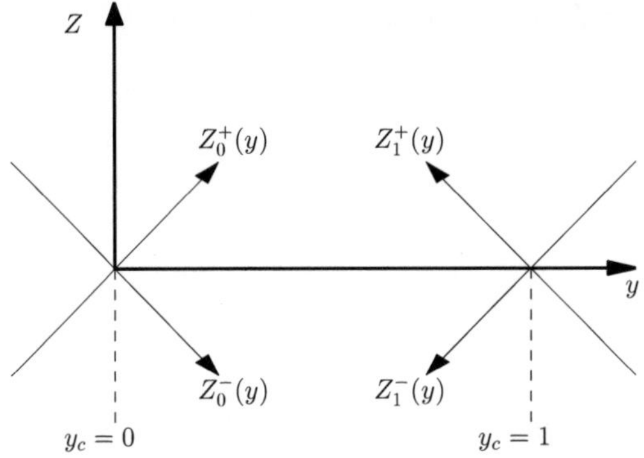

Fig. 10.7.   Saddle point diagram.

The negative solution from $y_c = 0$, in the direction of increasing y, $Z_0^-(y)$, can be calculated either from a mapping scheme similar to that used to find $Z_0^+(y)$ (though now $0 \le y \le s$), or by noting that, through equations (10.6),

$$Z_0^-(y) = H(y) - Z_0^+(y). \tag{10.18}$$

Two further solutions, emanating from $y_c = 1$ in the direction of decreasing y, can similarly be constructed on $[0, 1]$ by calculating the Taylor approximations to the two solutions near $y_c = 1$. We denote these by $Z_1^\pm(y)$, with $Z_1^+(y) > 0$ near $y_c = 1$. If the piston motion has the symmetry $H(y) = -H(1-y)$, then (see Theorem 10.2) when $Z(y)$ is a solution, $-Z(1-y)$ is also. By symmetry $Z_1^-(y) = -Z_0^+(1-y)$, hence, see Theorem 10.3, $Z_1^\pm(y)$ can also be written in terms of $Z_0^+(y)$ as follows:

$$Z_1^+(y) = H(y) + Z_0^+(1-y), \quad Z_1^-(y) = -Z_0^+(1-y). \tag{10.19}$$

These results follow from Theorems 10.1–10.3.

Theorem 10.1 *Let $G(y)$ be a solution of (10.6), then $Z(y) = H(y) - G(y)$ is also a solution and $Z(y) = -G(s(y))$.*

Proof. Let $Z(y) = H(y) - G(y)$ and substitute into (10.6). Then $Z(s) = Z(y) + H(s)$, $s = y + Z(y)$, since $y = s + H(s) - Z(s) = s - Z(y)$. By the definition of Z, $Z(y) = -G(s)$ and $y = s + G(s)$. i.e., $Z(y) = -G(s(y))$. $\qquad\square$

Theorem 10.2 *If $G(y)$ is a solution of (10.6), then $Z(y) = -G(1-y)$ is also solution when $H(y) = -H(1-y)$.*

Proof. Define $G(y) = -Z(1-y)$, and substitute $G(y)$ in (10.6). Then $Z(1-y) = Z(1-s) + H(1-y)$, with $1-y = 1-s+Z(1-s)$, which is (10.6). $\qquad\square$

Theorem 10.3 *$Z_1^{\pm}(y)$ can be written in terms of $Z_0^+(y)$ as follows: $Z_1^+(y) = H(y) + Z_0^+(1-y)$, $Z_1^-(y) = -Z_0^+(1-y)$.*

Proof. By the symmetry of $H(y)$ and Theorem 10.2, $Z_1^-(y) = -Z_0^+(1-y)$, hence, by Theorem 10.2, $-Z_1^+(1-y) = -H(1-y) - Z_0^+(y)$. $\qquad\square$

For definiteness we consider the piston motion $H(y) = A\sin(2\pi y)$. The structure of the four solution curves $Z_0^{\pm}$, $Z_1^{\pm}$ depends strongly on the magnitude of the similarity parameter A. In the small rate range, $0 < A < A_1$ the curves Z_0^+ and Z_1^+ (and also Z_0^- and Z_1^-) are indistinguishable, i.e., there are two invariant curves connecting the saddle points $y = 0, 1$. These are direct analogues of the separatrices for the differential Eq. (9.38). From numerical investigations we find that A_1 is approximately 0.01. However, it is not until $A > 0.08$ that Z_0^+ and Z_1^+ are distinguishable on a scale of physical interest; hence we take $A_1 = 0.08$, corresponding to $\varepsilon = 0.014$ for the first mode and $\varepsilon = 0.005$ for the second mode, see Fig. 10.8. When $A_1 < A < A_2$, where A_2 is approximately 0.8, we say A is in the finite-rate range. In this range the curve $Z_0^+(y)$ (and hence by (10.19), Z_1^+) is multivalued on $[0,1]$; $Z(y)$ becomes multivalued whenever $|Z'(y)| \to \infty$, which, by Eqs. (10.6), occurs whenever $1 + Z'(s) \to 0$. Then the 'breaking time' of some part of the propagating signal is less than its travel time in the tube. The finite-rate theory copes with this distortion, which is clearly outside the scope of any small-rate (differential equation) theory. Multivalued solutions $Z_0^+(y)$ are illustrated in Fig 10.8 for $A = 0.1, 0.2$ and 0.5. For $A = 0.1$ the multivalued loops are quite small, but become larger as A increases.

In the finite rate range, $Z_0^+(y)$ and $Z_1^+(y)$ are distinct and intersect an infinite number of times on $[0,1]$. This is illustrated in Fig. 10.9 for $A = 0.3$, where the multivalued loops of $Z_0^+(y)$ on $[\frac{1}{2}, 1]$

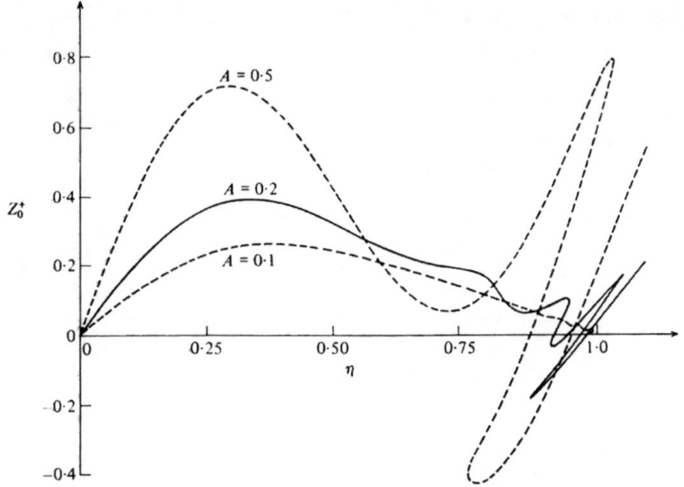

Fig. 10.8.   Solution curves Z_0^+.

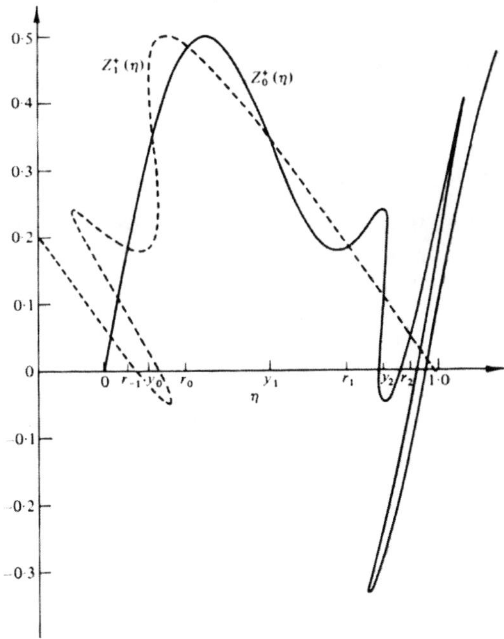

Fig. 10.9. $Z_0^+(y)$, $Z_1^+(y)$; $A = 0.3$.

are intersected by $Z_1^+(y)$. From Theorem 10.3, $Z_1^+(y) = Z_0^+(1 - y)$ when $H(y) = 0$. Therefore $Z_0^+(y)$ and $Z_1^+(y)$ intersect at $y = y_1 = \frac{1}{2}$. But

$$y_1 = y_0 + Z_0^+(y_0) = y_0 + Z_0^+(y_1), \qquad (10.20)$$

see Theorem 10.2 and note $H(y_0) = 0$. Hence $y_0 = y_1 - Z_0^+(y_1)$. There is a further point of intersection at $y = r_0$, $y_0 < r_0 < y_1$. Then

$$Z_0^+(r_i) = Z_1^+(r_i), \quad r_{i+1} = r_i + Z_0^+(r_i)$$

and

$$Z_0^+(y_i) = Z_1^+(y_i), \quad y_{i+1} = y_i + Z_0^+(y_i), \quad -\infty < i < \infty, \quad (10.21)$$

define two sequences of points of intersection. These are the only points common to $Z_0^+(y)$ and $Z_1^+(y)$. The limit points of these points of intersection are the saddle points at $y = 0$ and $y = 1$. Note that the algorithm for generating solutions is not affected by the multivaluedness since it is a purely algebraic process, independent of $Z'(y)$, in the multivalued region.

10.2.2. *Construction of Periodic Solutions*

In the finite-rate range, the invariant curves must be made single-valued before constructing the periodic solution to (10.1) and (10.2). The observation that multi-valued solutions correspond to breaking waves indicates that single-valued solutions should be obtained by inserting shocks using an equal-area rule. This choice of discontinuities is consistent with the mappings $T1$ and $T2$, which are both area preserving, and the weak shock conditions. A composite of the two discontinuous 'separatrices' on $[0, 1]$ joined by a further shock is chosen to satisfy the mean condition (10.2). It is then demonstrated that, on using the equal-area rule, this final discontinuous function is a solution of the functional Eq. (10.1). The direction of the jumps is inherent in the functional equation, which includes the cumulative distortion over a cycle, and hence no appeal to an entropy condition is necessary.

It is convenient here to note that the mapping (10.6) is area-preserving. To prove this, write (10.6), or equivalently (10.5), as

$$x_{n+1} = X(x_n, G_n) = x_n + G_n, \quad G_{n+1} = Y(x_n, G_n) = G_n + H(x_{n+1}) \tag{10.22}$$

or

$$X(x, y) = x + y, \quad W(x, y) = y + H(x + y).$$

Then the Jacobian of the mapping is

$$J = \det \begin{vmatrix} X_x & X_y \\ W_x & W_y \end{vmatrix} = \det \begin{vmatrix} 1 & 1 \\ H' & 1 + H' \end{vmatrix} = 1.$$

10.2.3. *Discontinuous Invariant Curves*

In 10.2.1 it was shown that, for $A > A_1$, the curve $Z_0^+(y)$ has an infinite number of multivalued loops on $[\frac{1}{2}, 1]$, each of which is intersected by the curve $Z_1^+(y)$. By Eqs.(10.6), $|Z'(y)| \to \infty$ as $y \to y_1^*$ whenever $1 + Z'(s) \to 0$ as $s \to s_1^*$, where $y_1^* = s_1^* + Z(s_1^*)$. The multivalued loop on $[y_2^*, y_1^*]$ of $Z_0^+(y)$, illustrated in Fig. 10.10 for $A = 0.5$, is the image of $Z_0^+(y)$ on $[s_1^*, s_2^*]$. A discontinuity is inserted at $y = x_1$, $y_2^* < x_1 < y_1^*$. The points marked a and e are the images of a' and e' at s^- and s^+ under the mapping (10.6). Thus, by Eqs.(10.6),

$$x_1 = s^- + Z_0^+(s^-) = s^+ + Z_0^+(s^+), \tag{10.23}$$

equivalent to two of the weak shock conditions. The equal-area rule (see Whitham, [1974]) is the third shock condition and can be

$$\int_{s^-}^{s^+} Z(y)dy - \frac{1}{2}\{Z(s^+) + Z(s^-)\}(s^+ - s^-) = 0. \tag{10.24}$$

Equations (10.6) and (10.23) then imply that the area enclosed by the loop $abcde$ is

$$\int_{abcde} Z_0^+(y)dy = \int_{s^-}^{s^+} Z_0^+(s)ds + \frac{1}{2}\{Z_0^{+2}(s^+) - Z_0^{+2}(s^-)\} = 0, \tag{10.25}$$

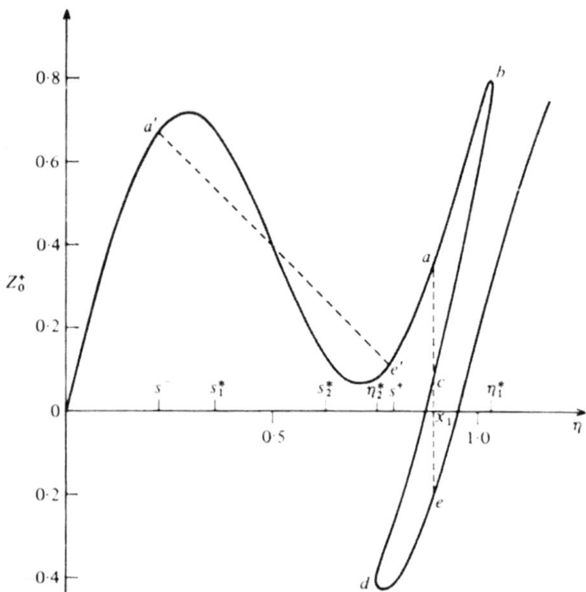

Fig. 10.10.   Discontinuities inserted in loops.

by Eqs. (10.23) and (10.24). Thus, each multivalued loop in Z_0^+ is made single-valued by inserting a discontinuity which cuts off lobes of equal area. It can additionally be shown that the areas bounded by Z_0^+ and Z_1^+ between successive points of intersection are all equal. Hence, there is a discontinuity in Z_0^+ in each interval $[r_i, r_{i+1}]$, $i \geq j$, when Z_0^+ first becomes multivalued in the jth interval. Thus, in the finite-rate range, Z_0^+ contains an infinite number of shocks on $[\frac{1}{2}, 1]$ whose strengths tend monotonically to zero and their locations tend to $y = 1$ as $i \to \infty$. This is illustrated in Fig. 10.11 for $A = 0.2$, which also exhibits the function $\hat{Z}(y) = Z_0^+(s(y))$, which is the distortion of $Z_0^+(y)$ under the simple wave mapping $T1$. Under this mapping the ith shock maps onto the $(i+1)$st shock, on using the equal-area rule. Thus, $\hat{Z}(y)$ and $Z_0^+(y)$ have shocks of identical strengths at the same locations. Since, by $T2$, $\hat{Z}(y) = Z_0^+(y) - H(y)$, and the continuous function $H(y)$ is added to $Z_0^+(y)$, the discontinuous function $Z_0^+(y)$ maps onto itself under Eqs. (10.6) and the equal-area rule, and hence may be considered a discontinuous invariant curve.

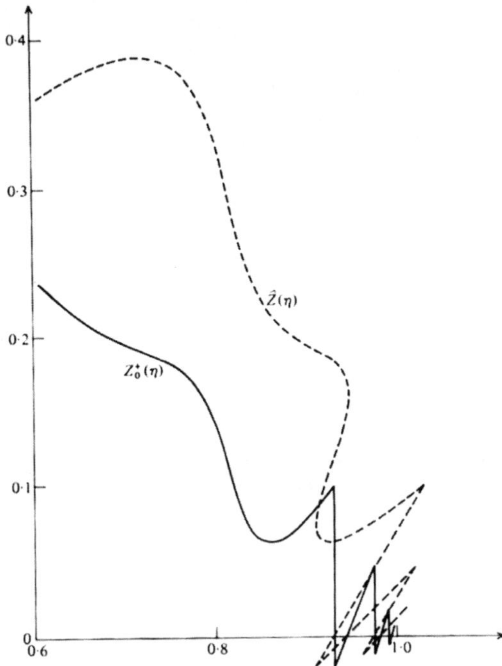

Fig. 10.11. $Z_0^+(y)$ and $\hat{Z}(y)$ for $A = 0.2$.

In a similar manner, we construct discontinuous functions from the multivalued solutions Z_0^- and $Z_1^\pm$. However, it is easily seen that, of these three functions, only that derived from Z_1^- is a discontinuous solution of (10.6) and (10.24). Any discontinuity introduced in Z_0^+ or Z_1^- is compressional and is maintained, while a discontinuity in Z_0^- or Z_1^+ produces a straight-line segment, corresponding to an expansion fan, under the mapping $T1$. The direction of the jumps in Z_0^+ and Z_1^- is consistent with that demanded by the entropy condition, Whitham [1974], as seen earlier in this chapter. We have thus constructed two single-valued separatrices, Z_0^+ and Z_1^-, on $[0, 1]$ in both the small and finite-rate ranges. The solutions derived from these are illustrated in Figs. 10.12 and 10.13 for $A = 0.05$ and $A = 0.3$ respectively, i.e., for the continuous and discontinuous separatrices. In the small-rate range the separatrices are the limiting curves of the continuous periodic curves, which do not contain fixed points

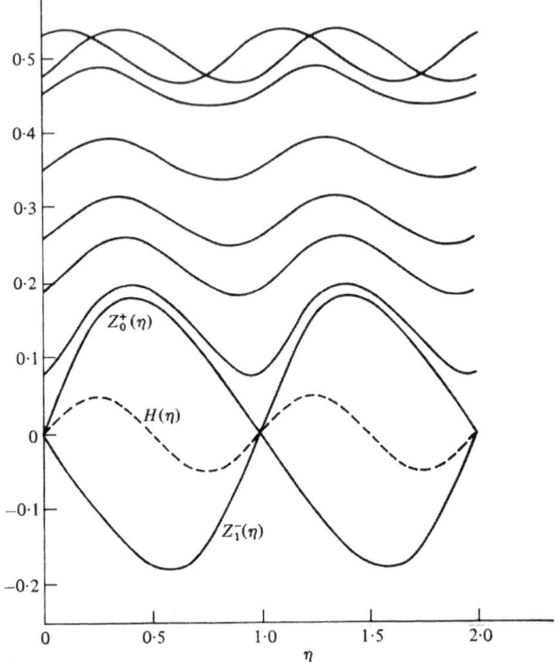

Fig. 10.12.   Small rate $Z - y$ plane; $A = 0.05$.

of Eqs. (10.6). Several of these continuous curves are included in Fig. 10.12, which also includes the separatrices associated with the subharmonic resonance at $\Delta = \frac{1}{2}$. The analysis for these latter curves is given later in this chapter. Further analysis shows that there are also separatrices about $\Delta = \frac{1}{3}, \frac{1}{4}, \frac{1}{5}$, so that the continuous curves do not fill the whole region between the separatrices associated with $\Delta = 0$ and $\Delta = \frac{1}{2}$. In both the small- and finite-rate ranges, the separatrices Z_0^+ and Z_1^- correspond to the piston frequencies

$$\omega^\pm = \omega_n \pm \frac{1}{2} \int_0^1 Z_0^+(y) \, dy,$$

$$= \omega_n \pm \frac{1}{2} \Delta_l(A),$$

(10.26)

on using Eqs. (10.1) and (10.2). If the applied frequency is such that

$$|\Delta| < \Delta_l(A)$$

(10.27)

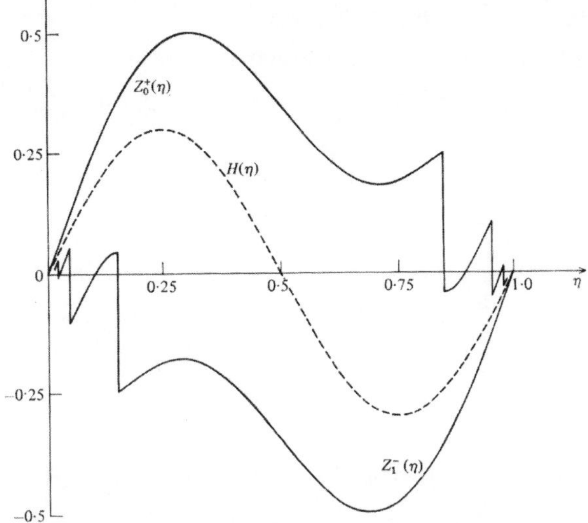

Fig. 10.13.   Finite rate separatrices: $A = 0.3$.

there is no continuous solution of (10.1) which satisfies the mean condition (10.2). When $|\Delta| > \Delta_l(A)$ there is no longer an amplitude of the signal leaving $x = 1$ which completes a cycle in the tube in a multiple of the period of H. Hence, $\Delta = \Delta_l(A)$ defines the edge of the linear resonance region and is equivalent to the statement (10.12).

The first branch of the transition curve $\Delta = \Delta_t(A)$, $0 \leq A \leq$ 0.12, passing through $A = \Delta = 0$, defines the boundary of the linear resonant region, Region I in Fig. 10.3. The parabolic nature of $\Delta_t(A)$ in the small rate limit, $A \ll 1$, is confirmed by the formula (9.47) for $\Delta_e(A)$. The region $0 \leq \Delta \leq \Delta_l(A)$ is the linear resonant region, and coincides with $\Delta = \Delta_t(A)$ for $0 \leq A \leq 0.8$. It is illustrated in Fig. 10.3, which indicates that in the small-rate range

$$\Delta_l(A) \simeq \Delta_e(A) \simeq \Delta_t(A)$$

coinciding with the transition curve separating the regions in the $A - \Delta$ plane corresponding to continuous and discontinuous periodic solutions of Eqs. (10.1) and (10.2). In the finite-rate range the edge of the linear resonance region does not correspond to the transition curve.

The area $\int_0^1 Z_0^+(y)dy$, defining the boundary of the linear resonance region, can be calculated, even in the finite-rate range, using
only continuous portions of $Z_0^+(y)$. On noting Eq. (10.26)

$$\Delta_l(A) = \int_0^1 Z_0^+(y)dy = \int_0^{r_1} Z_0^+(y)dy + \int_{r_1}^1 Z_1^+(y)dy$$

$$= \int_0^{r_1} Z_0^+(y)dy + \int_0^{1-r_1} [Z_0^+(y) - H(y)]dy, \text{ by (10.19)},$$

see Fig. 10.9 and note the lobes have equal area. The curve $\Delta = \Delta_l(A)$, calculated in this way, is plotted in Fig. 9.4.

10.2.4. *Solution in the Linear Resonance Region*

Single-valued solution curves of Eqs.(10.6) have been constructed
which join the saddle points at $y = 0$ and $y = 1$. When the applied
frequency, ω, is in the range $|2(\omega - \omega_n)| = |\Delta| < \Delta_l(A)$ there is no
single invariant curve, $Z(y)$, of Eqs.(10.6) which satisfies the mean
condition (10.2). Consequently we define the discontinuous function

$$G(y) = \left\{ \begin{array}{ll} Z_0^+(y), & 0 \le y < x_s, \\ Z_0^-(y), & x_s \le y < 1 \end{array} \right\}. \tag{10.28}$$

The location of the discontinuity, $x_s(\Delta)$, is chosen to satisfy the mean
condition (10.2); for example,

$$x_s(0) = \frac{1}{2}, \quad x_s(\Delta_l) = 1 \quad \text{and} \quad x_s(-\Delta_l) = 0.$$

This additional discontinuity must also satisfy the weak shock conditions. By the construction of Z_0^+ and Z_1^-, Eqs.(10.6) imply that
for any $0 < x_s < 1$ there exist numbers s^- and s^+, where $0 < s^- < x_s < s^+ < 1$, such that

$$x_s = s^- + Z_0^+(s^-) = s^+ + Z_1^-(s^+). \tag{10.29}$$

Hence, since y varies continuously on $[0, 1)$, from Eqs.(10.6)

$$G(x_s)^+ - G(x_s)^- = G(s^+) - G(s^-) = -(s^+ - s^-),$$

see Fig. 10.10 where $G(x_s)^+ = \lim_{y \searrow x_s} G(y) = Z_1^-(x_s)$.

$$G(x_s)^- = Z_0^+(s^+), \quad s^- - s^+ = Z_1^-(s^+) - Z_0^+(s^-). \qquad (10.30)$$

Integrating the first of Eqs.(10.6) with respect to y on $[0, 1]$, using Eq. (10.30) and the fact that H has zero mean value, again yields the equal-area rule (10.24) with Z replaced by G. Thus, the discontinuity at x_s introduced by the mean condition is consistent with the weak shock conditions.

It remains to show that the composite function $G(y)$ defined by Eq. (10.28) is a solution of the functional Eq. (10.6). We define a second composite function

$$\widehat{G}(y) = \begin{cases} Z_0^+(r(y)), & y = r + Z_0^+(r), & 0 \le y \le y_1 \\ Z_1^-(t(y)), & y = t + Z_1^-(t), & y_1 \le y < 1, \end{cases} \qquad (10.31)$$

which is the distortion of $G(y)$ by the simple wave mapping T_1. Then

$$\int_0^1 \widehat{G}(y)dy = \int_0^1 G(y)dy = \Delta,$$

since $\int_0^1 H(y)dy = 0$, as required by Eqs.(10.1), only if $y = x_s$. Hence, the position and strength of the shock at $y = x_s$ are preserved under the full mapping (10.1), since H is continuous. Thus, the composite function $G(y)$ defined by Eq. (10.28) is a discontinuous invariant curve. Note that the direction of the jump form Z_0^+ to Z_1^- results from the evolutionary nature of Eqs.(10.6). A composite function with a jump from Z_0^- to Z_1^+ does not satisfy Eqs.(10.6).

The function $\widehat{G}(y)$ represents the distorted signal returning to the piston after reflection from $x = 0$, but before it has been reinforced by the piston motion.

10.3. Finite Rate Subharmonic Oscillations

In Sec. 9.5 we described small rate subharmonic resonant oscillations in a closed tube when the applied frequency is near to half the fundamental, so that $|\omega - \Omega_0| < (1/2 - \Delta_q(A))/2$ where $\Omega_0 = 1/4$ is the "subharmonic" or "quadratic" resonant frequency. The boundary of the quadratic resonance region, $\Delta = \Delta_q(A)$, is Region II in Fig. 10.3, and for small A is given by $\Delta_q(A) = \frac{1}{2} - \frac{1}{\pi}A$. Here we extend these results to finite rate oscillations, using the full difference equations and not approximating them by differential equations for small rates.

To describe a consistent nonlinear theory of subharmonic resonance, we begin with the basic difference equation derived in 9.4, for a general forcing function $h(y)$ with unit period and $\int_0^1 h(y)dy = 0$:

$$g(y) = g(s) + Mh(y), \quad y = s + 2\omega[1 - Ng(s)]. \tag{10.32}$$

Define

$$G(y) = -2\omega Ng(y) + \Delta, \quad H(y) = -2\omega Nh(y), \quad \Delta = 2(\omega - \Omega_0), \tag{10.33}$$

and now $\Delta/2$ is the detuning from the quadratic resonant frequency $\Omega_0 = 1/4$. Then Eqs. (10.32) become

$$G(y) = G(s) + H(y), \quad y = s + G(s), \tag{10.34}$$

where G has unit period and satisfies the mean condition $\int_0^1 G(\eta)d\eta = \Delta$. For the sinusoidal piston velocity $h(y) = 2\pi\varepsilon\omega \sin(2\pi y)$ normally used in experiments

$$H(y) = A\sin(2\pi y), \tag{10.35}$$

where $A = -4\pi N\varepsilon\omega^2$ is a similarity parameter, and for a polytropic gas $N = -(1+\gamma)/2$.

The defining property of subharmonic resonance is that there is always an amplitude in the propagating signal that completes *two* cycles in the tube in one period of the forcing function $H(y)$. As before, we write the canonical equation (10.34) as the product of two mappings

$$T_1 : (s, G(s)) \to (y, \widehat{G}(y)), \quad \text{where } \widehat{G}(y) = G(s), \ y = s + G(s);$$

$$T_2 : (y, \widehat{G}(y)) \to (y, G(y)), \quad \text{where } G(y) = \widehat{G}(y) + H(y).$$

As described in Sec. 10.2.1, T_1 is the simple wave mapping that measures the distortion of the signal over a transversal of the tube, while T_2 represents the effect of the impact of the piston. Loosely speaking, the standard mapping (10.34) is a periodically forced simple wave map.

A given function is a solution of (10.34) if it maps onto itself under the product $T_1 T_2$. It is shown in Sec. 10.2.1 that solutions in the linear resonance region are constructed by using fixed points of $T_1 T_2$, i.e., the linear resonance region consists of these values of A and Δ for which $T_1 T_2$ has a fixed point. Thus, the different resonant regions in the $A - \Delta$ strip can be characterized by the existence of fixed points of the mapping $(T_1 T_2)^n$, $n = 1, 2, 3, \ldots$. We are concerned here with the case $n = 2$ for finite A. The procedure for $n = 3$, 4 in the small rate case $A \ll 1$ is given in Mortell and Seymour [1988].

Following the procedure in Sec. 9.5 we arrive at the mapping $(T_1 T_2)^2$, i.e., (9.79) and (9.80), in the form

$$Z(y) = Z(s) + \frac{1}{2}H(s) + \frac{1}{2}H(y) + H(r),$$
$$y = s + 1 + 2Z(s) + H(s) + H(r), \qquad (10.36)$$
$$r = s + \frac{1}{2} + Z(s) + \frac{1}{2}H(s),$$

where now

$$Z(y) = G(y) - \frac{1}{2}H(y) - \frac{1}{2} \qquad (10.37)$$

and the inclusion of $-\frac{1}{2}$ is explained following Eq. (10.48).

The problem of solving the governing nonlinear partial differential equations with applied frequencies in the quadratic resonance region has been reduced to finding solutions $Z(y)$ of (10.36) subject to the mean condition on G: $\int_0^1 G(t)dt = \Delta$ and the periodicity condition $Z(y + 1) = Z(y)$.

10.3.1. *Invariant Curves*

A curve that is mapped onto itself by the Eqs. (10.36) is called an invariant curve. The key to constructing invariant curves lies in recognizing that a fixed point of an invariant curve is analogous to a critical point of an ordinary differential equation. This observation was exploited in Seymour and Mortell [1980] to construct invariant curves in the linear resonance region.

A fixed point (y_c, Z_c) of an invariant curve is defined by

$$y = s = y_c, \quad \text{where } Z(y) = Z(s) = Z_c. \tag{10.38}$$

For Eqs.(10.36), the fixed points are located at y_c and r_c, which are the roots of

$$H(y) + H(r) = 0, \tag{10.39}$$

by the first of (10.36), $Z(y) = 0$ using (10.39) and $y = s$ in the second of (10.36), and the third of (10.36) yields

$$y = r - \frac{1}{2} + \frac{1}{2}H(r), \tag{10.40}$$

using (10.39).

The result $Z(y) = 0$ is one reason for the introduction of Z through Eq. (10.37). We take $H(y)$ to have the symmetry

$$H(y) = -H\left(y + \frac{1}{2}\right) \tag{10.41}$$

like the sinusoidal piston velocity (10.35) which is used in experiments. Then the location of the fixed points has a simple geometric construction given by Eqs. (10.39) and (10.40)

$$H(y) = H(t), \quad t = y + \frac{1}{2}H(y), \quad Z(y) = 0, \tag{10.42}$$

where $t = r - \frac{1}{2}$, see Fig. 10.14, taken from Mortell and Seymour [1981] with y here replacing η in the original, and $H(y)$ replacing $A\sin(2\pi\eta)$. Note that $\widehat{H}(y) = H(t(y))$ is the distortion of $H(y)$ under the simple wave map $t = y + \frac{1}{2}H(y)$. The construction for finding the critical points y_c and the corresponding $t_c = r_c - \frac{1}{2}$ is given in Fig. 10.14. The four distinct fixed points per period are located at $y_c = 0, \frac{1}{2}$, where $t_c = y_c$, and at $y_c = y_1, y_2$, where $0 < y_1 < \frac{1}{4}$, $\frac{3}{4} < y_2 < 1$. We note that $y_1 + y_2 = 1$, that y_1 and $t_1(= y_1 + \frac{1}{2}H(y_1))$ tend to $\frac{1}{4}$, and that y_2 and $t_2(= y_2 + \frac{1}{2}H(y_2))$ tend to $\frac{3}{4}$ as $A \to 0$. As $A \to \infty$, $y_1 \to 0$, $y_2 \to 1$ and $t_1, t_2 \to \frac{1}{2}$. By the symmetry of H, $r_1 = y_2$, $r_2 = y_1$ and $H(y_1) = -H(y_2)$.

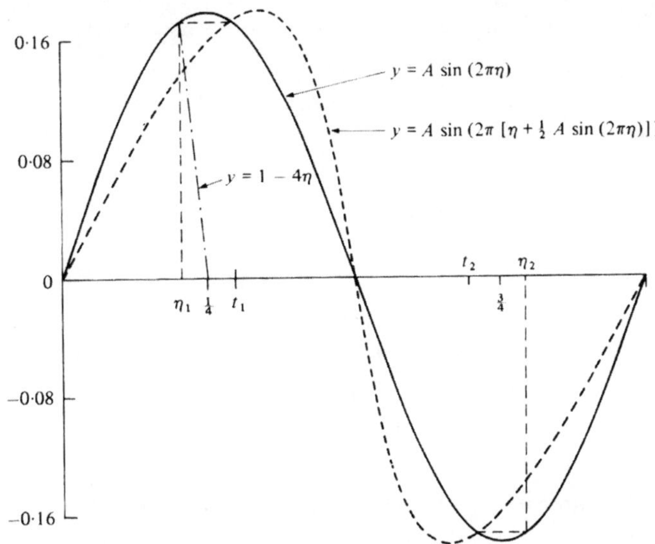

Fig. 10.14.   Location of fixed points of (10.36).

A fixed point of Eqs. (10.36) may be interpreted physically as a resonating wavelet in the propagating signal carrying the value $Z = 0$. Thus, a resonating wavelet carries a value of $g = g_c$ such that

$$g_c = \frac{1}{2\omega M}\left(\frac{1}{2} - \Delta\right) + \frac{1}{2}h(y_c),$$

see (10.32) and (10.33).

A solution containing a fixed point describes a signal which contains at least one amplitude, $g = g_c$, which completes two cycles in the tube in an odd multiple of the piston period. The mathematical significance of a fixed point is that the local structure of the functional Eq. (10.36), near the fixed point, can be described by an ordinary differential equation. When $H(y)$ and $H(r)$ are expanded about $y = y_c$ using Eqs. (10.36) and (10.42), we obtain

$$H(y) = H(y_c) + \mu_1(1 + \mu_1 + \alpha_1\mu_2)y + 2\mu_1\alpha_2 Z + O(y^2, Z^2),$$
$$H(r) = -H(y_c) + \alpha_1\mu_2 y + \mu_2 Z + O(y^2, Z^2),$$

$$(10.43)$$

where $\mu_1 = H'(y_c)$, $\mu_2 = H'(r_c)$, $\alpha_i = 1 + \frac{1}{2}\mu_i$ and here we replace y by $y - y_c$, so that $y = y - y_c$ and $y = 0$ is a fixed point. Then the first of Eqs. (10.36), using the second of (10.36) and (10.43), yields the differential equation

$$\frac{dZ}{dx} = \frac{Z + \alpha_1 y}{cZ + y}, \tag{10.44}$$

where $c = 2\alpha_2(\mu_1 + \alpha_1\mu_2)^{-1}$ to describe the structure of the nonlinear difference equation (10.36) in the neighborhood of the fixed points $y = 0$, $Z = 0$. The critical points y_1, y_2, where $y = 0$ and $\mu_1 = \mu_2 > 0$, correspond to saddle points (or hyperbolic points).

There are four solution curves emanating from each saddle point (two positive and two negative) and in general two saddle points per period in the quadratic resonance region. The eight possible solution curves in the interval $[y_1, y_1 + 1]$ are labeled $Z_{1,2}^{\pm}(y)$, $W_{1,2}^{\pm}(y)$ with for example, Z_2^+ and W_2^+ being the positive solutions leaving $(y_2, 0)$, Z_2^+ with positive slope and W_2^+ with negative slope (see Fig. 10.15).

The slope at the saddle points can be found simply by differentiating (10.44):

$$(cZ' + 1)Z' + (cZ + y)Z'' = Z' + \alpha_1,$$

so that at $Z = 0$, $y = 0$.

$$Z'(y_i) = \pm \left(\frac{\alpha_i}{c}\right)^{\frac{1}{2}}, \quad i = 1, 2 \tag{10.45}$$

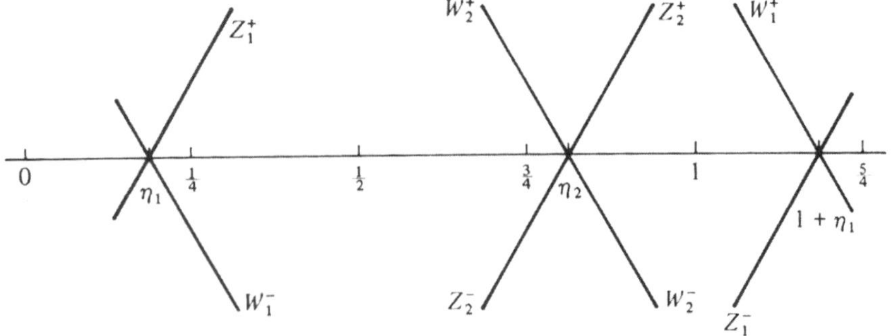

Fig. 10.15.   Notation for solutions from fixed points of (10.36).

are the possible slopes. The result (10.45) can also be found by differentiating (10.36) w.r.t. s and setting (y, s, r) to their values corresponding to y_1 or y_2. Repeated differentiations of (10.36) w.r.t s give the higher derivatives of $Z(y)$ at a saddle point, and hence the Taylor approximation to an invariant curve in the neighborhood of a saddle point can be computed to any accuracy. This procedure provides an initial segment containing a saddle point which may then be extended using the exact mapping (10.36). The only error introduced is in the truncation of the Taylor series.

Figure 10.15 illustrates the notation for solutions emanating from the fixed points of (10.36), again with y here replacing η.

Figure 10.16 shows typical positive solution curves $Z_1^+(y)$, $y_1(A) \le y \le y_2(A)$, and $Z_2^+(y)$, $y_2 \le y \le y_1 + 1$ of Eq. (10.36) with $H(y) = A\sin(2\pi y)$ when $A = 0.008$ and $A = 0.08$. The dashed curve $(A = 0.008)$ has been scaled by a factor $\frac{20}{3}$. The small rate case $A = 0.008$ shows the two lobes of the solution curves $Z_1^+(y)$ and $Z_2^+(y)$ being equal. This is not so for the case $A = 0.08$, just in the finite rate range.

In the finite rate range the invariant curves $Z = Z_1^+(y)$ and $Z = W_2^+(y)$ (and similarly $Z = Z_2^+(y)$ and $Z = W_1^+(y)$ are distinct and intersect an infinite number of times on $[y_1, y_2]$. This is illustrated in Fig. 10.17 for $A = 0.25$. The curve $Z_1^+(y)$ is intersected on $[y_1, y_2]$ by $W_2^+(y)$, and $Z_2^+(y)$ is intersected on $[y_2, y_1 + 1]$ by $W_1^+(y)$. The eight

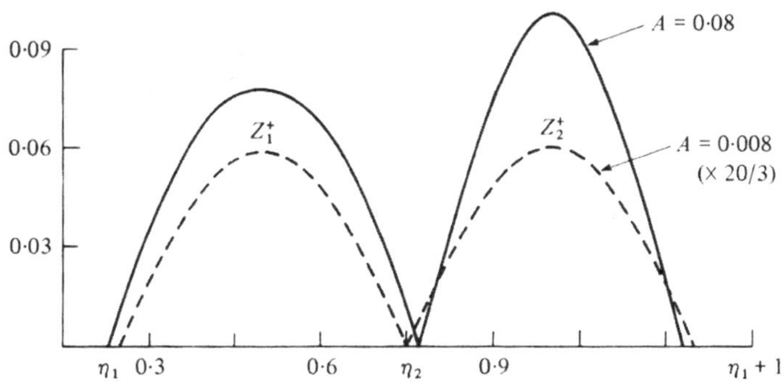

Fig. 10.16.   Solution curves $Z_1^+(y)$ and $Z_2^+(y)$.

possible solution curves of Eq. (10.36) on $[y_1, y_1 + 1]$ are denoted by

$$F_{1,2}^{\pm} = Z_{1,2}^{\pm} + \frac{1}{2}H + \frac{1}{2} \quad \text{and} \quad G_{1,2}^{\pm} = W_{1,2}^{\pm} + \frac{1}{2}H + \frac{1}{2}. \qquad (10.46)$$

The points of intersection at $y_0 = \frac{1}{2}$ and $x_0 = 1$ are recognized by noting that Eqs.(10.46) imply that

$$F_1^+\left(\frac{1}{2}\right) = G_2^+\left(\frac{1}{2}\right) \quad \text{and} \quad F_1^+(1) = G_2^+(1) \qquad (10.47)$$

as $H(\frac{1}{2}) = H(1) = 0$. Each generates a distinct sequence of intersection points on $[y_1, y_1 + 1]$

$$(y_i, G^+(y_i)) \quad \text{and} \quad (x_i, G^+(x_i)), \quad -\infty < i < \infty, \qquad (10.48)$$

under the mapping $T_1 T_2$ (Eq. (10.34)) and the periodicity of G. These are all of the intersection points on $[y_1, y_1 + 1]$, and the limit points of these sequences are the fixed points at $y = y_1, y_2$ and $y_1 + 1$. Thus, under the mapping $T_1 T_2$, $G_1^+(y)$ maps onto $F_2^+(y)$ and vice versa, on noting the presence of $\frac{1}{2}$ in the definition (10.37). However, as Eq. (10.36) is derived by using Eq. (10.34) twice, $Z_1^+(y)$ and $Z_2^+(y)$ map onto themselves under the mapping $(T_1 T_2)^2$. In fact, under $(T_1 T_2)^2$, $(y_i, Z_j^+(y_j)) \to (y_{i+2}, Z_j^+(y_{i+2}))$ and $(x_i, Z_j^+(x_i)) \to (x_{i+2}, Z_j^+(x_{i+2}))$. In addition, it can be shown that the areas bounded by curves such as Z_1^+ and W_2^+, between successive points of intersection, are all equal, and their common value is

$$\int_{y_0}^{x_1} [W_2^+(y) - Z_1^+(y)] dy. \qquad (10.49)$$

The proof of a similar result is given in Mortell and Seymour [1981]. The multivalued loops in Fig. 10.17 are made single valued by inserting shocks using the equal area rule so that the solutions are physically acceptable.

Figure 10.18 shows single-valued separatrices of (10.36) containing shocks, $G^{\pm}(y)$, $y_1 \leq y \leq y_2$, corresponding to second-order fixed points, for $A = 0.3$.

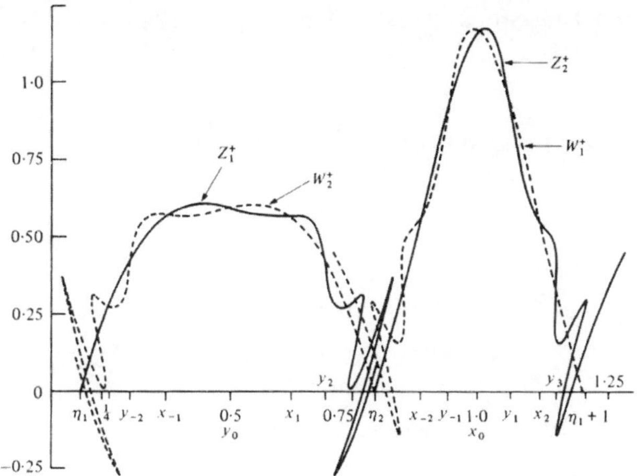

Fig. 10.17.   A shock is inserted in each loop.

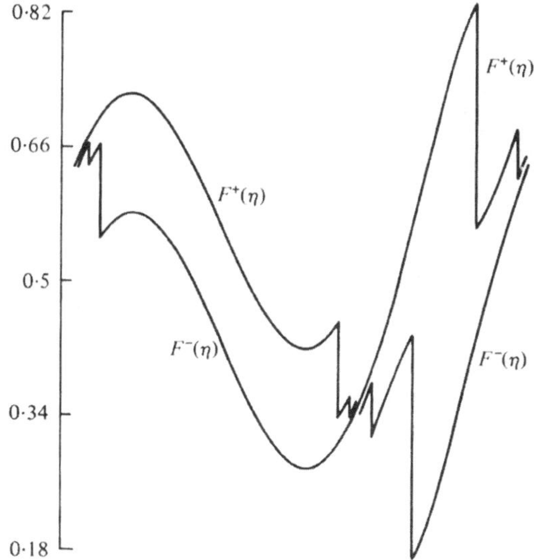

Fig. 10.18. $G^{\pm}(y)$ for $A = 0.3$ containing shocks.

In the small-rate theory the signal propagates undistorted as in
linear theory (although the signal shape is calculated from the nonlinear Eq. (9.87)). Hence, the shock strength on the closed end is the
same as that on the piston. This is not the case in the finite-rate

theory when the shock strengths change owing to the distortion of the wave form in traveling from the piston to the closed end.

10.4. Exact Discontinuous Solutions of An Area-Preserving Mapping

We have seen in Sec. 10.2.2, when dealing with the standard mapping, that an invariant curve Z_0^+ can have an infinite number of shocks on the interval $[1/2, 1]$. The strength of these shocks tends to zero monotonically as their locations tend to 1. Then we have discontinuous solutions of an area-preserving mapping — the areas bounded by Z_0^+ and Z_1^- between successive points of intersection are all equal — where the discontinuities are shocks. In this section, we follow the results in Seymour and Mortell [1981] and construct an exact discontinuous invariant curve for a particular simple forcing function $H(y)$ that is piecewise linear on $[0, 1]$, and has other properties similar to $\sin(2\pi y)$: it has zero mean, $H(0) = H(1/2) = H(1) = 0$ and is antisymmetric about the zero at $y = 1/2$, so that $H(y) = -H(1-y)$. The essential element of the piecewise linear forcing function is that the $y(x)$ relation in the transport equation, i.e., the second of (10.50), can be inverted. Then we have explicit expressions for the position and strength of the shocks and demonstrate the limit to the fixed point $(1, 0)$.

The area-preserving mapping is

$$F(y) = F(x) + H(y), \quad y = x + F(x). \tag{10.50}$$

Using Theorems 10.1–10.3 we can construct new solutions from known solutions by the following two properties.

P(i) If $F = Z_1(y)$ is a solution of (10.50), then $F = Z_2(y) = H(y) - Z_1(y)$ is also a solution and $Z_2(y) = -Z_1(x(y))$, $y = x + Z_1(x)$.

P(ii) If $F = Z_1(y)$ is a solution of (10.50), then $F = Z_3(y) = -Z_1(1 - y)$ is also a solution.

P(i) and P(ii) also imply that if $F = Z_1(y)$ is a solution of (10.50), then $Z_4(y) = H(y) + Z_1(1 - y)$ is also a solution. Hence, if $Z_1(y)$ is a solution of (10.50) then three further solutions, $Z_2(y), Z_3(y), Z_4(y)$ can be constructed.

10.4.1. *Construction of Invariant Curve $F_0^+(y)$, $0 \le y \le 1$*

The critical points of (10.50) are given by $F = H = 0$. Since $H(y)$ is antisymmetric about the middle zero at $y = \frac{1}{2}$, the critical points are at $y = 0, \frac{1}{2}, 1$. Differentiating (10.50) with respect to x yields

$$F'(y)[1 + F'(x)] = F'(x) + H'(y)[1 + F'(x)]. \qquad (10.51)$$

At a critical point $y = y_c = x_c$, since $F = 0$, then (10.51) yields

$$\mu^{\pm} = \frac{1}{2}[\lambda \pm \sqrt{\lambda^2 + 4\lambda}], \qquad (10.52)$$

where $\lambda = H'(y_c)$ and $\mu = F'(y_c)$. Now $y = 0, 1$ are saddles and $y = 0$ is a centre.

The piece-wise linear forcing is defined by

$$H(y) = \begin{cases} 4Ay, & 0 \le y \le \frac{1}{4} \\ 4A(\frac{1}{2} - y), & \frac{1}{4} \le y \le \frac{3}{4} \\ 4A(y - 1), & \frac{3}{4} \le y \le 1 \end{cases} \qquad (10.53)$$

with $H(y + 1) = H(y)$, see Fig. 10.19. This piece-wise linear $H(y)$ allows us to invert the $y - x$ relationship in the second of (10.50).

The slope of a solution at a critical point is $\mu^{\pm} = 2[A \pm \sqrt{A^2 + A}$. We take

$$A = \frac{\alpha^2}{4(1 + \alpha)}, \qquad (10.54)$$

so that

$$\mu^+ = \alpha \quad \text{and} \quad \mu^- = -\frac{\alpha}{1 + \alpha}. \qquad (10.55)$$

We now examine the cases

$$\alpha = 1, \quad \mu^+ = 1, \quad \mu^- = -\frac{1}{2},$$

$$\alpha = 2, \quad \mu^+ = 2, \quad \mu^- = -\frac{2}{3},$$

$$\alpha = 3, \quad \mu^+ = 3, \quad \mu^- = -\frac{3}{4}.$$

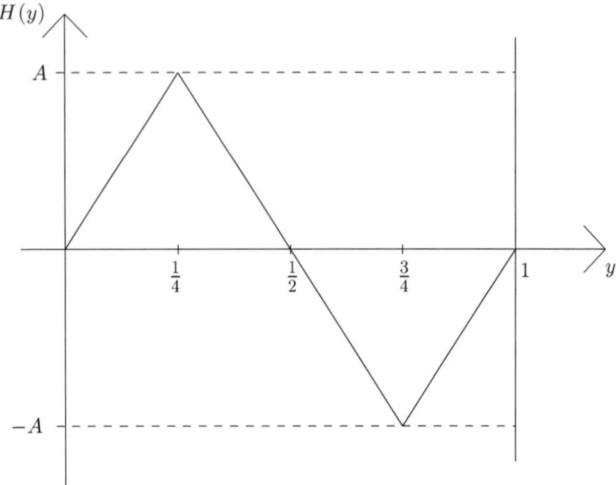

Fig. 10.19.   Forcing function $H(y)$.

Case $\alpha = 1$, $A = \frac{1}{8}$.

$$F_0^+(y) = y, \quad 0 \leq y \leq \tfrac{1}{4}, \tag{10.56}$$

since $\mu^+ = 1$. Recall $F_0^+(y)$ is the solution emanating from $y = 0$ with positive slope. Now $y = \frac{1}{4} = x + F(x) = 2x$ so that $x = \frac{1}{8}$. Also when $x = \frac{1}{4}$, $y = x + F(x) = 2x$ and $y = \frac{1}{2}$. Therefore $\frac{1}{8} \leq x \leq \frac{1}{4}$ implies $\frac{1}{4} \leq y \leq \frac{1}{2}$.

For $\frac{1}{4} \leq y \leq \frac{1}{2}$, $F_0^+(y) = F_0^+(x) + H(y)$, where $H(y) = \frac{1}{2}(\frac{1}{2} - y)$. Then $y = x + F_0^+(x) = 2x$, $x = \frac{1}{2}y$ and $F_0^+(y) = x + \frac{1}{4} - \frac{1}{2}y = \frac{1}{4}$. Therefore

$$F_0^+(y) = \tfrac{1}{4}, \quad \tfrac{1}{4} \leq y \leq \tfrac{1}{2}. \tag{10.57}$$

For $\frac{1}{4} \leq x \leq \frac{1}{2}$, $y = x + F_0^+(x) = x + \frac{1}{4}$ and $\frac{1}{2} \leq y \leq \frac{3}{4}$. Then $F_0^+(y) = F_0^+(x) + H(y) = \frac{1}{4} + \frac{1}{2}(\frac{1}{2} - y) = \frac{1}{2}(1 - y)$ and

$$F_0^+(y) = \tfrac{1}{2}(1 - y), \quad \tfrac{1}{2} \leq y \leq \tfrac{3}{4}. \tag{10.58}$$

Similarly for $\frac{1}{2} \leq x \leq \frac{3}{4}$, $\frac{3}{4} \leq y \leq 1$ and

$$F_0^+(y) = \tfrac{1}{2}(1 - y), \quad \tfrac{3}{4} \leq y \leq 1. \tag{10.59}$$

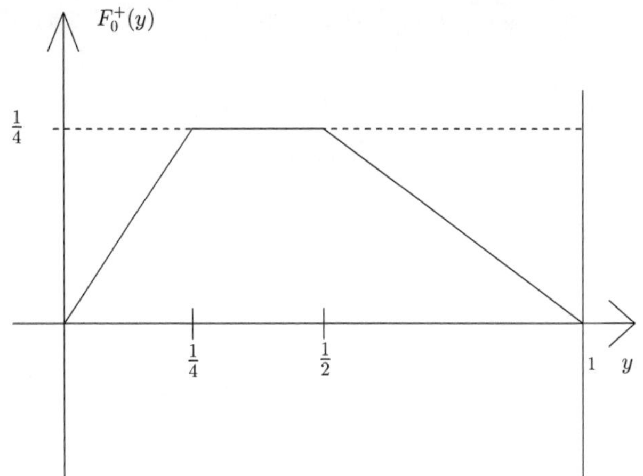

Fig. 10.20.   Continuous invariant curve $F_0^+(y)$ given by (10.56)–(10.59).

The invariant curve $F_0^+(y)$, given by (10.56)–(10.59), is depicted in Fig. 10.20. The invariant curve $F_0^+(y)$ is continuous on $[0,1]$ and connects the saddle points at $y = 0$ and 1. Three further invariant curves Z_2, Z_3, Z_4 may be constructed where $Z_1 = F_0^+$.

Case $\alpha = 2$, $A = \frac{1}{3}$ may be constructed in a similar way, to give the result

$$F_0^+(y) = \begin{cases} 2y, & 0 \le y \le \frac{1}{4} \\ \frac{2}{3}(1 - y), & \frac{1}{4} \le y \le 1. \end{cases} \tag{10.60}$$

This invariant curve is again continuous on $[0,1]$ and connects the saddle points. However, it should be noted that these cases $\alpha = 1, 2$ are exceptional since a value of α in a neighborhood of 1 or 2 will not yield a continuous invariant curve connecting the saddle points.

Case $\alpha = 3$, $A = \frac{9}{4}$ presents an entirely different situation. The single-valued invariant solution $F_0^+(y)$ is discontinuous, containing shocks, but still connects the saddle points. We give part of the solution $F_0^+(y)$ first, before giving some details of the calculations. The full solution involves an infinite number of straight lines of slopes α, joined by shocks of decreasing strength, and heading for the saddle point at $y = 1$.

The first five parts of $F_0^+(y)$ are given by (the details are given below)

$$F_0^+(y) = \begin{cases} 3y, & 0 \le y \le \frac{1}{4}, & 0 \le x \le \frac{1}{16} & (1) \\ \frac{9}{8} - \frac{3}{2}y, & \frac{1}{4} \le y \le \frac{3}{4}, & \frac{1}{16} \le x \le \frac{3}{16} & (2) \\ 3y - \frac{9}{4}, & \frac{3}{4} \le y \le 1, & \frac{3}{16} \le x \le \frac{1}{4} & (3) \\ \frac{21}{4}y - \frac{9}{2}, & \frac{3}{4} \le y \le 1, & \frac{1}{4} \le x \le \frac{3}{4} & (4) \\ 3y - \frac{45}{16}, & \frac{3}{4} \le y \le \frac{7}{4}, & \frac{3}{4} \le x \le 1, & (5) \\ \vdots \end{cases} \qquad (10.61)$$

The parts of $F_0^+(y)$ given by (10.61) are marked on Fig. 10.21 by $(1), \ldots, (5)$, and we note that from (3) onwards the solution is multi-valued. Figure 10.21 shows the solution $F_0^+(y)$ as described by (10.61), i.e., $\alpha = 3$ and we note that the slopes of (1), (3) and (5) are $\alpha = 3$.

Construction of $F_0^+(y)$, $\alpha = 3$.

$F_0^+(y) = 3y$, $0 \le y \le \frac{1}{4}$, since $\mu^+ = 3$. Then $y = x + F_0^+(x) = 4x$ or $x = \frac{1}{4}y$, so that $0 \le x \le \frac{1}{16}$. This is (1) in Fig. 10.21.

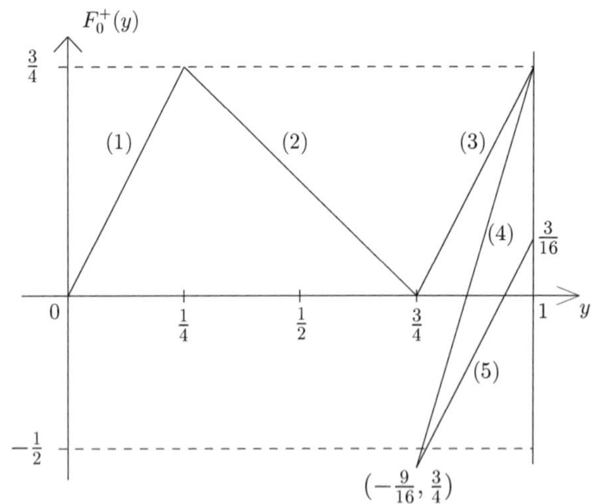

Fig. 10.21.   Parts of $F_0^+(y)$ given by (10.61).

For $\frac{1}{4} \le y \le \frac{3}{4}$, $F(y) = 3x + \frac{9}{4}(\frac{1}{2} - y)$ and $y = x + 3x = 4x$ or $x = \frac{1}{4}y$, then $F(y) = \frac{3}{4}y + \frac{9}{4}(\frac{1}{2} - y) = \frac{9}{8} - \frac{3}{2}y$. Now $F_0^+(y) = \frac{9}{8} - \frac{3}{2}y$, $\frac{1}{4} \le y \le \frac{3}{4}$, $\frac{1}{16} \le x \le \frac{3}{16}$.

We note $F_0^+(\frac{3}{4}) = 0$. This is (2) in Fig. 10.21.

For $\frac{3}{4} \le y \le 1$, $y = x + F(x) = 4x$ or $x = \frac{1}{4}y$. Therefore $\frac{3}{16} \le x \le \frac{1}{4}$.

$F(y) = 3x + \frac{9}{4}(y - 1)$ and $F_0^+(y) = 3y - \frac{9}{4}$. We note the slopes is $\alpha = 3$ and this is (3) in Fig. 10.21.

For $\frac{1}{4} \le x \le \frac{3}{4}$, $\frac{3}{4} \le y \le 1$ we have $F_0^+(y) = -\frac{9}{2} + \frac{21}{4}y$. This is (4) in Fig. 10.21 and the solution is now multivalued.

For $\frac{3}{4} \le x \le 1$, $\frac{3}{4} \le y \le \frac{7}{4}$ we have $F_0^+(y) = 3y - \frac{45}{16}$, which is (5) in Fig. 10.21, and the slope is 3.

Construction of $F_1^-(y)$, $\alpha = 3$.

For $\alpha = 3$, $\mu^- = -\frac{3}{4}$, hence $F_0^-(y) = -\frac{3}{4}y$. Then $F_1^-(y) = \frac{3}{4}(1 - y)$, see Property 2. This is the positive invariant curve starting at the saddle point at $y = 1$ with negative slope and heading towards $y = 0$. We need to check that the formula $F_0^-(y) = -\frac{3}{4}y$ holds for $0 \le y \le 1$. $F_0^-(y) = -\frac{\alpha}{1+\alpha}y$ and $H(y) = \frac{\alpha^2}{1+\alpha}y$, $\alpha = 3$.

Then $F_0^-(x) = F_0^-(y) - H(y) = -\alpha y$ and $x = (\alpha + 1)y$. With $0 \le y \le \frac{1}{4}$ we have $0 \le x \le 1$ and $x \ge y$. Hence, $F_0^-(x) = -\frac{\alpha}{1+\alpha}x$, $0 \le x \le 1$. Now $F_1^-(y) = \frac{3}{4}(1 - y)$, $0 \le y \le 1$.

Intersections of $F_0^+(y)$ with $F_1^-(y)$. The intersection of $F_1^-(y) = \frac{3}{4}(1 - y)$ with $F_0^+(y) = 3y$ (i.e., $\alpha = 3$) gives the intersection point $\left(\frac{1}{5}, \frac{3}{5}\right)$ on line (1) in Fig. 10.21. The intersection point on line (2) is $\left(\frac{1}{2}, \frac{3}{8}\right)$ and on (3), (4) and (5) they are $\left(\frac{4}{5}, \frac{3}{20}\right)$, $\left(\frac{7}{8}, \frac{3}{32}\right)$ and $\left(\frac{19}{20}, \frac{3}{80}\right)$, respectively.

Let the images of $\left(\frac{1}{5}, \frac{3}{5}\right)$ or $\left(\frac{1}{2+\alpha}, \frac{\alpha}{2+\alpha}\right)$ for $\alpha = 3$ on $F_1^-(y) = \frac{3}{4}(1 - y)$ be P_{i2}, $i = 1, 2, 3, \ldots$. Then we can easily conclude that

$$P_{i2} = (p_{i2}, \alpha q_{i2}),$$

where $q_{i2} = \frac{1}{(2+\alpha)(1+\alpha)^2}$ and $p_{i2} = 1 - (1 + \alpha)q_{i2}$, $i = 1, 2, 3, \ldots$. For example with $\alpha = 3$, $p_{12} = \frac{4}{5}$ and $\alpha q_{12} = \frac{3}{20}$, i.e., P_{12} lies on (3) and $p_{22} = \frac{19}{20}$. Also $\alpha p_{22} = \frac{3}{80}$, i.e., P_{22} lies on (5).

We still need the sequence $\left(\frac{1}{2}, \frac{3}{8}\right)$, $\left(\frac{7}{8}, \frac{3}{32}\right)$. Let the images of $\left(\frac{1}{2}, \frac{3}{8}\right)$ on $F_1^-(y) = \frac{3}{4}(1-y)$ be Q_{i1} where $Q_{i1} = (p_{i1}, \alpha q_{i1})$, and where $q_{i1} = \frac{1}{2(1+\alpha)^i}$, $p_{i1} = 1 - \frac{1}{2(1+\alpha)^{i-1}}$.

Then $p_{11} = \frac{1}{2}$, $q_{11} = \frac{1}{8}$ with $Q_{11} = \left(\frac{1}{2}, \frac{3}{8}\right)$ which lies on (2), while $p_{21} = \frac{7}{8}$, $q_{21} = \frac{1}{32}$ with $Q_{21} = \left(\frac{7}{8}, \frac{3}{32}\right)$ which lies on (4). Now $\left(\frac{1}{2+\alpha}, \frac{\alpha}{2+\alpha}\right)$ plus Q_{i1} and P_{i2}, $i = 1, 2, 3, \ldots$ constitute all the intersection points on $F_1^-(y) = \frac{3}{4}(1-y)$. Q_{i1} and P_{i2} are interlaced and tend to $(1, 0)$ as $i \to \infty$. Now $p_{i1} = \frac{1}{2}, \frac{7}{8}, \frac{31}{32}, \frac{127}{128}, \ldots$ and $p_{i2} = \frac{4}{5}, \frac{19}{20}, \frac{79}{80}, \frac{319}{320}, \ldots$ and it is easily checked that $p_{i2} < p_{i+1,1} < p_{i+1,2}$, $i = 1, 2, 3, \ldots$.

There is a similar sequence of intersections of F_1^- and F_0^+, beginning at $\left(\frac{1}{2+\alpha}, \frac{\alpha}{2+\alpha}\right)$, which limits to the fixed point $(0, 0)$.

An important consequence of the mapping (10.50) being area-preserving, see Sec.10.2.2, is that the area bounded by $F_0^+(y)$ and $F_1^-(y)$ between consecutive intersection points on $F_1^-(y) = \frac{3}{4}(1-y)$ are all equal. For example, the area bounded by F_0^+ and F_1^- between p_{12} and p_{21} is equal to the area bounded by these curves between p_{21} and p_{22}, see Fig. 10.22. Consequently, if a shock is inserted at p_{21} to make F_0^+ single valued the equal area rule for shocks is automatically satisfied.

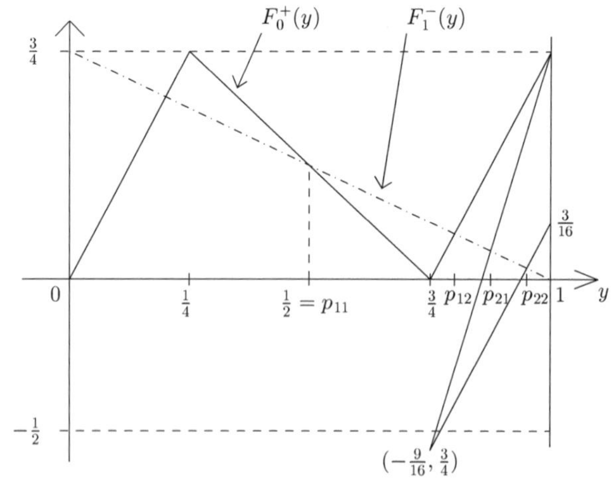

Fig. 10.22.   Intersection of $F_1^-(y)$ with $F_0^+(y)$.

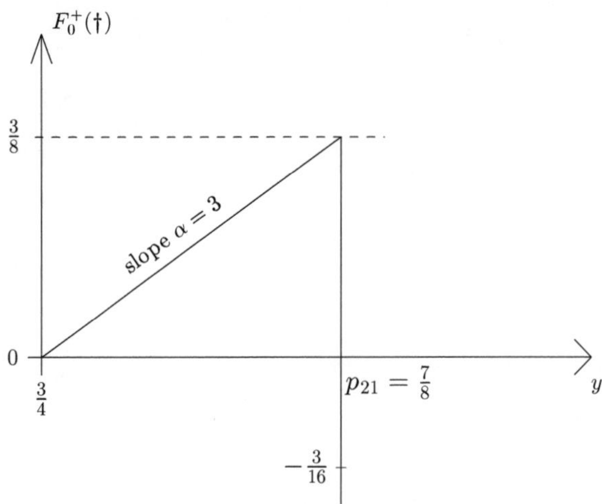

Fig. 10.23.   The first shock for $\alpha = 3$.

For a shock at p_{21}, the top of the shock is given by

$$\left(3y - \frac{9}{4}\right)\Big|_{y=p_{21}=\frac{7}{8}} = \frac{3}{8}.$$

Then the bottom of the shock lies on $F_0^+(y) = 3y - \frac{45}{16}$ and $(3y - \frac{45}{16})|_{y=\frac{7}{8}} = -\frac{3}{16}$, see in Fig. 10.23.

We now consider $F_0^+(y)$, $\frac{19}{20} = p_{22} \le y \le p_{32} = \frac{79}{80}$, and we expect a shock at $p_{31} = \frac{31}{32}$. The slope of $F_0^+(y)$ is again $\alpha = 3$, since

$$y = x + F(x) = x + 3x + C_1, \qquad \text{i.e., } x = y/4 + C_2.$$

Then $F(y) = F(x) + H(y) = 3x + \frac{9}{4}(y - 1) + C_1 = 3y + C_3$, since $H(y)$ lies in $\frac{3}{4} \le y \le 1$, and C_1, C_2, C_3 are constants. By repeating this argument, the slope of $F_0^+(y)$ between shocks is always $\alpha = 3$.

The bottom of the shock, $F_0^+ = -\frac{3}{16}$, at $y = \frac{7}{8}$ is connected to the top of the next shock at $y = \frac{31}{32}$ by a line with slope $\alpha = 3$. In this interval $F_0^+(y) = 3(y - \frac{7}{8}) - \frac{3}{16}$ and evaluating this at $y = \frac{31}{32}$ gives the top of the shock as $\frac{3}{32}$. Also $F_0^+(y) = 0$ at $y = \frac{15}{16}$.

We note that the distance between a zero of $F_0^+(y)$ and the following shock is given by $\frac{1}{8}$, $\frac{1}{16}$, $\frac{1}{32}$ ... so the zero after p_{31} is at $y = \frac{63}{64}$

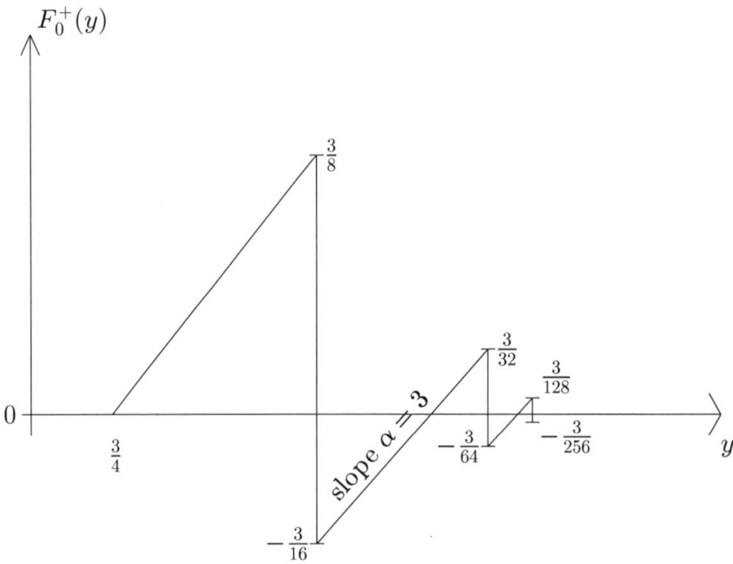

Fig. 10.24.   The first three shocks $\alpha = 3$.

and the following shock is at $y = p_{41} = \frac{63}{64} + \frac{1}{128} = \frac{127}{128}$. The line with slope $\alpha = 3$ connecting the bottom of the shock at $y = p_{31}$ to the top of the shock at $y = p_{41}$ is given by $F_0^+(y) = 3(y - \frac{63}{64})$. Then the bottom of the shock at p_{31} is $-\frac{3}{64}$ and the top of the shock at p_{41} is $\frac{3}{128}$. The diagram for the first three shocks for $\alpha = 3$ is shown in Fig. 10.24. The zeros of $F_0^+(y)$, $\frac{3}{4} \le y \le 1$, are at $y_i = (1 - \frac{1}{4^i})$, $i = 1, 2, 3, \ldots$. The shocks in $F_0^+(y)$ are at $y = p_{i1}$. The invariant solution on $[\frac{3}{4}, 1]$ is given by

$$F_0^+(y) = 3(y - y_i), \quad p_{i1} \le y \le p_{i+1,1}, \quad i = 2, 3, 4, \ldots$$

with

$$F_0^+(y) = 3\left(y - \frac{3}{4}\right), \quad \frac{3}{4} \le y \le p_{21},$$

where $p_{i1} = 1 - \frac{1}{2(1+\alpha)^{i-1}}$, with $\alpha = 3$, $i = 1, 2, 3 \ldots$.

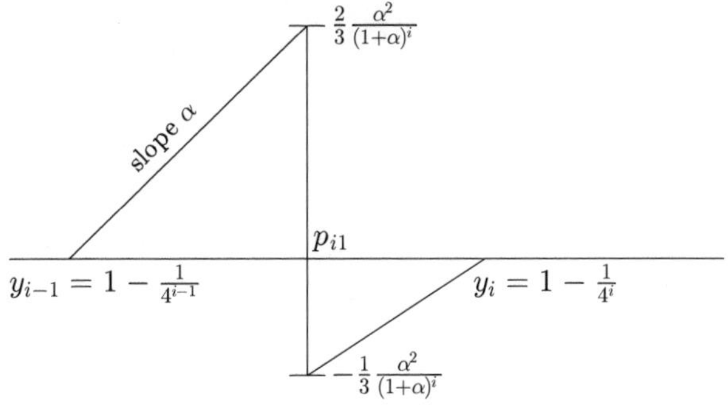

Fig. 10.25.   Invariant curve for $\frac{3}{4} \le y \le 1$ and $i = 2, 3, 4,$.

The general formulae for the top and bottom of the shocks are

$$\text{Top:} \qquad \frac{1}{2}\alpha(1+\alpha)^{1-i}, \quad i = 2, 3, 4, \ldots$$

$$\text{Bottom:} \qquad \frac{\alpha(1-\alpha)}{2(1+\alpha)^i}, \quad i = 2, 3, 4, \ldots.$$

These formulae also hold for $\alpha > 3$, see Seymour and Mortell [1981]. Also note $\frac{\alpha}{2(1+\alpha)^{i-1}} = 2\frac{\alpha(1-\alpha)}{(1+\alpha)^i}$ for $\alpha = 3$; i.e., the top of the shock is two times the bottom of the shock. The shock strength is $\frac{\alpha^2}{(1+\alpha)^i}$ for the shock at $y = p_{i1} = 1 - \frac{1}{2(1+\alpha)^{i-1}}$, $i = 2, 3, 4, \ldots$.

The shock position $y = p_{i1} \to 1$ as $i \to \infty$ and the corresponding shock strength $\frac{\alpha^2}{(1+\alpha)^i} \to 0$ as $i \to \infty$. The bottom of one shock is connected to the top of the next shock by a line of slope $\alpha = 3$. We have constructed an invariant curve of the area-preserving mapping (10.50), with $H(y)$ given by (10.53), which contains an infinite number of shocks on $[0, 1]$. Figure 10.25 shows the structure on the discontinuous invariant curve for $\frac{3}{4} \le y \le 1$ and $i = 2, 3, 4, \ldots$. This invariant curve has been found by a constructive procedure based on (10.50).

Chapter 11

The Evolution of Resonant Oscillations

The periodic response of a gas in a closed straight tube is described in detail in Chapters 9 and 10. In the small rate limit, when the piston frequency is in a band about a resonant frequency shock waves appear in the tube. Outside this frequency band the gas motion is continuous. For the periodic problem, the motion is characterized by the similarity parameter $A = 2\pi\varepsilon(\gamma + 1)\omega^2$, where ε is the ratio of the piston amplitude to the length of the tube. The small rate limit is defined as $A \ll 1$, when there is negligible distortion of a signal in traveling the length of the tube and the disturbance in the body of the gas is an acoustic oscillation. In the finite rate case, $A = O(1)$, the piston acceleration is no longer small and there is significant distortion of the signal in a traversal of the tube.

In this chapter, we consider the evolution of the motion from a state of rest to the final periodic state. The motion is again characterized by the similarity parameter A. There are two time-scales in the problem — the travel time in the tube (or, equivalently the period of the piston) and the time for significant distortion of the signal. When the ratio of travel time to distortion time is small, corresponding to $A \ll 1$, multiscale and nonlinearization methods can be used to construct the signal function in terms of the solution to a forced p.d.e. involving Elliptic functions. In the finite rate case, $A = O(1)$, the full difference equation must be solved directly.

276

11.1. Small Rate Evolution

In Chapter 9, we considered small amplitude, small rate periodic solutions to (9.6) and (9.7) subject to the boundary conditions (9.2). We now add the *initial conditions* of the gas at rest:

$$u(x,t) = 0 \quad \text{and} \quad c(x,t) = 1, \quad 0 \le x \le 1,\ t \le 0 \qquad (11.1)$$

and consider the evolution of the disturbance to a periodic state. In the small rate limit, $A \ll 1$, the ratio of travel time to distortion time is small, $O(\varepsilon)$

11.1.1. *A Perturbation Approach*

We first consider the evolution of the linear problem by considering the first term in a regular expansion of the form

$$u(x,t) = \varepsilon u_1(x,t) + \varepsilon^2 u_2(x,t) + \cdots \qquad (11.2)$$
$$c(x,t) = 1 + \varepsilon c_1(x,t) + \varepsilon^2 c_2(x,t) + \cdots,$$

in Eqs. (9.6) and (9.7). At $O(\varepsilon)$ the linear equations have the general solution

$$u_1 = f(t+x-1) - g(t-x) \quad \text{and} \quad c_1 = -\frac{\gamma-1}{2}[f(t+x-1) + g(t-x)].$$

The boundary condition $u(0,t) = 0$ implies

$$u_1(x,t) = f(t+x-1) - f(t-x-1)$$

and the second condition, $u(1,t) = 2\pi\varepsilon\omega\sin(2\pi\omega t)$, that

$$f(t) - f(t-2) = 2\pi\omega\sin(2\pi\omega t), \quad t > 0. \qquad (11.3)$$

The periodic solution to (11.3) for $\omega \ne \frac{1}{2}n$ is

$$f(t) = -\frac{\pi\omega\cos(2\pi\omega(t+1))}{\sin(2\pi\omega)}.$$

Thus, the resonant frequencies are $\omega = \omega_n = \frac{1}{2}n$. For the first resonance, $\omega = \frac{1}{2}$, (11.3) becomes:

$$f(t) - f(t-2) = \pi\sin(\pi t)$$

for which a particular solution with initial condition $f(t) = 0$, $t \leq 0$ is

$$f(t) = \frac{1}{2}\pi t \sin(\pi t).$$

This shows the initial linear growth in the signal $f(t)$ that becomes unbounded as $t \to \infty$.

To correct this growth we use a two-time, large time expansion with the first term $O(\varepsilon^{\frac{1}{2}})$

$$
\begin{aligned}
u(x,t) &= \varepsilon^{\frac{1}{2}} u_1(x,t;\tau) + \varepsilon u_2(x,t;\tau) + \cdots \\
c(x,t) &= 1 + \varepsilon^{\frac{1}{2}} c_1(x,t;\tau) + \varepsilon c_2(x,t;\tau) + \cdots ,
\end{aligned}
\tag{11.4}
$$

where $\tau = \varepsilon^{\frac{1}{2}} t$. This means that the basic approximation u_1 is a linear standing wave, and u_2 is an iteration about it. An expansion in powers of ε is used to derive the linear approximation, since the linear periodic solution, when valid, is $O(\varepsilon)$. However, the long-time nonlinear periodic solution, containing shocks, is $O(\varepsilon^{\frac{1}{2}})$, see (9.18) where $M = 2\pi\varepsilon\omega$. A regular perturbation expansion in powers of $\varepsilon^{\frac{1}{2}}$ yields the nonlinear periodic solution, see Sec. 9.2, and we generalize this by including the slow variable $\tau = \varepsilon^{\frac{1}{2}} t$.

At $O(\varepsilon^{\frac{1}{2}})$ the solution is

$$u_1 = f(t + x - 1, \tau) - f(t - x - 1, \tau) \tag{11.5}$$

and

$$c_1 = -\frac{\gamma - 1}{2}[f(t + x - 1, \tau) + f(t - x - 1, \tau)], \tag{11.6}$$

where

$$f(t - 2, \tau) = f(t, \tau) \tag{11.7}$$

on using the boundary conditions $u(0, t; \tau) = u(1, t; \tau) = 0$.

At $O(\varepsilon)$ the equations are

$$
\begin{aligned}
\left[\frac{\partial}{\partial t} + \frac{\partial}{\partial x}\right]&\left[u_2 + \frac{2}{\gamma - 1}c_2\right] \\
&= -\frac{\gamma + 1}{\gamma - 1}c_1\frac{\partial}{\partial x}\left[u_1 + \frac{2}{\gamma - 1}c_1\right] - \frac{\partial}{\partial \tau}\left[u_1 + \frac{2}{\gamma - 1}c_1\right], \quad (11.8)
\end{aligned}
$$

$$\left[\frac{\partial}{\partial t} - \frac{\partial}{\partial x}\right]\left[u_2 - \frac{2}{\gamma - 1}c_2\right]$$

$$= -\frac{\gamma + 1}{\gamma - 1}c_1\frac{\partial}{\partial x}\left[u_1 - \frac{2}{\gamma - 1}c_1\right] - \frac{\partial}{\partial \tau}\left[u_1 - \frac{2}{\gamma - 1}c_1\right]. \quad (11.9)$$

Inserting (11.5) and (11.6) into (11.8) and (11.9) and integrating yields

$$u_2 = \frac{\gamma + 1}{2}\left[xf(t + x - 1, \tau)\frac{\partial}{\partial x}f(t + x - 1, \tau)\right.$$

$$- xg(t - x, \tau)\frac{\partial}{\partial x}g(t - x, \tau)$$

$$+ \frac{1}{2}\int^{t-x} g(s, \tau)ds \cdot \frac{\partial}{\partial x}f(t + x - 1, \tau)$$

$$+ \frac{1}{2}\int^{t+x-1} f(s, \tau)ds \cdot \left.\frac{\partial}{\partial x}g(t - x, \tau)\right]$$

$$+ x\left[\frac{\partial}{\partial \tau}g(t - x), \tau) + \frac{\partial}{\partial \tau}f(t + x - 1, \tau)\right]. \quad (11.10)$$

In deriving (11.10) it is helpful to note

$$\left[\frac{\partial}{\partial t} \pm \frac{\partial}{\partial x}\right][xh(t \mp x - 1, \tau)] = \pm h(t \mp x - 1, \tau).$$

Now $g(t, \tau) = f(t - 1, \tau)$ so that $g(t - x, \tau) = f(t - x - 1, \tau)$. Then

$$u_2 = \frac{\gamma + 1}{2}x\left[f(t + x - 1, \tau)f'(t + x - 1, \tau)\right.$$

$$+ f(t - x - 1, \tau)f'(t - x - 1, \tau)$$

$$+ \left.\frac{\partial}{\partial \tau}\left(f(t + x - 1, \tau) + f(t - x - 1, \tau)\right)\right]$$

$$+ \frac{\gamma + 1}{4}\left[f'(t + x - 1, \tau)\int^{t-x} g(s, \tau)ds\right.$$

$$- g'(t - x, \tau)\int^{t+x-1} f(s, \tau)ds\right], \quad (11.11)$$

where the prime denotes differentiation w.r.t. the first term in the brackets. For example,

$$\frac{\partial}{\partial x} g(t - x, \tau) = -g'(t - x, \tau).$$

The boundary condition $u_2(0, t; \tau) = 0$ on $x = 0$ is satisfied by (11.11).

The condition on $x = 1$ implies for resonance at $\omega = \frac{1}{2}$

$$\frac{\pi}{2} \sin(\pi t) = \frac{\gamma + 1}{2} f(t, \tau) f'(t, \tau) + \frac{\partial f}{\partial \tau}(t, \tau), \tag{11.12}$$

since the integral terms in (11.11) reduce to

$$-\frac{\gamma + 1}{4} f'(t, \tau) \int_{t-2}^{t} f(s, \tau) ds$$

and this is zero once we take the mean of f as the reference state, which is consistent with the initial conditions.

We now observe that if the nonlinear term in (11.12) is discarded, the solution to (11.12) is

$$f = \frac{\pi}{2} \varepsilon^{\frac{1}{2}} t \sin(\pi t)$$

and then at $O(\varepsilon^{\frac{1}{2}})$ in the perturbation procedure

$$\varepsilon^{\frac{1}{2}} f = \frac{\pi}{2} \varepsilon t \sin(\pi t)$$

which agrees with the linear solution.

The steady state solution to (11.12) when $\frac{\partial f}{\partial \tau} = 0$ implies

$$(\gamma + 1) f(t) f'(t) = \pi \sin(\pi t) \tag{11.13}$$

which agrees with (9.17) when we note that $g(t, \tau) = f(t - 1, \tau)$.

If we introduce $F(t, \tau) = \frac{\gamma + 1}{2} f(t, \tau)$, then (11.12) takes the form

$$\frac{\partial F}{\partial \tau} + F \frac{\partial F}{\partial t} = \tilde{A} \sin(\pi t), \tag{11.14}$$

where $\tilde{A} = \frac{\pi}{4}(\gamma + 1)$. Equation (11.14) is essentially the same as Eq. (3.14) in Cox and Mortell [1983], with exact solutions that involve Elliptic functions. The initial condition is $F(t, 0) = \Delta$ and the mean

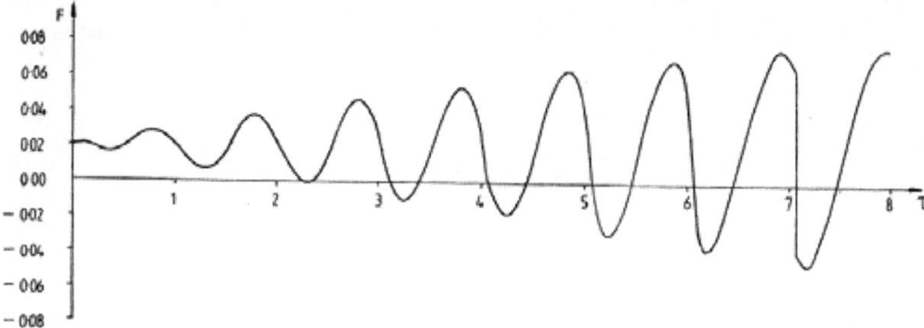

Fig. 11.1.   Evolution to shock; $\Delta = 0.02$.

condition at constant τ is $\int_0^1 F(s,\tau)ds = \Delta$. It should be noted that to relate the solution of (11.14) to the physical solution, $F(t,\tau)$ must be evaluated along the line $\tau = t/\omega$ in the (t,τ) plane. We need only evaluate $F(t,\tau)$ on $0 \le t \le 2$ due to the periodicity condition (11.7). Hence, $F(t,\tau)$ is solved for $0 \le t \le 2$, $\tau \ge 0$ along the lines $\tau = [t+(n-1)]/\omega$, where n is the piston cycle in which F is evaluated. Solutions evaluated on constant τ and extended periodically lead to spurious discontinuities.

Figure 11.1 shows that the growth and distortion of F given by Eq. (11.14) with $\Delta = 0.02$ until a shock forms in cycle 8 of the piston. Figure 11.2 shows that shocks are located at the zero of the forcing term for $\Delta = 0$. The shock forms in cycle 8 and the steady state is closely approximated at cycle 21. Figure 11.3 shows that for sufficiently large $\Delta = 0.06$, i.e., outside the resonant band, the evolving signal forms a shock that decays as $t \to \infty$ to leave a periodic linear solution. These figures are taken from Cox and Mortell [1983].

11.1.2. *Nonlinearization*

The linear equation for the signal f is (11.3) so that

$$u(x,t) = \varepsilon u_1(x,t) = f(t+x-1) - f(t-x-1),$$

and the Mach number is $2\pi\varepsilon\omega$. The question is, how can the linear difference equation be nonlinearized to result in the equation for the

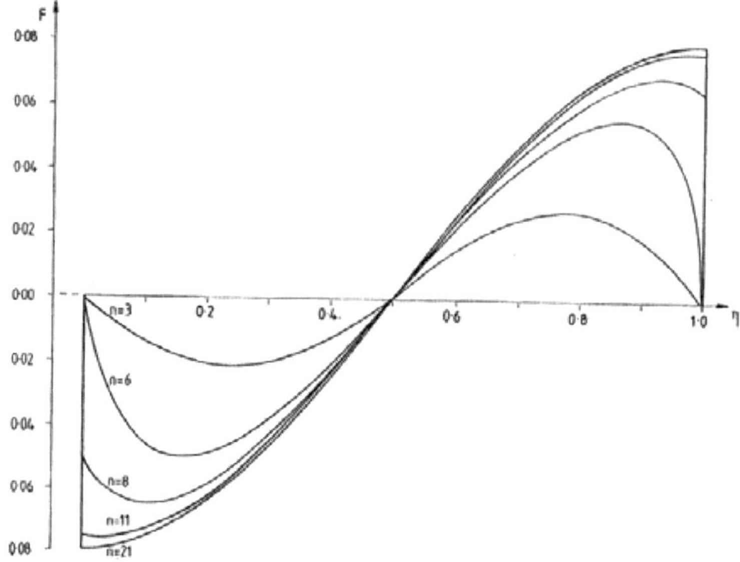

Fig. 11.2.   Steady state cycle 21; $\Delta = 0$.

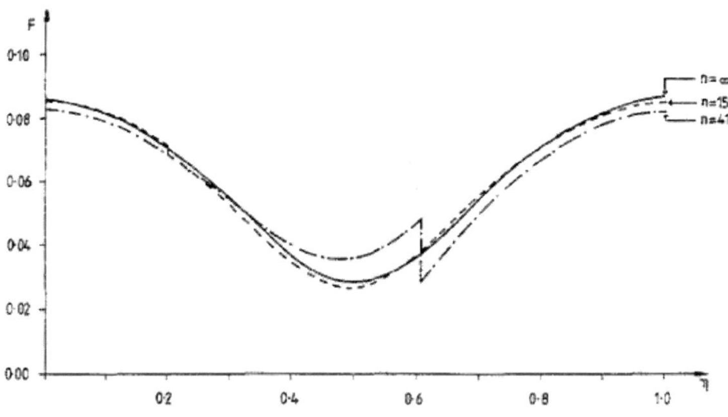

Fig. 11.3.   Decay of shock outside resonant band; $\Delta = 0.06$.

evolution of the resonant motion? The solution is to correct the linear travel time in the linear difference equation by including the first nonlinear correction. Then an expansion in powers of M, the square root of the Mach number, yields the small rate evolution equation.

We use Eqs. (9.31) and (9.32) in the form

$$G(\eta) - G(s) = M^2 \sin(\pi\eta),$$
$$\eta = s + 2 - 2NG(s),$$

(11.15)

so that comparing with (11.3) and noting that $u = \varepsilon u_1$, $M^2 = 2\pi\varepsilon\omega$. Here $\eta = \omega t$, $\omega = \omega_1 = \frac{1}{2}$, and $N = -\frac{\gamma+1}{2}$. M^2 is the Mach number of the input and we recall that the periodic output is $O(M)$. Hence, we have an expansion in powers of M and the input enters at second order.

Linear theory gives

$$G(\eta) - G(\eta - 2) = M^2 \sin(\pi\eta)$$

with the particular integral

$$G(\eta) = \frac{1}{2}M^2\eta\sin(\pi\eta).$$

To get the resonant solution when $\omega = \frac{1}{2}$, we write

$$G(s) = MG_0(s,\tau) + M^2G_1(s,\tau) + \cdots,$$

where $\tau = Ms$, which is a similar expansion to (11.4). Then

$$\begin{aligned}
G(\eta) &= G(s + 2 - 2NG(s))\\
&= G(s+2) - 2NG(s)G'(s+2) + \cdots\\
&= MG_0(s+2,\tau+2) + M^2G_1(s+2,\tau)\\
&\quad - 2NM^2G_0(s,\tau)\frac{\partial G_0}{\partial s}(s+2,\tau) + \cdots\\
&= MG_0(s+2,\tau) + 2M^2\frac{\partial G_0}{\partial \tau}(s+2,\tau) + M^2G_1(s+2,\tau)\\
&\quad - 2NM^2G_0(s,\tau)\frac{\partial G_0}{\partial s}(s+2,\tau) + O(M^3).
\end{aligned}$$

At $O(M)$, (11.15) implies

$$G_0(s+2,\tau) - G_0(s,\tau) = 0$$

(11.16)

and at $O(M^2)$

$$G_1(s+2,\tau) - G_1(s) + 2\frac{\partial G_0}{\partial \tau}(s+2,\tau)$$

$$- 2NG_0(s,\tau)\frac{\partial G_0}{\partial s}(s+2,\tau) = \sin(\pi(s+2)). \quad (11.17)$$

We take $G_1(s+2,\tau) - G_1(s) = 0$ to be part of the standing wave as in (11.16), and then $G_0(s,\tau)$ satisfies

$$\frac{\partial G_0}{\partial \tau}(s,\tau) - NG_0(s,\tau)\frac{\partial G_0}{\partial s}(s,\tau) = \frac{1}{2}\sin(\pi s). \quad (11.18)$$

Equation (11.18) agrees with (11.12) on noting $N = -\frac{\gamma+1}{2}$ and the form of the condition on $x = 1$. Linear theory is recovered since

$$G(s) = MG_0(s) = \frac{1}{2}M^2 s \sin(\pi s),$$

on noting that $\tau = Ms$.

11.1.3. *Damped Resonance using Nonlinearization*

We consider a closed tube where damping is introduced by a reflection coefficient at the end $x = 0$. The equations are (9.51) which are now written on $x = 1$ as

$$\bar{G}(y) = k\bar{G}(s) + M\bar{H}(y), \quad y = s + 2\omega - \omega(1+k)N\bar{G}(s) \quad (11.19)$$

with $y = \omega t$, $\omega = (1+\Delta)/2$ and $M = 2\pi\varepsilon\omega$. The linear resonant frequencies are $\omega = \omega_n = \frac{1}{2}n$, $n = 1,2,3,\ldots$ and we confine our attention to the first of these $\omega = \frac{1}{2}$.

   In (11.19), $M\bar{H}(y)$ is the velocity of the driver, where $M \ll 1$ is the Mach number of the input, and $N = -\frac{\gamma+1}{2}$. Also $\bar{H}(y)$ is normalized so that $\bar{H}(y+1) = \bar{H}(y)$, and $\bar{H}(y)$ has zero mean: $\int_0^1 \bar{H}(y)dy = 0$. A consequence is that $\int_0^1 \bar{G}(y)dy = 0$, i.e., $\bar{G}(y)$ has zero mean.

The canonical form of (11.19) is

$$G(y) - kG(s) = MH(y), \quad y = s + 1 + G(s), \tag{11.20}$$

where

$$G(y) = \Delta + b\bar{G}(y), \quad MH = \Delta(1 - k) + bM\bar{H}, \quad b = -\omega(1 + k)N$$

with $\int_0^1 G(y)dy = \Delta$.

To obtain a slowly varying evolution equation in terms of the small parameter M we assume that $k = 1 - \mu M^{1/2}, \omega = \frac{1}{2}(1 + M^{1/2}\Delta)$ and $b = -N + O(M^{1/2})$. It is anticipated that the amplitude of the periodic signal is $O(M^{1/2})$. Then, Eq. (11.20) is written as

$$G(y) - (1 - \mu M^{1/2})G(s) = MH(y), \quad y = s + 1 + G(s). \tag{11.21}$$

We assume a long-term expansion for G, with $G = O(M^{1/2})$,

$$G(s) = M^{1/2}G_0(s, \tau) + MG_1(s, \tau) + \cdots, \quad \tau = M^{1/2}s,$$

where the solution grows from $O(M)$ to $O(M^{1/2})$. Then, (11.21) yields at $O(M^{1/2})$.

$$G_0(s + 1, \tau) - G_0(s, \tau) = 0 \tag{11.22}$$

and at $O(M)$

$$\begin{aligned}
G_1(s + 1, \tau) &- G_1(s, \tau) \\
&= -\frac{\partial G_0}{\partial \tau}(s + 1, \tau) - \mu G_0(s, \tau) - bG_0(s + 1, \tau)\frac{\partial G_0}{\partial s}(s + 1, \tau) \\
&\quad + H(s + 1).
\end{aligned} \tag{11.23}$$

If G_1 in (11.23) is part of the standing wave, then on using (11.22) and to exclude secular growth terms

$$\frac{\partial G_0}{\partial \tau}(s, \tau) + \frac{\gamma + 1}{2}G_0(s, \tau)\frac{\partial G_0}{\partial s}(s, \tau) + \mu G_0(s, \tau) = H(s) \tag{11.24}$$

with $H(s) = \frac{\gamma + 1}{2}\bar{H}(s) + \mu\Delta$ and

$$\int_0^1 G_0(y)dy = \Delta. \tag{11.25}$$

Equation (11.25) fixes the position of a shock in the steady state. If there is no damping, $\mu = 0$, then (11.24) agrees with (11.18) on noting the different periods of the inputs: $\bar{H}$ has period 1 and $\sin(\pi s)$ has period 2. If $\frac{\partial G_0}{\partial \tau} = 0$, then (11.24) agrees with (9.54). If $\mu = 0$, the linear terms in (11.24) give $G(s) = M^{1/2} G_0(s) = M^{1/2} \tau H(s) = M s H(s)$, as required. Thus, the long-term evolution Eq. (11.24) contains the early linear growth of the signal and the long-time steady state. It can then be considered to be uniformly valid in time, and the initial condition of rest gives

$$G_0(t, 0) = \Delta. \qquad (11.26)$$

This result can also be derived by formally matching the initial linear growth and the long-term response (11.24). Then (11.24) is solved subject to (11.26). The analysis of the evolution of these damped oscillations is given in Cox and Mortell [1989a].

11.1.4. *Evolution Near Half the Fundamental Frequency*

The nonlinear periodic resonant solution at half the fundamental frequency that contains shocks in a resonant band is derived in Sec. 9.5. The result is motivated by using a simple perturbation scheme and the boundary conditions (9.72), where the Mach number is $M = 2\pi \varepsilon \omega$. Then (9.73) gives the solution at second order provided $\omega \neq \Omega_n = \frac{1}{4}(2n + 1)$, $n = 0, 1, 2, \ldots$.

To derive the relevant small rate equations to describe the evolution at frequencies around $\omega = \frac{1}{4}$, we follow the procedure in Sec. 9.5 when $|Z|, |Z'| \ll 1$ and $H(y) = -H(y - \frac{1}{2})$. Then the basic difference Eqs. (9.84) and (9.85) give (9.86), and (9.83) gives

$$Z(y) - Z(y - 1) + 2Z(y - 1)Z'(y - 1) = -\frac{1}{2} H(y) H'(y)$$

$$= -\frac{1}{2} A^2 H_0(y) H_0'(y), \qquad (11.27)$$

where $H(y) = AH_0(y)$ and $A = -4\pi N\varepsilon\omega^2$, see (9.78). For $|A| \ll 1$ we introduce the expansion

$$Z(y) = AZ_0(y,\widetilde{y}) + A^2 Z_1(y,\widetilde{y}) + \cdots, \qquad (11.28)$$

where $\widetilde{y} = Ay$. At $O(A)$, (11.27) gives

$$Z_0(y,\widetilde{y}) - Z_0(y-1,\widetilde{y}) = 0, \qquad (11.29)$$

i.e., Z_0 is a standing wave, modulated on the long time-scale $\widetilde{y}$, and at $O(A^2)$

$$\frac{\partial Z_0}{\partial \widetilde{y}}(y,\widetilde{y}) + 2Z_0\frac{\partial Z_0}{\partial y} = -\frac{1}{2}H_0(y)H_0'(y). \qquad (11.30)$$

This, with the standing wave condition (11.29) is the evolution equation. The initial condition for (11.30) is

$$f(r) = \begin{cases} 0, & -2\omega \leq r \leq 0 \\ 2\pi\varepsilon\omega \sin(2\pi r), & 0 \leq r \leq 2\omega, \end{cases} \qquad (11.31)$$

when $h(r) = 2\pi\varepsilon\omega \sin(2\pi r)$. Using (11.30), (9.76) and (9.82), the initial condition on Z is

$$Z(y) = \Delta \mp \frac{1}{2}A\sin(2\pi y), \quad \begin{cases} -2\omega \leq y \leq 0, \\ 0 \leq y \leq 2\omega \end{cases} \qquad (11.32)$$

or at subharmonic resonance, $\Delta = 0$,

$$Z(y) = \mp A\sin(2\pi y), \quad \begin{cases} -\frac{1}{2} \leq y \leq 0 \\ 0 \leq y \leq \frac{1}{2}. \end{cases} \qquad (11.33)$$

The initial condition for (11.30) is

$$Z_0(y,0) = \Delta \mp \frac{1}{2}A\sin(2\pi y), \quad \begin{cases} -2\omega \leq y \leq 0 \\ 0 \leq y \leq 2\omega. \end{cases}$$

With $H_0 = \sin(2\pi y)$, the steady state equation for Z_0 is

$$Z_0(y)Z_0'(y) = -\frac{\pi}{4}\sin(4\pi y), \qquad (11.34)$$

on using (9.76), and this integrates to give the separatrices

$$Z^{\pm}(y) = AZ_0^{\pm}(y) = \pm\frac{A}{2}\left|\cos(2\pi y)\right|, \quad 0 \le y \le 1.$$

The mean condition on F, $\int_0^1 F(y)dy = \Delta$, implies, on using (9.76), that

$$\frac{A}{\pi} = \int_0^1 Z^{+}(y)dy = |\Delta|.$$

If $|\Delta| > \frac{A}{\pi}$, then the solution $Z(y)$ is continuous and periodic. If $|\Delta| < \frac{A}{\pi}$, discontinuities connecting the separatrices must be inserted. Thus, $|\Delta| = \frac{A}{\pi}$ is the edge of the resonant band where shocks occur, and marks out the boundary between continuous and discontinuous solutions. It also marks the boundary between continuous and stochastic solutions.

Figure 11.4 shows the difference between small and finite rate separatrices in the steady state solution for $A = 0.008$ and $A = 0.08$, taken from Mortell and Seymour [1981].

Figure 11.5, taken from Cox and Mortell [1992], is an example of the evolution of quadratic resonance for $\Delta \neq 0$. The parameter j is a measure of the number of cycles in the tube. The shocks are of equal strength and half a forcing period apart. This is characteristic of small rate theory.

The experimental confirmation of shocks is given in Galiev, Ilgamov and Sadykov [1970], Zaripov and Ilgamov [1976] and Althaus

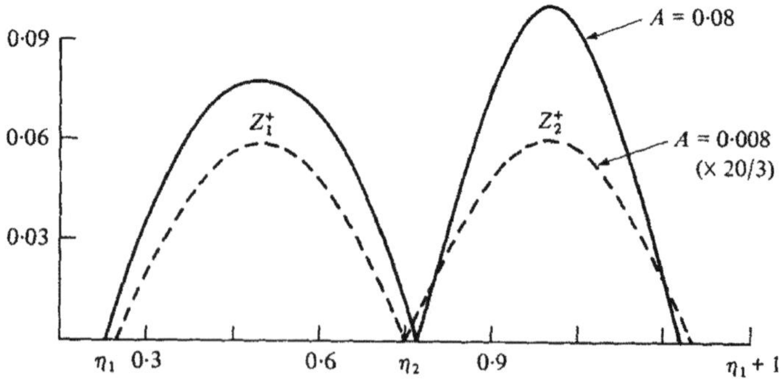

Fig. 11.4.   Steady state for $A = 0.008$ and $A = 0.08$.

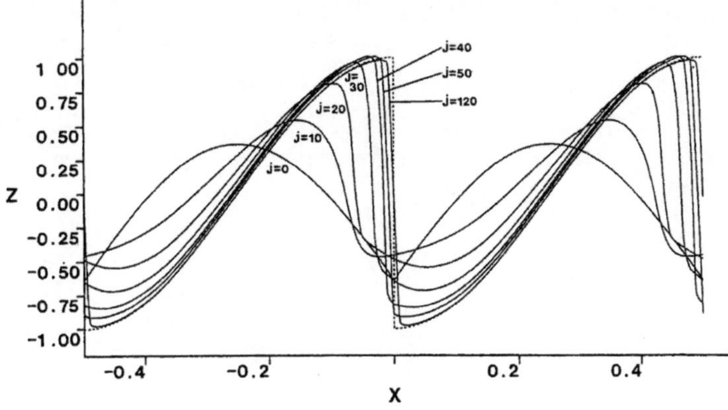

Fig. 11.5.   Evolution; quadratic resonance.

and Thomann [1987]. Keller [1975] gave a theoretical construction, for the periodic solution, based on Chester [1964].

11.2. Large Rate Evolution

In this section, we follow the results in Seymour and Mortell [1985] to examine the evolution to a periodic state of the functional Eq. (10.34) for large values of the rate parameter A. In this case the functional equation cannot be approximated by a differential equation. Hence, the basic equation is (10.34),

$$G(y) = G(s) + H(y), \quad y = s + G(s), \tag{11.35}$$

where $H(y)$ has unit period and $\int_0^1 H(y)dy = 0$. The initial condition corresponding to a state of rest is

$$G(y) = \Delta, \quad y \leq 0 \tag{11.36}$$

and $\Delta/2 = \omega - 1/2$ is the detuning from the fundamental resonance at $\omega = 1/2$.

Equations (11.35) are now rewritten so that the time variable in the current cycle is $t_0(r)$, $0 \leq r \leq 1$, and n cycles previously was $t_n(r)$. Then (11.35) implies

$$G(t_n(r)) = G(t_{n+1}(r)) + H(t_n(r)), \quad t_n(r) = t_{n+1}(r) + G(t_{n+1}(r))$$
$$(11.37)$$

and the parameter r denotes time within a cycle. Hence, $t_n(0) = t_{n+1}(1)$. Now, if $n = N$ indicates the initial cycle, then the first of (11.37) implies

$$G(t_i) = G(t_N) + \sum_{j=i}^{N-1} H(t_j). \tag{11.38}$$

This is seen by noting that

$$G(t_i) = G(t_{i+1}) + H(t_i),$$
$$G(t_{i+1}) = G(t_{i+2}) + H(t_{i+1}),$$
$$\cdots\cdots$$
$$G(t_{N-1}) = G(t_N) + H(t_{N-1})$$

and addition produces (11.38). Similarly

$$t_i = t_N + (N - i) + \sum_{j=i+1}^{N} G(t_j) \tag{11.39}$$

and

$$\sum_{j=i+1}^{N} G(t_j) = (N - i)G(t_N) + \sum_{j=0}^{N-1-i} jH(t_{i+j}) \tag{11.40}$$

on using (11.38). So finally

$$t_i = t_N + (N - i) + (N - i)G(t_N) + \sum_{j=0}^{N-1-i} jH(t_{i+j}), \tag{11.41}$$

$i = 0, 1, 2, \ldots, N-1$. Equations (11.38) and (11.41) are the equivalent of (11.35).

Now conditions in the current cycle are related to values of H in previous cycles by (11.38) and (11.41) setting $i = 0$, i.e.,

$$G(t_0) = G(t_N) + \sum_{j=0}^{N-1} H(t_j) \qquad (11.42)$$

and

$$t_0 = t_N + N + NG(t_N) + \sum_{j=0}^{N-1} jH(t_j) \qquad (11.43)$$

with $t_j = t_j(r))$ and $t_N(r) = r$, $0 \leq r \leq 1$. Equations (11.42) and (11.43) are the finite rate evolution solutions.

Before analyzing (11.42) and (11.43) we refer to Fig. 11.6 which shows the evolution of the solution of Eqs. (11.35) for $A = 0.05$ with $\Delta = 0$. The solid curve represents the current cycle, $N = 0$. We also indicate how conditions in the current cycle, $G(t_0(r))$, depend only on that part of the initial state in the neighborhood of the fixed point $t_N(0) = 0$. For example, with $A = 0.05$, $t_0(0.05) - 16 = 0.75$, i.e., $t = 16.75$ in the current cycle is the image of $t = 0.05$ in the initial cycle. In general, $G(t_0(r))$ is multivalued and can be

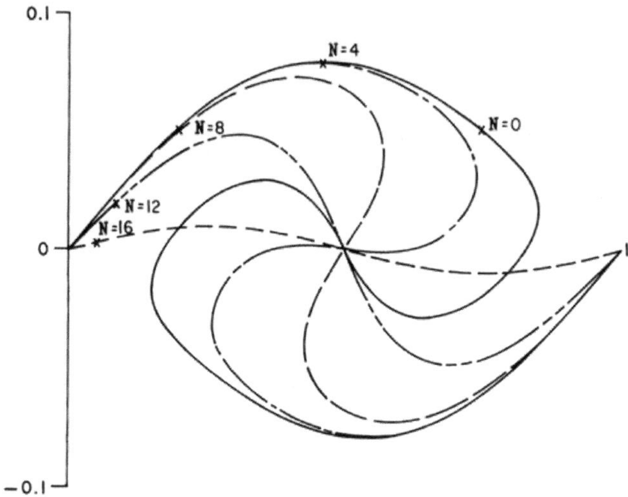

Fig. 11.6.   Evolution with $A = 0.05$.

made single valued by the insertion of shocks using the equal area rule. Then, of course, most of the information in the initial state is absorbed into the shock. For example, for $\Delta = 0$, by symmetry the shock remains at $i + 1/2$ in the ith cycle. Whenever $t_i(r) - i > 1/2$ the corresponding information in the initial cycle is absorbed by the shock in the ith cycle. If shocks are not inserted then $G(t_0(r))$ limits to the separatrices of the periodic state, Z_0 and Z_1 as $N \to \infty$, i.e., the $N = 0$ curve in Fig. 11.6 connects the two saddle points at 0 and 1.

Motivated by these observations we expect that in the periodic state as $N \to \infty$ in (11.42) and (11.43)

$$t_N(r) + N \to 0, \quad NG(t_N(r)) \to 0 \quad \text{and} \quad NH(t_N(r)) \to 0.$$

Then, the only value of G in the initial state that is maintained outside the shocks is $G = 0$.

The *periodic* solution is now given, from (11.42) and (11.43), by

$$G(t_0(r)) = \sum_{j=0}^{\infty} H(t_j(r)), \quad t_0(r) = \sum_{j=0}^{\infty} jH(t_j(r)), \qquad (11.44)$$

where

$$G(t_i(r)) = \sum_{j=i}^{\infty} H(t_j(r)), \quad t_i(r) = \sum_{j=0}^{\infty} jH(t_{i+j}(r)). \qquad (11.45)$$

If we set $y = t_n$ and $s = t_{n+1}$ in (11.35), then the second of (11.45) shows that the first of (11.35) is satisfied. The second of (11.35) now reads

$$t_n = t_{n+1} + 1 + G(t_{n+1})$$

and using (11.45) we require

$$\sum_{j=0}^{\infty} jH(t_{n+j}) = \sum_{j=0}^{\infty} jH(t_{n+1+j}) + 1 + \sum_{j=n+1}^{\infty} H(t_j)$$

which is clearly satisfied on writing out the terms. Therefore Eqs. (11.44) and (11.45) provide a periodic solution to (11.35).

Chapter 12

Shaped and Stratified Resonators

We have seen in Secs. 8 and 9 that in a nonlinear, homogeneous, finite elastic panel fixed at one end, or a polytropic gas in a straight closed tube of finite length, both longitudinal free and forced vibrations eventually exhibit shocks, see Mortell and Varley [1970], Chester [1964] and Seymour and Mortell [1973a]. It is also known that 'weak' inhomogeneities, satisfying the conditions of geometrical acoustics, do not prevent shock formation in a finite panel, see Mortell and Seymour [1972]. However, it has been shown in Lawrenson, Lipkens, Lucas, Perkins and Van Doren [1998], Mortell and Seymour [2004] and Mortell and Seymour [2005] that introducing strong nonuniformities either in the material or tube shape can prevent the formation of shocks, and hence allow a continuous (shockless) harmonic response to a harmonic input. Both cases are controlled by three parameters: an input amplitude or Mach number, M, a measure of the nonuniformity, k and the applied frequency, ω.

The difference between the two situations is summarized as follows.

- *Uniform panel or tube $(k = 0)$*

  (1) Shocks are present for frequencies in a band about the fundamental linear resonant frequency, ω_1.

  (2) Shocks are present for all frequencies for large enough M.

  (3) Resonant frequencies are all multiples of the fundamental; i.e., modes are commensurate, $\omega_n = n\omega_1$.

  (4) For an input $O(M)$ the resonant response is larger at $O(M^{\frac{1}{2}})$.

293

(5) The input energy at the boundary is dissipated through shocks as heat and limits the amplitude.

- *Nonuniform panel or tube ($k \neq 0$)*

   (1) For large enough k with $M \ll 1$, no shocks are present at any frequency.
   (2) Modes are incommensurate, meaning that, for higher modes, $\omega_n \neq n\omega_1$.
   (3) The motion is dominated by the first harmonic and associated eigenfunction.
   (4) The amplitude of the first harmonic is determined by a cubic equation.
   (5) For an input $O(M)$ there is a larger resonant response at $O(M^{\frac{1}{3}}) > O(M^{\frac{1}{2}})$, $M \ll 1$.

Here we first outline the very different mathematics involved through the simple example of an oscillating, inhomogeneous, elastic panel, [from Mortell and Seymour, 2005] and then consider the related, but more practical, problem of resonant macrosonic synthesis (RMS), see Lucas [1993].

12.1. Vibrations of an Inhomogeneous Elastic Panel

For this illustration we consider an example similar to that in Sec. 7.2.1, the forced oscillations of an elastic panel when the inhomogeneity scales both density and Young's modulus in the same way, so that $\rho(x) = E(x) = s(x)$. The variables are nondimensionalized in the same way as in Sec. 2.1 with $L = D = \|\frac{s(x)}{s'(x)}\|$. Then the dimensionless equations of motion relating stress σ, strain e and particle velocity u are

$$s(x)u_t = \sigma_x \quad \text{and} \quad u_x = e_t, \tag{12.1}$$

$$\sigma(e, x) = s(x)e(1 + be^2/6 + O(e^3)) = s(x)T(e), \tag{12.2}$$

where $T(e) = e + be^3/6 + O(e^4))$. Note that the e^2 term in the equation of state $T(e)$ has been omitted to simplify the calculations.

Eliminating σ, e and u satisfy

$$s(x)u_t = (s(x)T(e))_x \quad \text{and} \quad u_x = e_t. \tag{12.3}$$

The boundary conditions for a panel forced at one end and kept rigid at the other are

$$u(0,t) = 0 \quad \text{and} \quad u(1,t) = \varepsilon^3 K \sin(2\pi\omega t), \tag{12.4}$$

where $0 < \varepsilon \ll 1$, and $K = O(1)$ is a constant. Since we are considering near-resonant oscillations, the input amplitude is $O(\varepsilon^3)$, while the output response is anticipated to be $O(\varepsilon)$. We seek periodic solutions for $u(x,t)$ that have the same period as the input at $x = 1$:

$$u(x, t + 1/\omega) = u(x,t). \tag{12.5}$$

The problem of calculating the periodic resonant response is now completely determined by Eqs. (12.1)–(12.5).

12.1.1. *Homogeneous Panel*

To introduce the procedure and notation for the inhomogeneous case, we first consider the homogeneous case with $s(x) \equiv s_0 = \text{constant}$. It also illustrates where the procedure breaks down for the homogeneous case. The momentum Eq. (12.3) becomes

$$\frac{\partial u}{\partial t} = \frac{\partial T(e)}{\partial x} = \frac{\partial e}{\partial x}\left(1 + \frac{1}{2}be^2\right) + O(e^4) \tag{12.6}$$

and perturbation solutions are sought in the form:

$$e(x,t) = \varepsilon e_1(x,t) + \varepsilon^3 e_3(x,t) + \cdots, \tag{12.7}$$
$$u(x,t) = \varepsilon u_1(x,t) + \varepsilon^3 u_3(x,t) + \cdots, \tag{12.8}$$

where $|e_i|, |u_i| = O(1)$, $i = 1, 3 \ldots$. We also introduce the new time variable $\theta = 2\pi\omega t$, and expand the frequency in the series

$$2\pi\omega = \lambda_1 + \varepsilon^2 \delta + \cdots, \tag{12.9}$$

where $\lambda_1/2\pi$ is the fundamental frequency for the linear problem. Thus, $\delta \neq 0$ implies a detuning from resonance. Note that the expansions (12.7)–(12.9) are those usually associated with the Duffing equation when (12.4) applies, see Stoker [1950].

At $O(\varepsilon)$ the equations and boundary conditions are:

$$\lambda_1 \frac{\partial u_1}{\partial \theta} - \frac{\partial e_1}{\partial x} = 0, \quad \lambda_1 \frac{\partial e_1}{\partial \theta} - \frac{\partial u_1}{\partial x} = 0, \tag{12.10}$$

$$u_1(0,\theta) = 0, \quad u_1(1,\theta) = 0. \tag{12.11}$$

Equations (12.10) imply that u_1 satisfies

$$\lambda_1^2 \frac{\partial^2 u_1}{\partial \theta^2} - \frac{\partial^2 u_1}{\partial x^2} = 0. \tag{12.12}$$

A periodic solution, satisfying (12.11), is

$$u_1(x,\theta) = A\phi(x)\sin\theta, \quad \lambda_1 e_1(x,\theta) = -A\phi'(x)\cos\theta, \tag{12.13}$$

where A is arbitrary and the eigenfunction is

$$\phi(x) = \sqrt{2}\sin(\lambda_1 x), \quad \text{with } \lambda_1 = \pi. \tag{12.14}$$

Thus, $\lambda_1 = \pi$ ($\omega = 1/2$) is the fundamental eigenvalue and $\phi(x)$ is the corresponding normalized eigenfunction, *viz.*, $\int_0^1 \phi^2(x)dx = 1$. The solution given by (12.13) and (12.14) is a linear standing wave of arbitrary amplitude.

At $O(\varepsilon^3)$ the elimination of e_3 between the momentum and compatibility equations implies that

$$\lambda_1^2 \frac{\partial^2 u_3}{\partial \theta^2} - \frac{\partial^2 u_3}{\partial x^2} = \lambda_1 \frac{\partial}{\partial \theta}\left[\frac{1}{2}be_1^2 \frac{\partial e_1}{\partial x} - \delta\frac{\partial u_1}{\partial \theta}\right] - \delta\frac{\partial^2 e_1}{\partial \theta \partial x}, \tag{12.15}$$

with now nonhomogeneous boundary conditions

$$u_3(0,\theta) = 0, \quad u_3(1,\theta) = K\sin\theta. \tag{12.16}$$

Using (12.13), Eq. (12.15) becomes

$$\lambda_1^2 \frac{\partial^2 u_3}{\partial \theta^2} - \frac{\partial^2 u_3}{\partial x^2} = \left[-\frac{3}{8}bA^3\phi(\phi')^2 + 2\lambda_1\delta A\phi\right]\sin\theta$$
$$+ \left[-\frac{3}{8}bA^3\phi(\phi')^2\right]\sin 3\theta. \tag{12.17}$$

The form of (12.17) implies that we seek solutions of the form

$$u_3(x,\theta) = P(x)\sin\theta + Q(x)\sin 3\theta. \tag{12.18}$$

Then $P(x)$ is determined by

$$P'' + \lambda_1^2 P = \frac{3}{8}bA^3\phi(\phi')^2 - 2\lambda_1\delta A\phi, \tag{12.19}$$

$$P(0) = 0, \ \ P(1) = K, \tag{12.20}$$

while $Q(x)$ is determined by

$$Q'' + 9\lambda_1^2 Q = \frac{3}{8}bA^3\phi(\phi')^2, \tag{12.21}$$

$$Q(0) = 0, \ \ Q(1) = 0. \tag{12.22}$$

If $\phi_3(x)$ is an eigenfunction of the linear operator in (12.21) and (12.22), then the existence of a solution to (12.21) and (12.22) requires that

$$\frac{3}{8}bA^3 \int_0^1 \phi_3\phi(\phi')^2 dx = 0 \tag{12.23}$$

which implies $A = 0$. With $A = 0$ in (12.19), the existence of a solution to (12.19) and (12.20) requires

$$K \int_0^1 x\phi\,dx = 0 \tag{12.24}$$

which implies $K = 0$, see Boyce and DiPrima [2012]. Thus, the only free vibration with the same period as the forcing function is the zero solution, and there is no continuous solution to the resonant forced problem within the theory.

If we denote the eigenvalues of the basic linear problem (12.10) and (12.11) by λ_n, $n = 1, 2, 3\ldots$, then in this case $\lambda_n = n\lambda_1$. Due to nonlinear interactions, the right-hand side of (12.21) is resonant and leads to secular terms in the solution, thus inhibiting a harmonic solution. If, however, $3\lambda_1$ is not an eigenvalue of the linear operator in (12.21) and (12.22) then resonance does not arise. Thus, if we can find an inhomogeneity such that the eigenvalues of the linear problem λ_n have the property that $\lambda_n \neq n\lambda_1$, then resonance does not arise at

$O(\varepsilon^3)$ and a harmonic solution becomes possible. We say that such eigenvalues are *incommensurate*.

12.1.2. *Inhomogeneous Panel*

We now consider Eqs. (12.1)–(12.5) when the inhomogeneity, $s(x)$, is a specified function of x. To use the procedure outlined in Varley and Seymour [1988], we need to put the linear variable coefficient equation into the Webster horn format, and hence we define the new variable $f(x,t) = s(x)e(x,t)$. Then (12.1)–(12.3) become

$$s(x)u_t - f_x = \frac{b}{6}\left(\frac{f^3}{s^2}\right)_x \quad \text{and} \quad f_t = su_x. \tag{12.25}$$

We use the same form of expansion as (12.7)–(12.9), again with $\theta = 2\pi\omega t = (\lambda_1 + \varepsilon^2\delta + \ldots)t$, where λ_1 is the lowest eigenvalue of the linear problem, but this is now dependent on the form of $s(x)$ and is to be determined. Again δ measures the detuning.

At $O(\varepsilon)$ the homogeneous boundary conditions are (12.11), while eliminating f_1 from (12.25), u_1 satisfies the Webster horn equation:

$$\lambda_1^2\frac{\partial^2 u_1}{\partial\theta^2} - \frac{1}{s}\frac{\partial}{\partial x}\left(s\frac{\partial u_1}{\partial x}\right) = 0. \tag{12.26}$$

The solution can again be written in the form (12.13), where A is constant, but now the eigenfunction $\phi(x)$ is a solution to the variable coefficient problem

$$\frac{1}{s}\frac{d}{dx}\left(s\frac{d\phi}{dx}\right) + \lambda_1^2\phi = 0, \quad \phi(0) = 0, \quad \phi(1) = 0, \tag{12.27}$$

where $\phi(x)$ is normalized so that $\int_0^1 s(x)\phi^2(s)ds = 1$.

We saw in Sec. 5.6 that, from the results described in Varley and Seymour [1988], there are several families of inhomogeneities, $s(x)$, for which we can construct the *exact* general solution to (12.27). Here we use a simple form of inhomogeneity given by (5.88), $s(x) = \frac{l}{k}\tan^2(\sqrt{kl}(x-X))$, where l, k and X are arbitrary constants. Then, following the procedure in Sec. (5.6), if $F(x)$ is defined as

$$F(x) = \cos(\lambda_1 x) - \frac{\alpha_0}{\lambda_1} \sin(\lambda_1 x), \qquad (12.28)$$

where $\alpha_0 = k\sqrt{s(0)}$, then the exact solution of (12.27) is

$$\phi(x) = \gamma \left(\frac{1}{\sqrt{s(x)}} \frac{dF}{dx} + kF \right), \qquad (12.29)$$

where γ is chosen to normalize the amplitude and the boundary conditions imply that the eigenvalue equation for λ is

$$\tan \lambda = \frac{\lambda(\alpha_1 - \alpha_0)}{\lambda^2 + \alpha_0 \alpha_1}, \qquad (12.30)$$

where $\alpha_1 = k\sqrt{s(1)}$. The constants l, k and X completely determine α_0 and α_1 and hence the solutions of (12.30). For $\alpha_1 > \alpha_0$ and $\alpha_1 \alpha_0 > 0$ the solutions of (12.30) are positive, discrete and incommensurate, so that in particular $\lambda_3 \neq 3\lambda_1$, where λ_1 is the smallest eigenvalue.

Using the exact linear variable coefficient solution given by (12.13), (12.29) and (12.30), we now construct the nonlinear solution of (12.1)–(12.5) that represents a continuous harmonic oscillation. At $O(\varepsilon^3)$ the equations for u_3 and f_3 are

$$\lambda_1 \frac{\partial u_3}{\partial \theta} - \frac{1}{s} \frac{\partial f_3}{\partial x} = \frac{b}{6s} \frac{\partial}{\partial x} \left(\frac{1}{s^2} f_1^3 \right) - \delta \frac{\partial u_1}{\partial \theta}, \qquad (12.31)$$

$$\frac{\partial u_3}{\partial x} - \frac{\lambda_1}{s} \frac{\partial f_3}{\partial \theta} = \frac{\delta}{s} \frac{\partial f_1}{\partial \theta} \qquad (12.32)$$

with the nonhomogeneous boundary conditions (with $K \neq 0$ we can always set $K = 1$) $u_3(0, \theta) = 0$, $u_3(1, \theta) = \sin \theta$.

Using (12.29) and noting that $f_1(x, \theta) = -\frac{1}{\lambda_1} As(x)\phi'(x) \cos \theta$, on eliminating f_3 from (12.31) and (12.32), the equation for $u_3(x, \theta)$ becomes

$$\lambda_1^2 \frac{\partial^2 u_3}{\partial \theta^2} - \frac{1}{s} \frac{\partial}{\partial x} \left(s \frac{\partial u_3}{\partial x} \right) = U(x) \sin \theta + V(x) \sin 3\theta, \qquad (12.33)$$

where

$$U(x) = V(x) + 2\lambda_1\delta A\phi, \quad V(x) = \frac{bA^3}{8\lambda_1^2}\frac{1}{s}\frac{d}{dx}(s(\phi')^3). \tag{12.34}$$

We seek a solution to (12.33) of the form

$$u_3(x,\theta) = P(x)\sin\theta + Q(x)\sin 3\theta, \tag{12.35}$$

then $P(x)$ is determined by

$$\frac{1}{s}\frac{d}{dx}\left(s\frac{dP}{dx}\right) + \lambda_1^2 P = -U(x), \quad P(0) = 0, \quad P(1) = 1. \tag{12.36}$$

Since λ_1 is an eigenvalue, a solution to (12.36) exists only if the orthogonality condition holds:

$$\int_0^1 \left[U(x) + \left\{\frac{s'}{s} + \lambda_1^2 x\right\}\right]s(x)\phi(x)dx = 0. \tag{12.37}$$

We made the substitution $P(x) = \widehat{P}(x) + x$ to get homogeneous boundary conditions for $\widehat{P}(x)$ and then use the orthogonality condition to obtain (12.37) as the required *amplitude–frequency relation*. On using the definition of $U(x)$, (12.37) implies that A satisfies the cubic equation

$$NA^3 - 2\delta\lambda_1 A = M, \tag{12.38}$$

where

$$N = \frac{b}{8\lambda_1^2}\int_0^1 s(x)(\phi')^4 dx, \quad M = \lambda_1^2\int_0^1 xs(x)\phi dx + \int_0^1 s'(x)\phi dx. \tag{12.39}$$

For a specific inhomogeneity $s(x)$, Eqs. (12.38) and (12.39), will yield an *amplitude-frequency curve* relating the amplitude A and the detuning, δ, from the linear resonant frequency λ_1. For an excitation exactly at resonance (i.e., $\delta = 0$) the amplitude A is given by $A = (\frac{M}{N})^{1/3}$.

We illustrate with a specific example for the simple form of the inhomogeneity: $s(x) = \tan^2(\psi x + \pi/4)$, with $s(0) = 1$. When $\psi = 0.6$, $s(1) = 28$, but even with such large changes we can still construct exact solutions. In this case the smallest eigenvalue is $\lambda_1 = 3.69$. Note

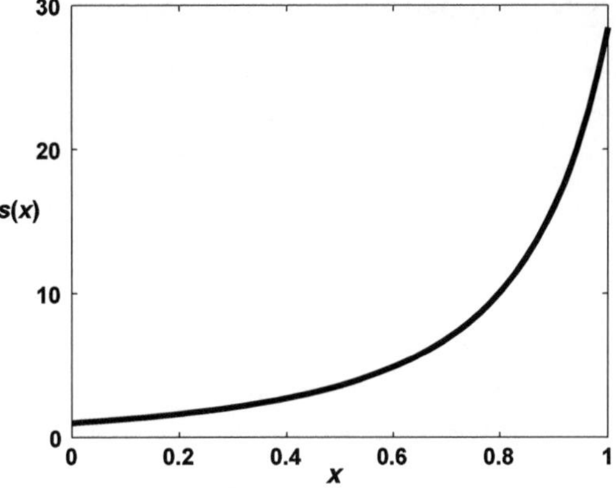

Fig. 12.1. $s(x) = \tan^2(0.6x + \pi/4)$.

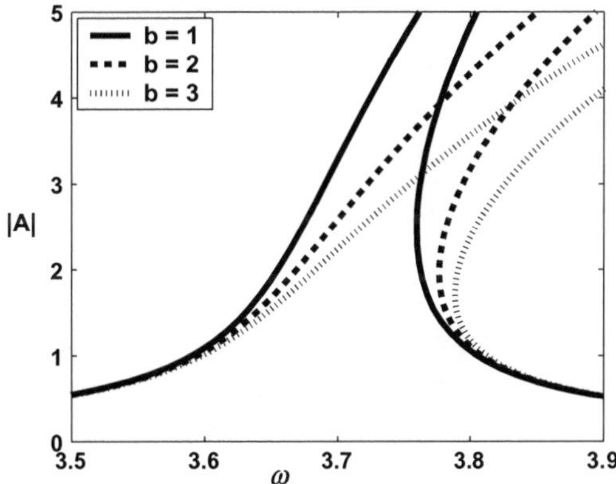

Fig. 12.2.   Typical amplitude–frequency curve.

that the larger the change in $s(x)$, the further λ_1 moves from π, the eigenvalue of the homogeneous problem. Figure 12.1 shows the profile of $s(x)$; there is a very large change over unit distance. Figure 12.2 shows the corresponding amplitude-frequency curves for three values of $b > 0$. The magnitude of b is a measure of the nonlinearity, while the sign of b determines the sign of N, see (12.39), which determines whether the curve bends to the right or left, see (12.38). A negative value of b or N gives a softening response.

12.2. Shaped Resonators

Following the earlier experiments of Lucas [1993], Lawrenson, Lipkens, Lucas, Perkins and Van Doren [1998] showed through a series of experiments that by appropriately designing the shape of a closed container, the resonant oscillations of a gas in the container could reach macrosonic pressures while remaining shockless. This result is in sharp contrast to resonant oscillations in closed uniform cylindrical tubes, where acoustic saturation due to shocks is a feature of the motion, see Sec. 9. Using expansion methods similar to those in the previous Sec. 12.1.2 it was shown analytically in Mortell and Seymour [2004] that for resonant cavities shaped like a cone, a horn or a bulb the nonlinear interaction between the cavity shape and the gas leads to a shockless resonant output. The deliberate shaping of a waveform by designing the shape of the resonator to give the desired output is the basis for Resonant Macrosonic Synthesis (RMS), see Lawrenson, Lipkens, Lucas, Perkins and Van Doren [1998]. The significance of RMS is that continuous waveforms can be synthesized to allow a large amount of energy to be added to the wave and extremely high dynamic pressure achieved while avoiding acoustic saturation due to shocks.

There are two significant numerical investigations of RMS: Ilinskii, Lipkens, Lucas, Van Doren and Zabolotskaya [1998] gave a numerical solution for the problem based on a one-dimensional model that included viscous damping and reproduced qualitatively the amplitude–frequency curves of the experiments in Lawrenson, Lipkens, Lucas, Perkins and Van Doren [1998]; Chun and Kim [2000] also gave numerical solutions of the one-dimensional nonlinear equations for resonant oscillations that include attenuation terms related to viscosity. The analytical study of Ockendon, Ockendon, Peake, and Chester [1993] considered geometric variations characterized by different orders of magnitude of the small geometric parameter. The regime where area variations are large enough to affect the linearized spectrum was considered briefly, but no explicit solution for any shape like a cone, horn or bulb was attempted. Hamilton, Ilinskii and Zabolotskaya [2001] examined the case of a tube with a slowly varying cross section. However, they also were limited to small variations in cross-sectional area of a cylinder with constant, but arbitrary, cross section. Again this has the limitation of not being able to deal with the shapes of interest for RMS. The analytic procedure given here and in Mortell and Seymour [2004] in no way depends on resonator shapes being small deviations from the cylindrical.

The critical factor in the formation of shock waves in a straight cylindrical tube is the excitation, via nonlinearity, of an infinite number of modes whose frequencies are integer multiples of the fundamental. The modes emerge from linear undamped acoustic theory. Thus, it makes sense to seek resonator shapes leading to eigenvalues of the linear problem that are not integer multiples of the lowest eigenvalue, when the eigenvalues are said to be *incommensurate*. Then we can expect that shocks will not form and a continuous, periodic output, dominated by the lowest eigenmode, will follow.

We note that these analytical solutions describing nonlinear resonant oscillations in tubes of variable cross section do not require any dissipation to produce bounded results, see Mortell and Seymour [2004].

12.2.1. *Expansion and Linear Equations*

We consider the one-dimensional motion of a polytropic gas contained in a tube of arbitrary axisymmetric shape, closed at both ends and oscillated along its axis by an external periodic driving force. This is essentially the experimental set-up in Lawrenson, Lipkens, Lucas, Perkins and Van Doren [1998].

If $s(x), 0 \leq x \leq 1$, is the cross-sectional area and $a(t)$ the imposed acceleration on the resonator, the dimensionless equations of conservation of mass and momentum relating the velocity (u) and density (ρ) in a polytropic gas are:

$$(s\rho)_t + (su\rho)_x = 0, \quad u_t + uu_x + \rho^{\gamma-2}\rho_x = -a(t). \qquad (12.40)$$

For later convenience we introduce the new variables $e(x,t)$, $f(x,t)$ by

$$f(x,t) = s(x)u(x,t), \quad \text{and} \quad e(x,t) = \rho(x,t) - 1, \qquad (12.41)$$

where $e(x,t)$ is the condensation. Since the resonator is closed at both ends, the boundary conditions are

$$f(0,t) = 0 = f(1,t). \qquad (12.42)$$

To examine the resonance problem, as formulated in Lawrenson, Lipkens, Lucas, Perkins and Van Doren [1998], we take the applied acceleration as

$$a(t) = \varepsilon^3 \cos\theta \qquad (12.43)$$

where $\theta = 2\pi\omega t$, and $0 < \varepsilon^3 \ll 1$ is the magnitude of the applied acceleration. We seek periodic solutions that have the same period as the external forcing:

$$f\left(x, t + \frac{1}{\omega}\right) = f(x,t) \qquad (12.44)$$

and, as in Sec. 12.1.2, assume a perturbation expansion of the form:

$$e(x,t) = \varepsilon e_1(x,t) + \varepsilon^2 e_2(x,t) + \varepsilon^3 e_3(x,t) + \cdots, \qquad (12.45)$$
$$f(x,t) = \varepsilon f_1(x,t) + \varepsilon^2 f_2(x,t) + \varepsilon^3 f_3(x,t) + \cdots, \qquad (12.46)$$

where $|e_i|, |f_i| = O(1)$, $i = 1, 2, 3 \ldots$. Equations (12.45), (12.46) and (12.43) imply we are iterating about a linear standing wave. Then at $O(\varepsilon)$ with zero boundary conditions, we obtain the linear problem on $0 \le x \le 1$:

$$\frac{\partial f_1}{\partial t} + s(x)\frac{\partial e_1}{\partial x} = 0 \quad \text{and} \quad s(x)\frac{\partial e_1}{\partial t} + \frac{\partial f_1}{\partial x} = 0, \qquad (12.47)$$
$$f_1(0,t) = 0, \quad f_1(1,t) = 0. \qquad (12.48)$$

Eliminating e_1 from (12.47), $f_1(x,t)$ satisfies

$$\frac{\partial^2 f_1}{\partial t^2} - s(x)\frac{\partial}{\partial x}\left(\frac{1}{s}\frac{\partial f_1}{\partial x}\right) = 0 \qquad (12.49)$$

which is the Webster horn equation, see Pierce [1989], and separating variables: $f_1(x,t) = X(x)T(t)$ leads to the eigenvalue problem on $0 \leq x \leq 1$

$$\frac{d}{dx}\left(\frac{1}{s(x)}\frac{dX}{dx}\right) + \frac{\lambda^2}{s(x)}X = 0, \quad X(0) = 0, \quad X(1) = 0. \qquad (12.50)$$

The Webster horn equation is a direct consequence of the change of variable (12.41) with no further assumptions.

Linear problems such as (12.47)–(12.50) are standard where material properties vary with position, see Ellermeier [1994b]. For our purposes here in showing the structure of the analytic solution, the crux of the matter arises in solving such problems analytically for physically interesting area functions $s(x)$. These solutions are implicit in the methods described in Sec. 5.6 and here we exploit these to solve the acoustic resonance problem in tubes of varying cross-sectional area that correspond approximately to experimental conditions. However, it should be noted that if, for a given shape $s(x)$, an analytic solution of the Webster horn equation is not available, a numerical calculation of the eigenfunction and eigenvalues can also be used to calculate the coefficients in the amplitude–frequency relation, see Fig. 12.6.

12.2.2. *Nonlinear Theory: Amplitude-Frequency Relation*

Motivated by (12.13) and (12.9), we *assume* that solutions $f_1(x,t)$ to (12.49) are of the form

$$f_1(x,\theta) = A\phi(x)\sin\theta, \qquad (12.51)$$

where $\theta = 2\pi\omega t = (\lambda_1 + \varepsilon^2\delta + \cdots)t$, $\varepsilon^2\delta$ measures the detuning and A is an arbitrary constant. Here $\phi(x)$ is the eigenfunction determined by (12.50) corresponding to the fundamental eigenvalue λ_1, and normalized so that $\int_0^1 s^{-1}(x)\phi^2(x)dx = 1$. On using (12.51), the corresponding representation for $e_1(x,\theta)$ is

$$e_1(x,\theta) = \frac{A}{\lambda_1 s(x)}\phi'(x)\cos\theta.$$

The form (12.51) is called the dominant first harmonic approximation. As in Sec. 12.1.2, we have expanded the frequency, ω, in a Duffing-like expansion about the fundamental frequency $\omega = \lambda_1/2\pi$. It will emerge at $O(\varepsilon^3)$ that the amplitude A, arbitrary within linear theory, is connected to the frequency ω through the amplitude–frequency relation. We emphasize that the fundamental requirement imposed here is that higher eigenvalues are not integer multiples of the fundamental i.e., $\lambda_n \neq n\lambda_1$, $n = 2, 3, \ldots$

The problem that determines $f_2(x, \theta)$ is

$$\frac{\partial}{\partial x}\left(\frac{1}{s}\frac{\partial f_2}{\partial x}\right) - \frac{\lambda_1^2}{s}\frac{\partial^2 f_2}{\partial \theta^2} = A^2 B_2(x)\sin 2\theta, \quad f_2(0,\theta) = 0, \quad f_2(1,\theta) = 0,$$

$$(12.52)$$

where

$$B_2(x) = \frac{\lambda_1}{s^2}\left[(\gamma + 1)\phi\phi' - 2\frac{s'}{s}\phi^2\right].$$

Then, setting $f_2(x, \theta) = A^2 B(x)\sin 2\theta$, $B(x)$ is determined by

$$\frac{d}{dx}\left(\frac{1}{s}\frac{dB}{dx}\right) + \frac{(2\lambda_1)^2}{s}B = B_2(x), \quad B(0) = 0, \quad B(1) = 0$$

and, since $2\lambda_1$ is not an eigenvalue, $B(x)$ exists.

$f_3(x, \theta)$ is determined by

$$\frac{\partial}{\partial x}\left(\frac{1}{s}\frac{\partial f_3}{\partial x}\right) - \frac{\lambda_1^2}{s}\frac{\partial^2 f_3}{\partial \theta^2} = C_1(x)\sin\theta + A^3 C_2(x)\sin 3\theta, \quad (12.53)$$

$$C_1(x) = \left[A^3 C_3(x) + A\frac{\delta\phi'}{\lambda_1 s}\right]'$$

$$+ \lambda_1\left[A^3 C_4(x) - 1 - A\frac{\delta\phi}{s}\right],$$

$$f_3(0,\theta) = 0, \quad f_3(1,\theta) = 0.$$

Note that, except for the detuning and forcing terms, all terms on the right-hand side of Eq. (12.53) are proportional to A^3. The details of the $C_i(x)$ are messy, but straightforward to compute; they involve terms like $[\lambda_1\phi^3 - \frac{2s'}{\lambda_1 s}\phi^2\phi']$.

Now we let

$$f_3(x, \theta) = P(x)\sin\theta + Q(x)\sin 3\theta,$$

where $Q(x)$ is determined from

$$\frac{d}{dx}\left(\frac{1}{s}\frac{dQ}{dx}\right) + \frac{(3\lambda_1)^2}{s}Q = A^3 C_2(x), \ \ Q(0) = 0, \ Q(1) = 0. \quad (12.54)$$

Since $3\lambda_1$ is not an eigenvalue, $Q(x)$ exists with no restriction on A.
Then $P(x)$ is determined by

$$\frac{d}{dx}\left(\frac{1}{s}\frac{dP}{dx}\right) + \frac{\lambda_1^2}{s}P = A^3[\lambda_1 C_4(x) + C_3'(x)] \quad (12.55)$$

$$-A\lambda_1\delta\left[\frac{\phi}{s} - \lambda_1^{-2}\left(\frac{\phi'}{s}\right)'\right] - \lambda_1,$$

$$P(0) = 0, \ \ P(1) = 0. \quad (12.56)$$

Since λ_1 is an eigenvalue, a solution to (12.55) and (12.56) exists only
if

$$\int_0^1 \{A^3[\lambda_1 C_4(x) + C_3'(x)] - A\lambda_1\delta\left[\frac{\phi}{s} - \lambda_1^{-2}\left(\frac{\phi'}{s}\right)'\right] - \lambda_1\}\phi(x)dx = 0$$

which simplifies to

$$\int_0^1 \{A^3[\lambda_1 C_4(x) + C_3'(x)] - \lambda_1\}\phi(x)dx = 2A\lambda_1\delta. \quad (12.57)$$

Equation (12.57) is the required **amplitude–frequency relation**
that implies that A satisfies the cubic equation

$$NA^3 - 2\delta\lambda_1 A = M, \quad (12.58)$$

where

$$N = \int_0^1 [\lambda_1 C_4(x) + C_3'(x)]\phi(x)dx, \ \ M = \lambda_1\int_0^1 \phi dx. \quad (12.59)$$

For a specific area variation $s(x)$, Eq. (12.58) will yield an
amplitude–frequency curve relating the amplitude A and the
detuning δ for the shaking of a closed tube in the neighborhood of
the fundamental resonance. We note that N and M depend only on

the basic eigenvalue λ_1 and eigenfunction $\phi(x)$, which are determined by the shape function $s(x)$.

If the tube is at rest, but the gas is driven by a piston at $x = 1$, the boundary condition (12.48) is replaced by

$$f(1,t) = \varepsilon^3 \cos\theta \tag{12.60}$$

with $a(t) \equiv 0$ and $P(x)$ is now determined by

$$\frac{d}{dx}\left(\frac{1}{s}\frac{dP}{dx}\right) + \frac{\lambda_1^2}{s}P = A^3[\lambda_1 C_4(x) + C_3'(x)] - A\lambda_1\delta\left[\frac{\phi}{s} - \lambda_1^{-2}\left(\frac{\phi'}{s}\right)\right],$$
$$P(0) = 0, \; P(1) = 1.$$

The amplitude–frequency relation is still (12.58) with N given by (12.59), but

$$M = \int_0^1 \left[\lambda_1^2\frac{x}{s} + \left(\frac{1}{s}\right)'\right]\phi\,dx.$$

It remains to examine the predictions of (12.58) for different $s(x)$ describing cone, horn and bulb resonators. Exact solutions to (12.50) for the corresponding eigenfunctions $\phi(x)$ were constructed in Sec. 5.6 using the methods described in Varley and Seymour [1988].

Cone Resonator: $s(x) = (1 + kx)^2$. A cone is the simplest example of a nonuniform shape for which there is an exact solution to (12.50) and this was constructed in example 5.6.3.2. To confirm the form of $\phi(x)$, let

$$\phi(x) = (1 + kx)F' - kF \tag{12.61}$$

for an arbitrary constant k, where $F(x)$ satisfies $F'' + k^2F = 0$, $F(0) = F(1) = 0$. Then direct substitution confirms that $\phi(x)$ satisfies (12.50) and yields the eigenvalue Eq. (5.108), here in the form

$$\tan\lambda = \frac{\lambda k^2}{k^2 + \lambda^2[1 + k]}. \tag{12.62}$$

The fundamental eigenfunction is given by

$$\phi(x) = \chi(\lambda_1 k^2 x \cos\lambda_1 x - [\lambda_1^2 + kx\lambda_1^2 + k^2]\sin\lambda_1 x), \tag{12.63}$$

where $\lambda = \lambda_1$ is the smallest positive eigenvalue of (12.62) and the constant χ is a normalizing factor so that $\int_0^1 s^{-1}\phi^2 dx = 1$. It is clear from (12.62) that for $k \neq 0$ the higher eigenvalues are not integer multiples of the fundamental. Since $3\lambda_1$ is not an eigenvalue of (12.62), $Q(x)$ exists and hence the terms in the perturbation expansion at $O(\varepsilon^3)$ are periodic and thus bounded.

The response curves of p_1/p_0, where $p = p_0 + \varepsilon p_1$ represent the ratio of the amplitude of the pressure signal in the fundamental mode to the ambient pressure at the small end of the resonator, $x = 0$. They are calculated from

$$\left|\frac{p_1}{p_0}\right| = \frac{A\gamma\varepsilon\phi'(0)}{\lambda_1 s(0)}.$$

For $k = 7$ (similar to the example in Lawrenson, Lipkens, Lucas, Perkins and Van Doren [1998]) the first three eigenvalues are $\lambda_1 = 3.98$ $(= 1.27\pi)$, $\lambda_2 = 6.95$ $(= 2.21\pi)$, $\lambda_3 = 9.95$ $(= 3.17\pi)$. The amplitude–frequency curves for the excess pressure ratio when $k = 7$ are shown in Fig. 12.3 for $\varepsilon = 0.05$, 0.08 and 0.1. We recover the case of a uniform circular cylinder by setting $k = 0$, so that $s(x) \equiv 1$. Then the positive eigenvalues of (12.62) are $\lambda_n = n\pi$, and (12.63) yields $\phi(x) = \sqrt{2}\sin n\pi x$. Now the higher eigenvalues are integer multiples

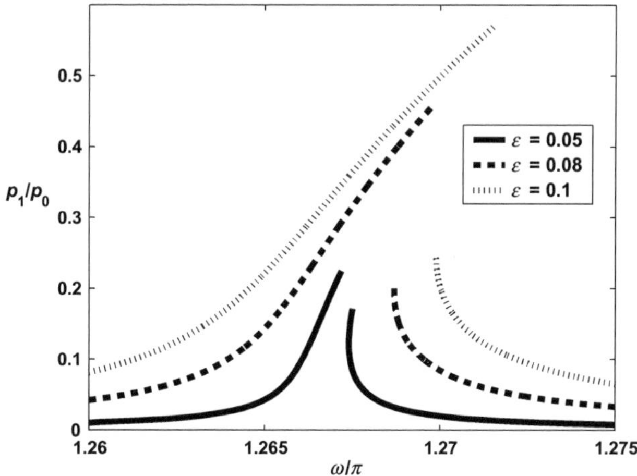

Fig. 12.3.   Excess pressure ratio for cone, $k = 7$.

of the fundamental eigenvalue, resonance in the higher modes cannot be avoided and shocks occur.

<u>Bulb Resonator.</u> Bulb-like resonators are described by cross sections of the form $s(x) = i(x)$ in (5.98) where $z_1(x)$, $z_2(x)$ are given by (5.88). The details of the eigenvalue equation for the bulb shape in Fig. 5.2 are given in Mortell and Seymour [2004]. The first three eigenvalues are given by $\lambda_1 = 4.43\ (= 1.41\pi)$, $\lambda_2 = 6.96\ (= 2.22\pi)$, $\lambda_3 = 9.87\ (= 3.14\pi)$. The corresponding amplitude–frequency curves for the excess pressure ratio are shown in Fig. 12.4 for $\varepsilon = 0.03$, 0.04 and 0.05. Note that in Fig. 12.3 the curves bend to the right (hardening), while in Fig. 12.4 they bend to the left (softening). The difference is the result of a change of sign of N in (12.58) from positive to negative. We have of course neglected all dissipative effects, but in our non-dissipative theory, the maximum amplitude of the peak output is achieved when $N \sim 0$. It was shown in Mortell and Seymour [2004] that, by varying the bulb shape, the response curve changes from hardening (bends right) to softening (bends left) and the amplification has increased by an order of magnitude around the fundamental resonant frequency as N passes through a zero. It may be that in an experiment this effect is masked by dissipation, but in

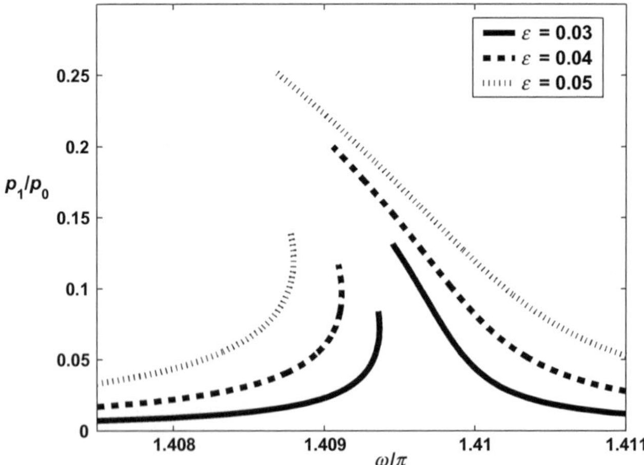

Fig. 12.4. Excess pressure ratio for bulb.

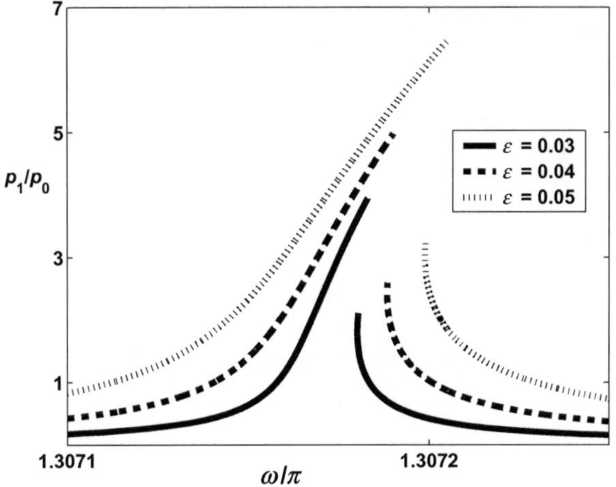

Fig. 12.5.   Excess pressure ratio for $N \backsim 0$.

designing a resonator for maximum amplitude, choosing a shape for which $N \sim 0$ may well be optimal. This possibility is illustrated in Fig. 12.5, where an excess pressure ratio in excess of 5 is achieved. The bulb described by the area

$$s(x) = \frac{1}{a_0^2}(a_0 + a_1 x + a_2 x^2 + a_4 x^4)^2, \quad 0 \leq x \leq 1$$

with $a_0 = 0.025$, $a_1 = -0.2$, $a_2 = 1.15$, $a_4 = -0.9$, is given in Hamilton, Ilinskii and Zabolotskaya [2009] as an approximation to the bulb used in the experiments in Lawrenson, Lipkens, Lucas, Perkins and Van Doren [1998]. Here $s(x)$ is normalized so that $s(0) = 1$, the bulb is closed at both ends and is oscillated parallel to its axis at or near resonance. This shape is not among those that allow an exact solution of the Webster horn Eq. (12.49). Hence, the eigenvalues λ_1, λ_2, λ_3 are calculated numerically to check they are incommensurate, and the fundamental eigenvalue ϕ is also calculated numerically. Then λ_1 and ϕ are substituted into the analytic amplitude–frequency relation (12.58) to yield the amplitude–frequency curve. In this case $\lambda_1 = 1.5850$, $\lambda_2 = 2.7267$ and $\lambda_3 = 3.1008$ and these are incommensurate. The response curve shown in Fig. 12.6 has small viscous damping and shows a not very pronounced softening that is

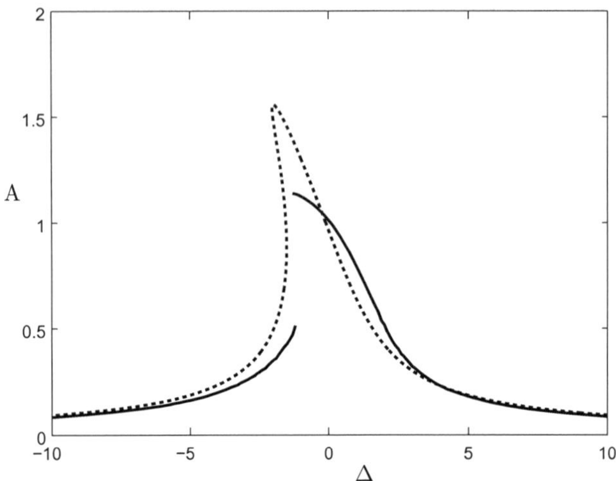

Fig. 12.6.   Numerical amplitude-frequency curve (solid), analytic (dashed).

in line with the experiment in Lawrenson, Lipkens, Lucas, Perkins and Van Doren [1998]. The detuning is given by $\omega = \lambda_1 + \varepsilon^2\Delta$ in Fig. 12.6.

12.2.3. *Spherical Geometry*

A particularly simple geometry for which a dominant first harmonic solution is possible is a full sphere. Chester [1991] considered the problem of a gas contained in a sphere, forced periodically on the surface with a bounded velocity at the centre. He showed that, for a sufficiently small input Mach number, radially symmetric resonant oscillations are continuous. He then raised questions about how the flow changes when contained between two concentric spheres and, in particular, what happens when the radii do not differ appreciably. A further complication that may be present in spherical geometries is the radial stratification of the ambient gas density. Results similar to those in Chester [1991], but for a cylindrical geometry, are given in Ellermeier [1997].

Here we follow the analysis of Seymour, Mortell and Amundsen [2011], [2012], and Amundsen, Mortell and Seymour [2015] and consider the resonant oscillations of a radially stratified gas between concentric spheres. The problem is defined by four dimensionless parameters: the applied frequency ω, the Mach number of the applied velocity M, the geometric parameter J, and the stratification parameter k.

The dimensionless geometric parameter that characterizes concentric spherical shells is J, the ratio of the inner radius to the distance between the spheres. When $J \gg 1$ and the flow is radially symmetric, the motion of the gas between the radial lines can be viewed as approximating the flow in a closed tube with slowly varying cross section, or the flow in the frustum of a cone. In this limit, as $J \to \infty$, the gas flow is similar to that in a closed-end straight tube and hence will contain shocks. Alternatively, when $J \to 0$ the problem is that of a full sphere as considered by Chester [1991], and when the Mach number is small enough the flow is continuous. However, when the Mach number is sufficiently large, the nonlinearity dominates and shocks appear in the flow. The latter flows are not covered by either the geometric acoustics or dominant first harmonic approximations and numerical solutions are required.

Introducing a stratification in the ambient gas density can also produce incommensurate eigenvalues. In Seymour, Mortell and Amundsen [2011] a specific class of inhomogeneity is used to put the linear equations in the Webster horn format for which there are exact solutions. Then the effects of, and interactions between, nonlinearity, geometry and inhomogeneity are examined. When modes are incommensurate, weakly nonlinear oscillations can be continuous. These continuous motions may become discontinuous, i.e., contain a shock, by increasing the Mach number of the input — a nonlinear effect, by changing the inhomogeneity, or by increasing the radius of the inner sphere while maintaining a fixed distance between the spheres — a geometric effect. The approximate solutions are confirmed by means of a numerical solution of the full equations.

Resonant oscillations between concentric spheres and coaxial cylinders when the flow contains shocks and the dominant first harmonic approximation is not valid are considered in Seymour, Mortell and Amundsen [2012]. Analytic approximate solutions for the shock strength and wave form are constructed using the nonlinear geometric acoustics techniques from Sec.7.2 that are validated by exact numerical solutions. The case where a gas stratification is introduced by the requirement for an equilibrium condition in the reference state is considered in Amundsen, Mortell and Seymour [2015]. Continuous and shocked solutions and the transition between them are examined numerically by varying the stratification parameter. The example of the resonant oscillations of a vertically propagating atmospheric wave is given.

Changes of variable: To cover large changes in geometry and stratification it is necessary to change variables to convert the linear equations into the Webster horn format. This is algebraically complicated, but ultimately the solution method in the dominant first harmonic limit is similar to that in Sec. 12.1. The governing equations are transformed into the standard form before a Duffing-like perturbation expansion is used to find the amplitude–frequency relation for a continuous motion. At linear theory there are exact solutions for certain forms of the inhomogeneity via the function $s(R)$, see (12.72). Both the geometry and inhomogeneity of the sound field are unified in the one function $s(R)$.

The gas is contained between two concentric spheres, $0 < r_0 \leq r \leq r_1$. The interior boundary is fixed, while the external boundary oscillates periodically at or near a resonant frequency, the fundamental frequency of a linear free vibration. The motion of the gas is assumed to be radially symmetric.

Using Eulerian coordinates, the undamped governing equations in dimensional variables are

$$\rho\left[\frac{\partial u}{\partial t} + u\frac{\partial u}{\partial r}\right] + \frac{\partial p}{\partial r} = 0, \quad \frac{\partial \rho}{\partial t} + \frac{\partial(u\rho)}{\partial r} + \frac{2u\rho}{r} = 0 \qquad (12.64)$$

with the equation of state for the isentropic flow of a polytropic gas

$$\frac{p}{p_s} = \left(\frac{\rho}{\rho_s(r)}\right)^\gamma = (1 + e)^\gamma = 1 + \gamma e + \frac{\gamma(\gamma - 1)}{2}e^2 + \cdots.$$

Pressure and density are measured from their values in a reference state $(p_s, \rho_s(r))$, where $e(r, t) = \rho/\rho_s - 1$ is the condensation, p_s is a constant, and $a_s(r) = \sqrt{\frac{\gamma p_s}{\rho_s(r)}}$ the associated sound speed. The reference state is isobaric, i.e., maintained by a temperature gradient. The subscripted variables p_s, ρ_s, a_s refer to the reference state.

The boundary conditions are

$$u(r_0, t) = 0, \quad u(r_1, t) = \omega_p l \sin(\omega_p t), \tag{12.65}$$

where l is the maximum boundary displacement with period τ_p and frequency $\omega_p = 2\pi/\tau_p$. Defining the reference sound speed at the inner boundary, $a_0 = \sqrt{\frac{\gamma p_s}{\rho_s(0)}}$ and $\rho_0 = \rho_s(0)$, velocity, pressure and density are nondimensionalized with respect to $(a_0, \rho_0 a_0^2, \rho_0)$ and (u, p, ρ) are considered as functions of dimensionless length, x, and time $\frac{t a_0}{r_1 - r_0}$. The shell thickness, $r_1 - r_0$, is fixed, and x is defined by

$$x = \frac{r - r_0}{r_1 - r_0} = \frac{r}{r_1 - r_0} - J, \tag{12.66}$$

where $J = \frac{r_0}{r_1 - r_0}$ is a dimensionless measure of the internal radius compared with the thickness of the shell. Hence, $0 \le x \le 1$. The dimensionless frequency is $\omega = \frac{r_1 - r_0}{a_0} \omega_p$.

The dimensionless form of Eqs. (12.64) and (12.65) are

$$\omega u_\theta + u u_x + \frac{1}{\rho_s(x)}(1 + e)^{\gamma - 2} e_x = 0, \tag{12.67}$$

$$\omega e_\theta + \frac{1}{\rho_s(x)}[\rho_s(x)(1 + e)u]_x + \frac{2(1 + e)u}{J + x} = 0, \tag{12.68}$$

$$u(0, \theta) = 0, \quad u(1, \theta) = M \sin\theta, \tag{12.69}$$

where $\theta = \omega t$ and the Mach number of the applied input velocity is $M = \varepsilon^3 = \frac{\omega_p l}{a_0} \ll 1$. Since l is the boundary displacement there is a further restriction, $\frac{l}{r_1 - r_0} \ll 1$; i.e., the boundary displacement is much less than the distance between the spheres. The solutions sought have the same period as the boundary forcing,

$$u(x, \theta) = u(x, \theta + 2\pi) \tag{12.70}$$

and are small deviations from the reference state, $\rho = \rho_s$, $p = p_s$, $u = 0$, $e = 0$, which is an exact solution of (12.67) and (12.68). It is noted that for $J \to \infty$ Eqs. (12.67) and (12.68) yield the equations for a straight tube.

A new dependent variable $w(x, \theta)$ and new length variable $R(x)$ are defined by

$$w(x, \theta) = (J + x)^2 \rho_s(x) u(x, \theta) \quad \text{and} \quad \frac{dR}{dx} = \sqrt{\rho_s(x)} \equiv c(R) \tag{12.71}$$

with $R(0) = 0$. With

$$s(R) = (J + x)^2 c(R), \tag{12.72}$$

Eqs. (12.67) and (12.68) become

$$w w_\theta + c w \left(\frac{w}{cs}\right)_R + s(1 + e)^{\gamma - 2} e_R = 0, \tag{12.73}$$

$$w e_\theta + \frac{1}{s}[(1 + e)w]_R = 0, \tag{12.74}$$

$$w(0, \theta) = 0, \quad w(R_1, \theta) = \varepsilon^n \sin \theta \tag{12.75}$$

for $0 < R < R(1) = R_1$, where $\varepsilon^n = M(J + 1)^2 \rho_s(1) \ll 1$ ($n = 2$ or 3) is a constraint limiting a combination of the Mach number of the input and the geometric parameter J to ensure a continuous motion. It is worth noting that the dimensionless parameter ε combines the Mach number of the input, the geometry of the shell and the value of the inhomogeneity at the outer sphere. In writing the condition at $R = R_1$ in (12.75) we anticipate that the solution is $O(\varepsilon)$, while the input is $O(\varepsilon^n)$. The requirement $\varepsilon^n \ll 1$ implies that M must become ever smaller as the J increases in order to ensure a continuous output. As $J \to \infty$, we are dealing with a straight closed tube and the resonant output is shocked for any nonzero M when there is no damping and $n = 2$. For small or finite J, when we anticipate a dominant first harmonic solution, $n = 3$.

Equations (12.73) and (12.74) are the canonical form of the governing Eqs. (12.67) and (12.68) and are the basis of the perturbation scheme for both continuous and shocked motions.

<u>Linear system and limits</u>: Nonlinear problems are usually solved for small amplitude disturbances and hence, perturbation expansions

typically begin with the equivalent linear form of (12.73) and (12.74):

$$\lambda\frac{\partial w_1}{\partial\theta} + s\frac{\partial e_1}{\partial R} = 0 \quad \text{and} \quad \lambda\frac{\partial e_1}{\partial\theta} + \frac{1}{s}\frac{\partial w_1}{\partial R} = 0, \tag{12.76}$$

$$w_1(0,\theta) = 0, \ w_1(R_1,\theta) = 0. \tag{12.77}$$

Eliminating e_1 from (12.76), $w_1(R,\theta)$ satisfies the Webster horn equation,

$$\lambda^2\frac{\partial^2 w_1}{\partial\theta^2} - s\frac{\partial}{\partial R}\left(\frac{1}{s}\frac{\partial w_1}{\partial R}\right) = 0, \tag{12.78}$$

that can be solved exactly for specific forms of $s(R)$. This can be done in many ways (see Sec. 5.6), but here one simple form is selected where the effect of the stratification can be calculated explicitly:

$$s(R) = J^2(1 + kR)^2. \tag{12.79}$$

Then (12.71) and (12.72) imply that $R(x)$ and the density $\rho_s(x)$ are

$$R(x) = \frac{Jx}{J + x(1 - kJ)} \quad \text{and} \quad \rho_s(x) = \frac{J^4}{[J + x(1 - kJ)]^4}. \tag{12.80}$$

Note that both the spherical geometry and the stratification of the gas are unified in $s(R)$. The effect of the geometry is contained in the term $(J + x)^2$ in (12.72), so that specifying a form of $s(R)$ defines a particular class of gas stratification through $c(R)$, see (12.71). From (12.79) and (12.80), the gas is homogeneous when $kJ = 1$, $\rho_s(x) \equiv 1$, $R(x) = x$, $R(1) = 1$ and $s(x) = J^2(1 + x)^2 = (\frac{r}{r_1 - r_0})^2$. Figure 12.7 shows several possible density variations when $J = 2$, although we will delay considering the effect of stratification until Sec. 12.2.3.3. On setting

$$w_1(R,\theta) = A\phi(R)\sin\theta, \quad \text{and} \quad e_1(R,\theta) = \frac{A}{\lambda s(R)}\phi'(R)\cos\theta, \tag{12.81}$$

where A is an arbitrary amplitude, $\phi(R)$ satisfies the eigenvalue problem

$$\frac{d}{dR}\left(\frac{1}{s(R)}\frac{d\phi}{dR}\right) + \frac{\lambda^2}{s(R)}\phi = 0, \ \phi(0) = \phi(R_1) = 0, \tag{12.82}$$

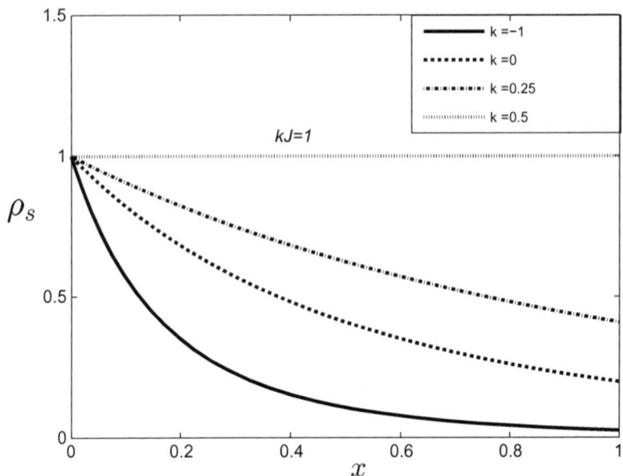

Fig. 12.7.   Density stratification for $J = 2$.

where $R_1 = R(1)$. The eigenfunction ϕ is normalized so that $\int_0^{R_1} s^{-1}(R)\phi^2(R)dR = 1$.

With $s(R)$ given by (12.79), the solution to (12.82) has the form

$$\phi(R) = (1 + kR)F'(R) - kF(R),$$

where $F(R)$ satisfies $F'' + \lambda^2 F = 0$. This can be confirmed by direct substitution. The boundary conditions imply that λ is determined by the eigenvalue equation

$$\tan(\lambda R_1) = \frac{\lambda R_1}{1 + \left(\frac{\lambda}{k}\right)^2 (1 + kR_1)}. \tag{12.83}$$

Thus, the eigenvalues are, in general, incommensurate, i.e., $\lambda_n \neq n\lambda_1$, $n = 2, 3, 4 \ldots$.

The eigenvalue Eq. (12.83) contains the effects of the geometry of the shell through (12.79) and the ambient density stratification of the gas, given by (12.80). Two limits are described in the context of an *homogeneous* gas: the result given in Chester [1991] for a full sphere and the 'plane-wave' case for a shell of large internal radius J and fixed thickness.

When $kJ = 1$ the ambient density, $\rho_s(x)$, is constant $(= 1)$, $c(x) \equiv 1$, $R(x) = x$ and hence $R_1 = 1$. The eigenvalue Eq. (12.83) now only contains the effect of the spherical geometry and becomes

$$\tan \lambda = \frac{\lambda}{1 + (\lambda J)^2(1 + J^{-1})}. \tag{12.84}$$

<u>Full sphere</u>: The case of a full sphere containing an homogeneous gas is recovered in the limit $J = \frac{r_0}{r_1 - r_0} \to 0$, since J is a measure of the internal radius of the shell. Then the eigenvalue Eq. (12.84) reduces to

$$\tan \lambda = \lambda. \tag{12.85}$$

This equation is given in Ellermeier [1997], is implicit in Chester [1991], and yields incommensurate eigenvalues and hence no shocks for sufficiently small ε.

<u>Cone</u>: When $J = 1/k$, (12.84) is the eigenvalue equation for the frustum of a cone with slope k containing an homogeneous gas, see Mortell and Seymour [2004], and is the theoretical underpinning for the experimental results of Lawrenson, Lipkens, Lucas, Perkins and Van Doren [1998]. The radially symmetric oscillations in a segment of a spherical shell give the appropriate context for oscillations in the frustum of a cone, thus obviating the need for the assumption (see Ilinskii, Lipkens, Lucas, Van Doren and Zabolotskaya [1998] and Mortell and Seymour [2004]) of a quasi one-dimensional motion.

<u>Plane-Wave</u>: The 'plane-wave' case arises from the limit as $J \to \infty$, when the internal radius of the shell becomes large compared with the fixed shell thickness and the radial lines are 'almost' parallel within the shell. Then the eigenvalue Eq. (12.84) becomes

$$\tan \lambda = 0 \tag{12.86}$$

with solutions $\lambda = \lambda_n = n\pi$, $n = 1, 2, 3, ...$; the eigenvalues are commensurate and shocks result. This is, of course, the case of axial resonance in a closed tube, see Chester [1964].

12.2.3.1. *Dominant first harmonic case*

In the dominant first harmonic range, an expansion of the form
(12.7)–(12.9) inserted in Eqs. (12.73) and (12.74) with boundary conditions (12.75) and $n = 3$ produces an amplitude–frequency equation
of the form (12.38). In Seymour, Mortell and Amundsen [2011], continuous solutions obtained by this perturbation procedure are compared with those found directly by a numerical procedure. Here we
give a summary of those results for an *homogeneous* gas.

Figure 12.8 shows the effect of increasing the input Mach number
for an homogeneous gas when $J = 2$, $k = 1/2$, $(kJ = 1)$ which is in
the continuous range for small ε. The solution curve changes from
being continuous for $\varepsilon = 0.03$ (the top graph) to being shocked at
$\varepsilon = 0.1$ (the bottom graph). The effect of the increasing nonlinearity
eventually dominates that of geometry to produce a shock. For the
$\varepsilon = 0.1$ case the presence of numerical viscosity gives the appearance
of an almost continuous solution.

The response curves for a continuous solution of an homogeneous
gas $J = 2$, $k = 1/2$, $kJ = 1$, $\varepsilon = 0.03$ and are given in Fig. 12.9 for
$\mu = 0$, 0.25 and 1.0, where μ is an introduced viscosity and $\nu = \varepsilon^2 \mu$.

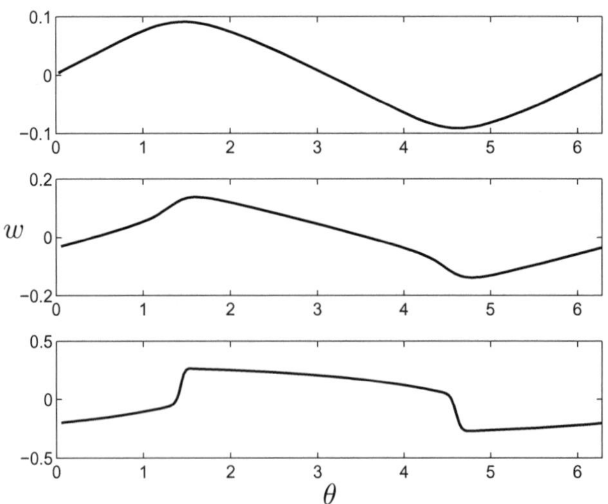

Fig. 12.8.   Top continuous, $\varepsilon = 0.03$, bottom shocked, $\varepsilon = 0.1$.

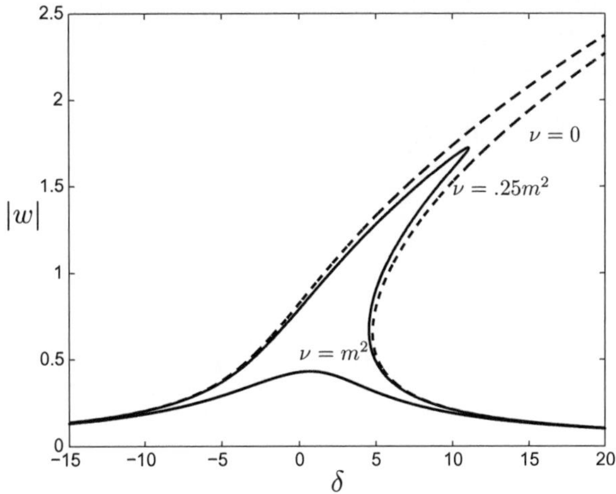

Fig. 12.9.　Response curves with viscosity.

The curves bend to the right (i.e., a hardening response) and there is no hysteresis for sufficiently large ν. In the figure m corresponds to ε in the text.

Figure 12.10 shows the effect of increasing Mach number on the full sphere $(J = 0)$ solution. The dominant first harmonic solution for $\varepsilon = 0.03$ (dashed) is compared with a full numerical solution for $\varepsilon = 0.4$ (solid) containing shocks, both at $x = 0.5$. The continuous solution corresponds to that in Chester [1991].

12.2.3.2.　*Geometric acoustics approximation and shocks*

When the radius of the inner sphere is large compared with the fixed distance between the spheres, $1 \ll J$, the radial lines converge slowly towards the origin. Then the eigenvalues of Eq. (12.84) are commensurate and are approximated by $\lambda = \lambda_n = n\pi$, $n = 1, 2, 3, \ldots$ and shocks result. The geometrical set-up is now equivalent to a closed cylinder, with a slowly varying cross section, containing a gas oscillating parallel to its axis at a resonant frequency.

To consider the propagation of waves in an homogeneous gas between the spheres in this limit it is convenient to change variables

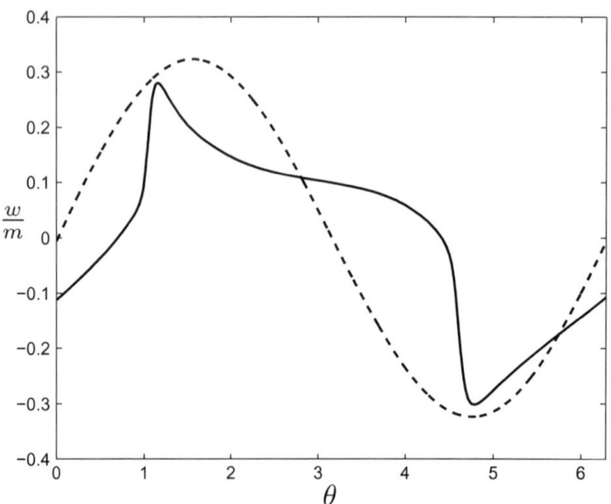

Fig. 12.10.   Full-sphere, $\varepsilon(=m) = 0.4$ and $\varepsilon = 0.03$.

from (x, θ) to (x, α), where $\alpha(x, T) = \text{constant}$ is the nonlinear characteristic variable and $T(x, \alpha)$ is the arrival time of the wavelet $\alpha = \text{constant}$ at the location $x = \frac{r - r_0}{r_1 - r_0}$. For the equivalent linear equations, $T(x, \alpha) = x + \alpha$, where $\alpha = T$ on $x = 0$. Then, if $w(x, \theta) = W(x, \alpha)$,

$$w_x = W_x - \alpha_T T_x W_\alpha$$

and Eqs. (12.73) and (12.74) become

$$\left(1 - T_x \frac{W}{s}\right) W_\alpha - s T_x (1 + E)^{\gamma - 2} E_\alpha = -\frac{s}{\alpha_T} E_x + \text{nonlinear terms},$$

$$\left(1 - T_x \frac{W}{s}\right) E_\alpha - \frac{T_x}{s}(1 + E) W_\alpha = -\frac{1}{s \alpha_T} W_x + \text{nonlinear terms}.$$

The characteristic condition, (2.32), see Seymour and Mortell [1975] and Sec. 5.5, is

$$\det \begin{vmatrix} 1 - T_x \frac{W}{s} & -s T_x (1 + E)^{\gamma - 2} \\ -\frac{T_x}{s}(1 + E) & 1 - T_x \frac{W}{s} \end{vmatrix} = 0,$$

i.e.,

$$T_x = \pm \left[(1+E)^{\frac{\gamma-1}{2}} \mp \frac{W}{s} \right]^{-1} = \pm \left[1 - \frac{\gamma-1}{2} E \pm \frac{W}{s} + O(E^2) \right].$$

In the geometrical acoustics format, T (or α) is considered a 'fast' variable, while the 'slow' variation occurs through the dependence on x. Following the methods of Sec. 7.2, for a small amplitude ($\varkappa$) outgoing wave the nonlinear geometrical acoustics expansion is: (see Sec. 7.2.1)

$$W = \sum_{n=0}^{n} \varkappa^n W_n(x) f_n(\alpha), \quad E = \sum_{n=0}^{n} \varkappa^n E_n(x) f_n(\alpha),$$

$$T = \alpha + x + \varkappa \sum_{n=0}^{n} \varkappa^n T_n(x, \alpha).$$

This yields the first order solution, the modulated simple wave approximation,

$$W = \sqrt{s(x)} f(\alpha), \quad E = \frac{f(\alpha)}{\sqrt{s(x)}}, \tag{12.87}$$

where $f_1(\alpha) = f(\alpha)$ and $f(\alpha)$ is specified on the boundary where $\alpha = T - x$ is the linear characteristic and $s(x) = (J+x)^2$, see (12.79) with $kJ = 1$. The nonlinear outgoing characteristic is

$$T = \alpha + x - \varkappa \frac{(\gamma+1)}{2} f(\alpha) \int \frac{dx}{\sqrt{s}} = \alpha + x - \varkappa \frac{(\gamma+1)}{2} f(\alpha) \ln \left(\frac{J+x}{J} \right),$$

$$\tag{12.88}$$

where the boundary condition is $x = 0$, $\alpha = T$ and $w(0,\theta) = \varkappa J f(\theta)$. For a wave traveling inwards, where the characteristic is tagged by $\beta = T$ at $x = 1$,

$$W = \sqrt{s(x)} g(\beta), \quad E = -\frac{g(\beta)}{\sqrt{s(x)}} \tag{12.89}$$

and the nonlinear characteristic is

$$T = \beta - (x - 1) - \varkappa \frac{(\gamma+1)}{2} g(\beta) \ln \left(\frac{J+x}{J+1} \right). \tag{12.90}$$

With the ansatz that the motion of the gas between the spheres can be approximated by two modulated simple waves traveling in opposite directions that do not interact to order $\varkappa$, we obtain from (12.87) and (12.89)

$$W(\alpha, \beta, x) = \sqrt{s(x)}[f(\alpha) + g(\beta)], \quad E(\alpha, \beta, x) = \frac{1}{\sqrt{s(x)}}[f(\alpha) - g(\beta)],$$

(12.91)

where α and β are given by (12.88) and (12.90). This is a generalization of nonlinearization to oppositely traveling nonlinear waves when the interaction is negligible.

The unknown functions f and g are determined from the boundary conditions at $x = 0$ and $x = 1$:

$$W(0) = 0, \quad \text{and} \quad W(1) = \varkappa \sin \theta.$$

(12.92)

It should be noted that, since we anticipate a shock, the amplitude of the forcing function is now taken as $\varkappa = \varepsilon^2$, i.e., $\varepsilon^2 = \frac{\omega l}{a_0}(J+1)^2 = M^2(J+1)^2$ and the period is 2π.

Equations (12.88)–(12.92) then imply that $g(\eta)$ has zero mean over a period and satisfies the nonlinear difference equation

$$g(\eta) - g(s) = \frac{\varepsilon^2}{(J+1)} \sin(\eta), \quad \eta = s + 2\omega + g(\eta)(\gamma + 1)\omega \ln\left(\frac{J+1}{J}\right),$$

(12.93)

where $2\omega = 2\pi + \Delta$, $\omega = \pi$ is the fundamental frequency and $\Delta/2$ is the detuning. $\eta - s$ in (12.93) is the nonlinear travel time, so including the term proportional to $g(\eta)$ in (12.93) is a form of the nonlinearization technique. Defining

$$G(\eta) = g(\eta)(\gamma + 1)\omega \ln\left(\frac{J+1}{J}\right) + \Delta,$$

(12.94)

$G(\eta)$ satisfies the difference equation (the standard mapping)

$$G(\eta) - G(s) = B \sin(\eta), \quad \eta = s + G(s),$$

(12.95)

where the similarity parameter is

$$B = \frac{(\gamma + 1)\varepsilon^2 \omega}{(J+1)} \ln\left(\frac{J+1}{J}\right).$$

(12.96)

The form of dependence of B on J through $\ln(\frac{J+1}{J})$ is specific to the spherical geometry.

Following the procedure in Sec. 9.4, for $|G|, |G'| \ll 1$ the shock strength of $e(\theta)$ at $x = 0$ at resonance for sufficiently large J, is

$$[e] = \frac{8M}{\sqrt{\pi\,(\gamma+1)}J}\sqrt{\frac{(J+1)}{\ln\,(1+1/J)}}. \qquad (12.97)$$

Then the limiting case, $J \gg 1$, gives

$$[e] \simeq 8M\,[\pi(\gamma+1)]^{-1/2}\left(1 + \frac{3}{4}J^{-1} + O(J^{-2})\right). \qquad (12.98)$$

The asymptote $8M\,[\pi(\gamma+1)]^{-1/2}$, $J \to \infty$, corresponds to the case of an homogeneous gas in a straight tube with a closed end.

In Seymour, Mortell and Amundsen [2012], for $J \gg 1$ the results of the geometric acoustics representation are compared with the exact numerical solution. Numerical solutions are generated using a variety of approaches as described in Seymour, Mortell and Amundsen [2012]. Figure 12.11 shows the comparison between the exact numerical solution of the full equations for $e(\theta)$ at $x = 0$ (top graph) and the velocity $u(\theta)$ at $x = 0.5$ (bottom graph) with the corresponding geometrical acoustics representations when $J = 30$ and the output Mach number is $M = 0.005$. Clearly the geometrical acoustics solution is a good approximation and thus in these circumstances the gas motion in the system of two concentric spheres behaves as though it were in a straight closed tube. That the motion consists of two oppositely travelling waves is seen from the two jumps in $u(\theta)$ at $x = 0.5$.

The effect of decreasing J is illustrated in Fig. 12.12 for $e(\theta)$ at $x = 0$ (top graph) and $u(\theta)$ at $x = 0.5$ (bottom graph) when $J = 8$ and the output Mach number is again $M = 0.005$. Even with the smaller value of J, the geometric acoustics approximation is in qualitative agreement with the exact solution, and the motion in the gas is still similar to that in a straight, closed-end tube.

Since the geometrical acoustics solution always contains a shock, it fails for those values of J and M for which the solution is continuous, e.g., $J = 2$, $M = 0.005$. However, the lower values of J for which

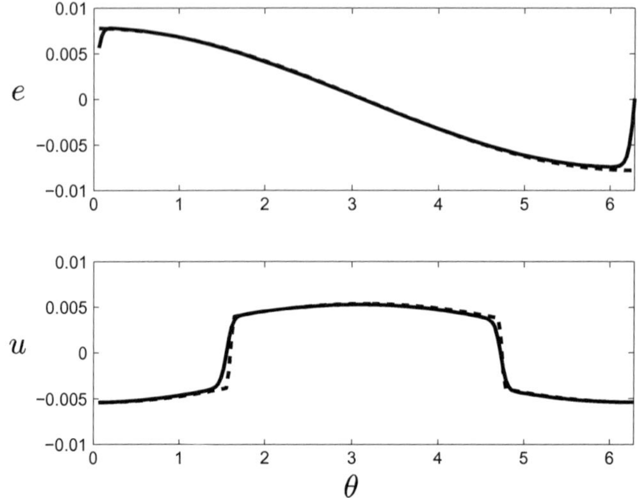

Fig. 12.11. Approximate (dashed) and exact (solid), $J = 30$.

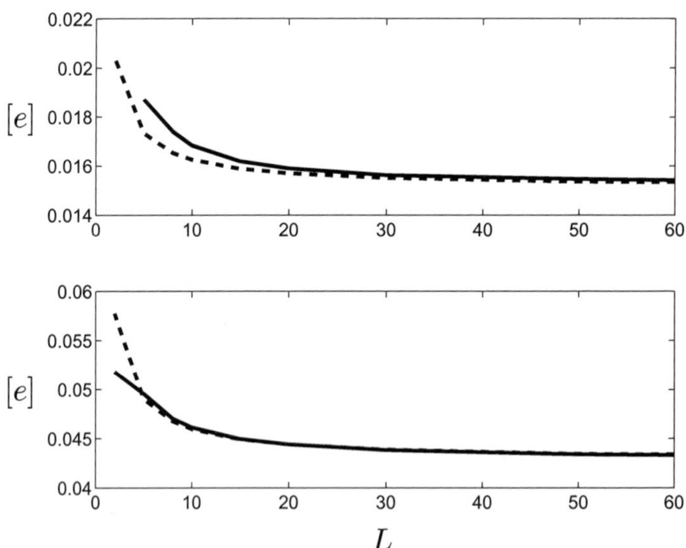

Fig. 12.12. Approximate (dashed) and exact (solid), $J = 8$.

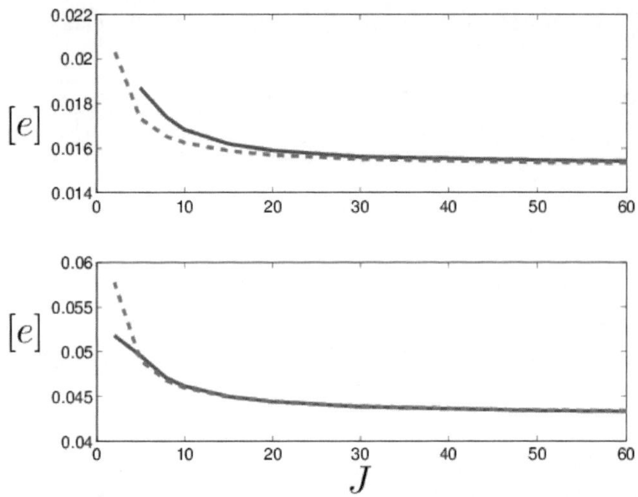

Fig. 12.13.   Shock strength: $M = 0.005$ (top), $M = 0.015$.

there is at least qualitative agreement are surprising. Figure 12.13 gives a comparison of the shock strength of $e(\theta)$ for $M = 0.005$ (top graph) and $M = 0.015$ (bottom graph) at $x = 0$ for different J, using the geometrical acoustics formula (12.97) and the exact numerical solution. For $J \geq 10$ and $M = 0.005$ the error from using the geometrical acoustics solution is $\leq 5\%$, while for $J \geq 5$ it is $\leq 10\%$. A shock forms near $J = 4$ and a single mode solution is valid for $J < 4$. So for an error $\leq 5\%$, with $M = 0.005$, a numerical solution is required for J in the range $4 < J < 10$. For $M = 0.015$ the corresponding range for a numerical solution is $0.25 < J < 0.5$.

12.2.3.3. *Inhomogeneous material*

Following Amundsen, Mortell and Seymour [2015], here we consider the effect of an inhomogeneity when a body force term is included in the equations of motion by the requirement of an equilibrium condition in the reference state. Most examples of waves in bounded media that include inhomogeneities do not include the effect of the body force, e.g., Mortell and Seymour [2005], Ellermeier [1993], [1994a], Seymour, Mortell and Amundsen [2011]. In the case of Ellermeier

[1993], the reference state is isobaric, i.e., maintained by a temperature gradient, so that, while the equilibrium density varies, the pressure is constant, and a body force does not occur. It is shown in
Amundsen, Mortell and Seymour [2015] how, when the equilibrium
pressure and density vary with distance, an approximating density
may be derived that leads to an eigenvalue equation with incommensurate eigenvalues and an approximate analytic solution. This
analytic solution compares well with the numerical solution of the
original exact equations.

In the region where the eigenvalues become commensurate,
shocks can again be a feature of the resonant motion. As in Sec. 12.2.3
these solutions are represented by the superposition of oppositely
traveling modulated simple waves.

Total pressure and density satisfy the equation of state for a polytropic gas, $P_T(r, t) = \mu[\rho(r, t)]^\gamma$, and are measured from the equilibrium state $P(r)$ and $\rho_0(r)$, where $P(r) = \mu\rho_0^\gamma(r)$, and μ is a constant
to be determined. The equilibrium is maintained by a body force
$P'(r)\rho_0^{-1}(r)$. The variations in pressure and density from equilibrium
are defined by

$$P_T(r,t) = P(r) + p(r,t), \quad \rho(r,t) = \rho_0(r)[1 + e(r,t)], \qquad (12.99)$$

where $p(r, t)$ is the excess pressure, $e(r, t)$ is the condensation.

The dimensionless radial variable, x, is defined in (12.66). Defining the ambient density and sound speed at the inner boundary as
$\rho_a = \rho_0(0)$, and $a_a = \sqrt{\mu\gamma\rho_a^{\gamma-1}}$, velocity, pressure and density are
nondimensionalized with respect to $(a_a, \rho_a a_a^2, \rho_a)$ and (u, p, ρ) are
considered as functions of x, and dimensionless time, $\frac{ta_a}{r_1-r_0}$. Note
that $\mu = P(0)$ since the density is normalized so that $\rho_a = 1$. The
boundary conditions in the new variables are again (12.69).

In terms of the dimensionless variables the equations of motion
are

$$u_t + uu_x + \frac{1}{\rho_0[1 + e]}\left(P' + \frac{\partial p}{\partial x}\right) = \frac{P'}{\rho_0}, \qquad (12.100)$$

$$e_t + \frac{1}{\rho_0}[\rho_0(1 + e)u]_x + \frac{2(1 + e)u}{J + x} = 0. \qquad (12.101)$$

The equivalent linear equations are

$$u_t + (\rho_0^{\gamma-1} e)_x = 0, \ \ e_t + \frac{1}{\rho_0}[\rho_0 u]_x + \frac{2u}{J+x} = 0, \qquad (12.102)$$

where $\mu\gamma = 1$ to ensure $\rho_0(0) = 1$, and note that the body force term disappears at this order. To convert (12.102) into divergence form, both independent and dependent variables are changed. However, because of the presence of the body force term, this is more complicated than the example in Sec. 12.2.3. Firstly we define $R(x)$ in terms of the density as

$$\frac{dR}{dx} = \rho_0(x)^{(1-\gamma)/2} \equiv c(R), \ \text{ with } R(0) = 0. \qquad (12.103)$$

Then we define new dependent variables $w(R,t)$ and $E(R,t)$ as

$$w(R,t) = (J+x)^2 \rho_0 u(x,t), \ \ E(R,t) = \rho_0^{\gamma-1} e(x,t). \qquad (12.104)$$

Finally we define $s(R)$ by

$$s(R) = (J+x)^2 \rho_0(x)^{(3-\gamma)/2}. \qquad (12.105)$$

Then the linear Eqs.(12.102) are

$$w_t + sE_R = 0, \ \ E_t + \frac{1}{s}w_R = 0 \qquad (12.106)$$

and, on eliminating E, w satisfies the Webster horn equation:

$$w_{tt} + s\left(\frac{1}{s}w_R\right)_R = 0.$$

The main reason for the change of variables is that the Webster horn equation and hence Eqs.(12.106) can be integrated exactly when $s(R) = J^2(1+kR)^2$ for some constant k. Then one form of solution is $w(R,\theta) = A\phi(R)\sin\theta$ with $\phi(R)$ satisfying (12.82) where $F(R)$

satisfies $F'' + \omega^2 F = 0$. Hence, we equate

$$s(R) = J^2(1 + kR)^2 = (J + r)^2 \rho_0(x)^{(3-\gamma)/2} \tag{12.107}$$

which implies that

$$\rho_0(x) = \left(\frac{1 + kR}{1 + x/J}\right)^{4/(3-\gamma)}. \tag{12.108}$$

Integration of (12.103) and (12.108) gives

$$(1 + kR)^{1-q} = 1 - kJ + kJ\left(1 + \frac{x}{J}\right)^{1-q}, \quad q = \frac{2(1 - \gamma)}{(3 - \gamma)}.$$

This determines $R(x)$ which, using (12.108), implies that

$$\rho_0(x) = \left(1 + \frac{x}{J}\right)^{q-2}\left(1 - Jk + Jk\left(1 + \frac{x}{J}\right)^{1-q}\right)^{\frac{q-2}{q-1}}. \tag{12.109}$$

This is the specific form of the equilibrium density for which the linear equations can be solved exactly. It contains the two arbitrary constants, J and k. We note that the homogeneous case, $\rho_0(x) \equiv 1$ when $R(x) = x$, is recovered when $Jk = 1$, and the region of interest is now $0 \le R \le R_1 = R(1)$.

Polytropic atmosphere: An example of an external forcing producing an inhomogeneity is a polytropic atmosphere when in (12.100) $P'\rho_0^{-1} = -g$, the gravitational acceleration. The density and temperature are functions of elevation, z, with the equilibrium density

$$\rho_0(z) = \rho_{00}\left(1 - \frac{\gamma - 1}{\gamma H}z\right)^{1/(\gamma-1)}, \tag{12.110}$$

where γ is the polytropic exponent, H is the isothermal scale height, and ρ_{00} is the surface density, see Lamb [1945] Sec. 310. The equilibrium pressure is given by

$$P(z) = \rho_{00}^\gamma\left(1 - \frac{\gamma - 1}{\gamma H}z\right)^{\frac{\gamma}{\gamma-1}}.$$

To use the previous results, we need to change from z to r as the independent variable. Note that with $z = r - r_0$, then $x = \frac{z}{r_1 - r_0}$. For

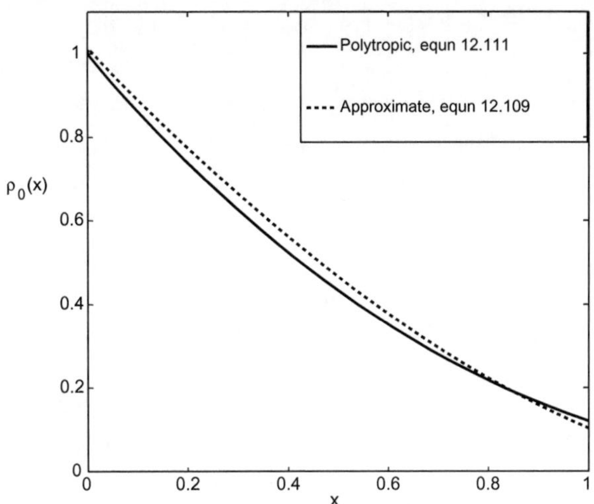

Fig. 12.14.   Comparison of densities.

the earth's radius of $r_0 = 6355$ km and a 17 km atmosphere $(= 2H)$, then $L = 374$ and $x = \frac{z}{17}$. Hence, corresponding to (12.110), we have

$$\rho_0(x) = \rho_{00}(1 - 0.57x)^{2.5}. \tag{12.111}$$

In Fig. 12.14 we compare the physical density profile given by (12.111) with the approximation given by (12.109) for $L = 374$, $\gamma = 1.4$ and $k = -0.5$.

Using (12.104) and (12.103) the full nonlinear equations become

$$\omega w_\theta + cw\left(\frac{c}{s}w\right)_R + sE_1{}^{\gamma-1}E_R = Y(s, c, E), \tag{12.112}$$

$$\omega E_\theta + \frac{1}{s}w_R + \frac{1}{s}(wc^2E)_R = 0, \tag{12.113}$$

where $\theta = \omega t$, $E_1 = 1 + c^2E$ and

$$Y(s, c, E) = 2sc'c^{-1}\{(\gamma - 1)^{-1}[c^{-1}E_1{}^{\gamma-1} - 1] - EE_1{}^{\gamma-1}\}. \tag{12.114}$$

These equations are exact and are the canonical forms of the nonlinear equations. In terms of $w(R, \theta)$ the boundary conditions are

$$w(0, \theta) = 0, \quad w(R_1, \theta) = M(J + 1)^2\rho_0(1)\sin\theta. \tag{12.115}$$

With $s(R) = J^2(1 + kR)^2$, the solution of the linear Eqs. (12.106) again yields the eigenvalue Eq. (12.84). This equation, has, in general, incommensurate eigenvalues and we can expect continuous solutions. It corresponds to the variable equilibrium density (12.109), which is a consequence of the body force term on the right-hand side of (12.100). For a given J, we can model a given density $\rho_0(r)$ by a choice of the parameter k in (12.109). If the reference state is isobaric, as in Sec. 12.2.3, the equilibrium density is variable but the pressure is constant. Then the right-hand side of (12.100) is zero, see Eq. (12.67).

The approximate analytic solutions in both the dominant first harmonic region and the shocked region can be calculated in a similar way to those in Secs. 12.2.3.1 and 12.2.3.2. Next these approximate solutions are compared with the exact numerical solution for various values of the parameters. In Fig. 12.15 there is a continuous solution for the parameter values $J = 8$, $\varepsilon = 0.03$, $\lambda_2 = 0$ and $k = -0.6$. Here λ_2 is the detuning. We see that the one-term dominant first harmonic approximation (dashed) for the motion in the stratified gas is in close agreement with the exact (solid) curve. When the forcing

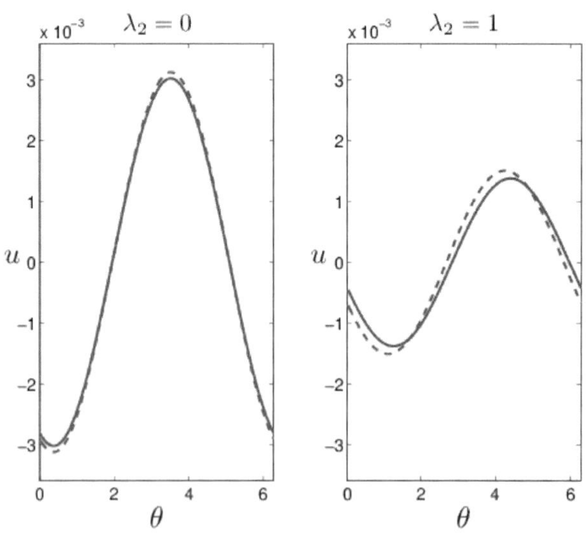

Fig. 12.15.   Continuous solutions, $k = -0.6$.

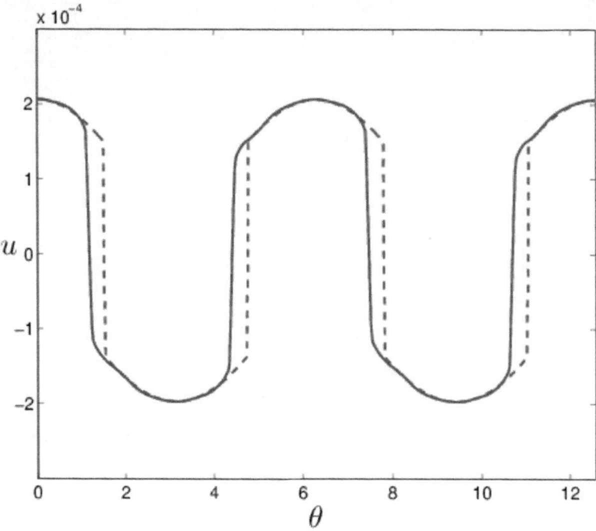

Fig. 12.16.   Shocked solutions for $k = 0$.

is off resonance, $\lambda_2 = 1$, the amplitude of the motion is significantly reduced. By comparison, in Fig. 12.16, the gas is inhomogeneous with $k = 0$ while the other parameters are fixed, the same as in Fig. 12.15. Now the motion is shocked and is reduced by an order of magnitude compared to the continuous case. The approximation is quite good, with a slightly bigger error in the phase of the shock. This could be due to the interaction of the waves, which has been neglected.

Figure 12.17 illustrates the usefulness of the approximate analytic solution. It shows the amplitude–frequency curve calculated numerically (solid) and that given by the dominant first harmonic approximation for the atmospheric density profile (12.111) (dashed) when the eigenfunction and eigenvalue are calculated numerically for $\varepsilon = 0.03$ and a dimensionless damping coefficient is $\nu = 0.25\varepsilon^2$. There is good quantitative agreement, except near the maximum. However, the computer time to produce the dominant first harmonic approximation is less than 1% of the time for the full numerical solution. Thus, the dominant first harmonic approximation is time efficient and gives an accurate result.

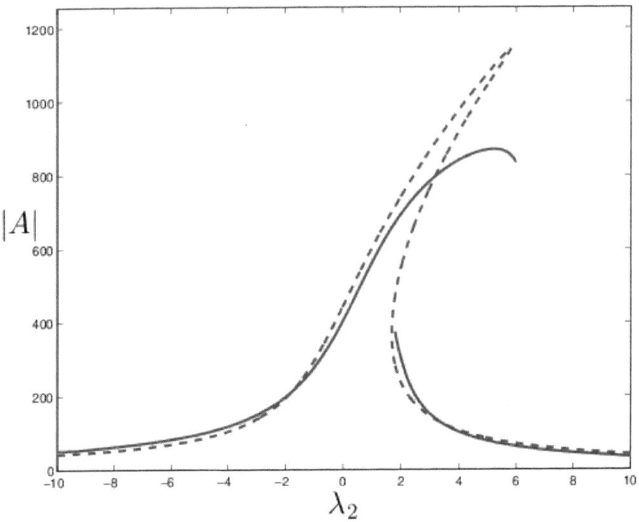

Fig. 12.17. Amplitude-frequency curve for $\varepsilon = 0.03$.

12.2.4. *Evolution for Cone and Bulb Shapes*

In Mortell, Mulchrone and Seymour [2009] the results in Mortell and Seymour [2004] are extended to consider the evolution of the resonant oscillations described in the experiments of Lawrenson, Lipkens, Lucas, Perkins and Van Doren [1998] for cone, horn and bulb resonators. The coupled equations governing the evolution of the slowly varying amplitude and phase are derived. The detail of much of the calculation here is similar to that in Sec. 12.2 except that the dependent variables are also assumed to depend a 'slow' time scale $\tau = \varepsilon^2 t$. The constants in the cubic amplitude–frequency equation now depend on τ and a pair of coupled equations for the amplitude $A(\tau)$ and phase $v(\tau)$ determine the evolution process. The effect of small viscous damping is also included.

The basic equations are (12.40) with a viscous term added to the right-hand side of the momentum equation of the form $\eta \frac{\partial}{\partial x}[\frac{1}{s}\frac{\partial(su)}{\partial x}]$, where η is a dimensionless damping coefficient that includes the coefficients of viscosity and bulk viscosity. The perturbation scheme is again of the form (12.45) and (12.46), but with all expansion terms also depending on τ, so that for example $f(x,\theta) = \varepsilon f_1(x,\theta,\tau) + \dots$.

In addition, it is assumed the damping coefficient also acts over the 'slow' time, so that $\eta = \varepsilon^2 \eta_1$. Hence, the forcing, detuning, $\varepsilon^2 \delta$, damping, $\varepsilon^2 \eta_1$ and the nonlinearity all act over the timescale $\tau = \varepsilon^2 t$.

A calculation similar to that in 12.2 produces the first-order solution

$$f_1(x, \theta, \tau) = A(\tau)\phi(x)\sin(\theta + \upsilon(\tau)), \tag{12.116}$$

where the amplitude and phase of the oscillation are functions of the slow timescale τ. Corresponding to f_1, we also have

$$e_1(x, \theta, \tau) = \frac{A(\tau)}{\lambda_1 s(x)}\phi'(x)\cos(\theta + \upsilon(\tau)). \tag{12.117}$$

$A(\tau)$ and $\upsilon(\tau)$ then satisfy

$$NA^3 \cos \upsilon - M = 2\delta\lambda_1 A \cos \upsilon \tag{12.118}$$
$$+2\lambda_1^2[A' \sin \upsilon + A\upsilon' \cos \upsilon]$$
$$+\lambda_1^3 \eta_1 A \sin \upsilon$$

and

$$NA^3 \sin \upsilon = 2\delta\lambda_1 A \sin \upsilon \tag{12.119}$$
$$-2\lambda_1^2[A' \cos \upsilon - A\upsilon' \sin \upsilon]$$
$$-\lambda_1^3 \eta_1 A \cos \upsilon,$$

where the coefficients M and N are those in the steady-state amplitude-frequency relation, and are given in Eq. (12.59) and in Eq. (40) in Mortell and Seymour [2004]. Now (12.118) and (12.119) may be put in the form

$$2\lambda_1^2 A' = -M \sin \upsilon - \lambda_1^3 \eta_1 A \tag{12.120}$$

and

$$2\lambda_1^2 A\upsilon' = NA^3 - M \cos \upsilon - 2\lambda_1 \delta A. \tag{12.121}$$

When $A(\tau)$, $\upsilon(\tau)$ are independent of τ, i.e., in the steady state, (12.120) and (12.121) reduce to

$$NA^3 - [M^2 - (\lambda_1^3 \eta_1 A)^2]^{1/2} - 2\lambda_1 \delta A = 0. \tag{12.122}$$

This is the steady-state amplitude–frequency relation including damping. When there is no damping, i.e., $\eta_1 = 0$, then (12.122) reduces to the amplitude–frequency relation in Eq. (12.58). Equations (12.120) and (12.121) are those given in Kurihara and Yano [2006] on identifying

$$\sigma = \frac{N}{2\lambda_1^2}, \quad \beta = -\frac{M}{2\lambda_1^2}, \quad \widehat{v} = v - \pi/2, \quad \Gamma = -\frac{1}{\lambda_1}, \quad \eta = \lambda_1 \eta_1.$$

We note that the constants M, N, λ_1 are determined by a steady-state calculation and thus the evolution follows easily by inserting the steady-state parameters into the evolution Eqs. (12.120) and (12.121). It is important to note that the constants M, N depend on the form of the eigenfunction $\phi(x)$. In particular, the coefficient, N, of A^3 is positive for a cone or horn and negative for a bulb, thus giving a hardening or softening response respectively.

<u>Numerical results</u>: Phase plots of the system dynamics were generated numerically in Mortell, Mulchrone and Seymour [2009] for parameter sets corresponding to a cone ($M = 0.3282$, $N = 16.688$, $\lambda_1 = 3.979$ and $\eta_1 = 0.006$) and to a bulb ($M = 17.9$, $N = -2.85$, $\lambda_1 = 4.427$ and $\eta_1 = 0.045$). In each case the detuning δ is varied to demonstrate the full range of system behavior. The cone corresponds to the experimental cone used in Lawrenson, Lipkens, Lucas, Perkins and Van Doren [1998] with response curves shown in Fig. 12.3. The bulb used is that shown in Fig. 1(c) of Mortell and Seymour [2004], and the response curves (for $\eta_1 = 0$) and comparison with experiment are shown in Fig. 12.4. A requirement for the validity of the expansion is that $\max |\varepsilon A| \ll 1$ and this is satisfied in both cases.

Figures 12.18–12.21 illustrate the dynamics of the system in the case of a cone shape. Figure 12.18 plots the tuning curve for A as a function of δ and is characterized by a right sloping spike. Points on the curve correspond to fixed points of A in the $A - v$ phase space. For $\delta = 0$, there is a single real value of $|A|$ and the phase plot is shown in Fig. 12.19. For positive A there is a stable spiral with an anticlockwise sense. As the value of δ is increased, a saddle point and clockwise stable spiral emerge in addition to the anticlockwise stable

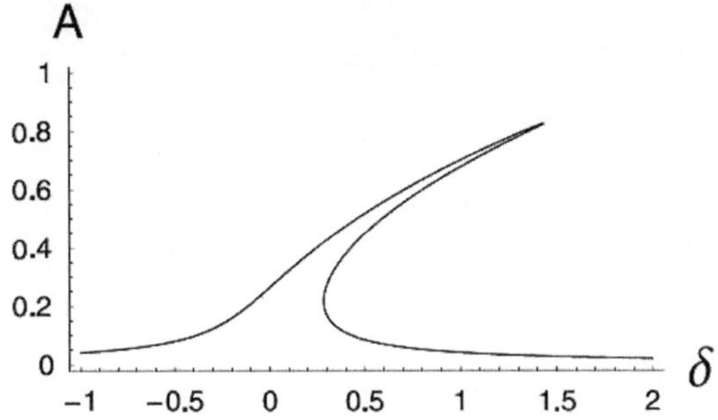

Fig. 12.18. $A - \delta$ tuning curve for cone.

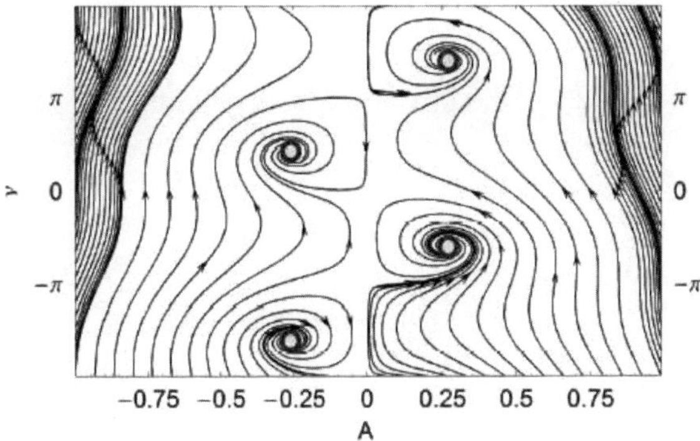

Fig. 12.19.   Phase plot for cone with $\delta = 0$.

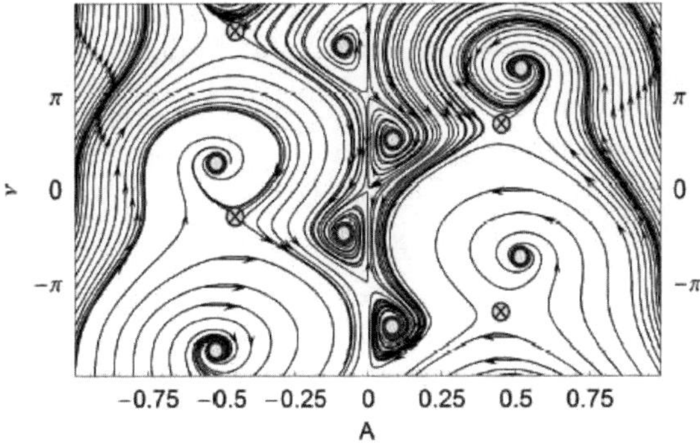

Fig. 12.20.   Phase plot for cone with $\delta = 0.5$.

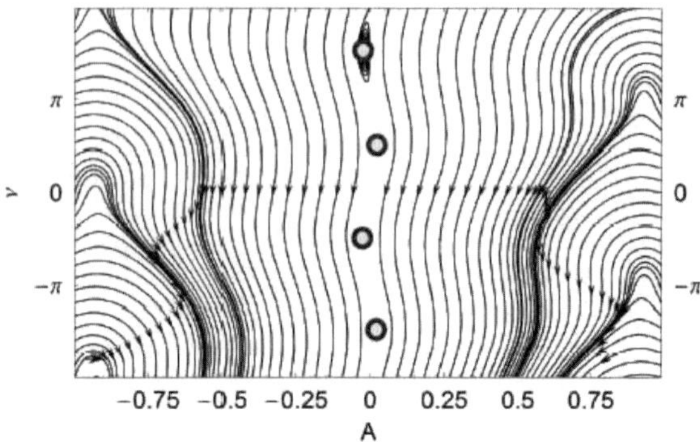

Fig. 12.21.   Phase plot for cone with $\delta = 1.8$.

spiral. Then at $\delta = 0.5$ there are three real values of $|A|$. There is a new stable spiral close to $A = 0$ with a clockwise sense, see Fig. 12.20. A saddle point and an anticlockwise stable spiral (corresponding to that of Fig. 12.19) exists near $A = 0.5$. As the value of δ is increased further, the original anticlockwise stable spiral joins together with the saddle point and mutually annihilate each other. Finally at $\delta = 1.8$, only the clockwise stable spiral remains, see Fig. 12.21. These spirals are close to $A = 0$ relative to the domain of the diagram and are not visually well developed. If the value of δ is now decreased, the reverse evolution occurs. For the bulb shape, the dynamics are similar to those of the cone but with a different sense. In the tuning curve there is a left sloping spike. At $\delta = 0$ there is a single clockwise stable spiral. Decreasing δ to -3, a saddle point and an anticlockwise stable spiral emerge, while the clockwise stable spiral remains. As δ decreases further the saddle point and the clockwise stable spiral mutually annihilate leaving only the stable anticlockwise spiral near $A = 0$. This behavior is illustrated in Mortell, Mulchrone and Seymour [2009]. The amplitude and phase may evolve to different steady states depending on the initial conditions.

Chapter 13

Resonant Sloshing in a Shallow Tank

In this chapter, we deal with waves that are both nonlinear and frequency dispersed. The result of nonlinearity is that the waveform steepens, while the result of frequency dispersion is to spread the waveform. Thus, we have competing physical effects, steepening and spreading, and we examine how a balance is achieved to give continuous solutions in the context of resonantly forced oscillations induced in a tank of shallow water.

13.1. Basics of Frequency Dispersion

If we substitute a solution of the form $Ae^{i(kx-\omega t)}$ in a linear p.d.e., the resultant condition for a nontrivial solution is an equation of the form $G(\omega, k) = 0$; ω is the frequency, $k = 2\pi/\lambda$ the wave number and λ is the wavelength. If $G(\omega, k) = 0$ is solved for ω with $\omega = W(k)$ real, then $W(k)$ is called the *dispersion relation*. A simple example is given by the beam equation

$$\varphi_{tt} + \gamma^2 \varphi_{xxxx} = 0.$$

Substituting $\varphi(x, t) = Ae^{i(kx-\omega t)}$ and solving,

$$\omega = W(k) = \pm\gamma k^2$$

are the dispersion relations representing waves traveling to left and right.

We define $\theta = kx - \omega t$ as the phase, and the phase speed as

$$c = \frac{dx}{dt}\big|_\theta = \frac{\omega}{k}.$$

For $\omega = W(k)$, the phase speed is $c(k) = \frac{W(k)}{k}$, and depends on k when $W''(k) \neq 0$. Then different modes with different wave numbers k have different phase speeds, i.e., the modes separate and the wave "disperses".

Example 1. The simple kinematic equation $\varphi_t + c_0\varphi_x = 0$ gives $G(\omega, k) = \omega - c_0 k$, hence $W(k) = c_0 k$, with $W''(k) = 0$. This wave is not dispersed.

Example 2. We consider a surface wave with wave number k in water with constant undisturbed depth H_0, where shallow water is defined as $kH_0 \ll 1$. The linear equation for the velocity potential $\varphi(x, y, t)$ is, see Sec. 8.2.3,

$$\varphi_{xx} + \varphi_{yy} = 0$$

with the boundary condition on the free surface $y = 0$, $\varphi_{tt} = -g\varphi_y$, and on the impermeable bottom surface $y = -H_0$, $\varphi_y = 0$.

Let $\varphi = \Phi(y)e^{i(kx-\omega t)}$, then

$$\Phi'' - k^2\Phi = 0,$$

so that $\Phi(y) = \Phi_0 \cosh(k(y + H_0))$ satisfies the o.d.e. for $\Phi(y)$ and the boundary conditions on $y = -H_0$. Then the boundary condition on $y = 0$ gives the dispersion relation

$$\omega^2 = gk\tanh(kH_0), \text{ or } W(k) = [gk\tanh(kH_0)]^{\frac{1}{2}}, \text{ with } W''(k) \neq 0.$$

The phase speed is

$$c(k) = \frac{W(k)}{k} = \left[\frac{g}{k}\tanh(kH_0)\right]^{\frac{1}{2}}.$$

For deep water, $kH_0 \to \infty$, $c(k) \to (\frac{g}{k})^{\frac{1}{2}}$, which is dispersive, while for shallow water, $kH_0 \to 0$, $c(k) \to c_0 = (gH_0)^{\frac{1}{2}}$, which is nondispersive.

The next approximation as $kH_0 \to 0$ gives

$$W(k) = (gH_0)^{\frac{1}{2}} k \left[1 - \frac{1}{6}(kH_0)^2 + \cdots \right]$$

or

$$W(k) = c_0 k - \gamma k^3, \qquad \gamma = \frac{1}{6} c_0 H_0^2.$$

It is easily checked that this dispersion relation arises from the linear KdV equation

$$\varphi_t + c_0 \varphi_x + \gamma \varphi_{xxx} = 0.$$

The corresponding nonlinear KdV equation is

$$\varphi_t + c_0 \left(1 + \frac{3}{2} \frac{\varphi}{H_0} \right) \varphi_x + \gamma \varphi_{xxx} = 0.$$

If $\gamma = 0$ we get the nonlinear equation for long waves in shallow water, see Sec. 8.2.5, and if we neglect the nonlinear term $\frac{3}{2} \frac{\varphi}{H_0} \varphi_x$ we get the linear KdV equation.

An exact solution of the nonlinear KdV equation is the solitary wave given by, see Whitham [1974],

$$\varphi = H_0 \zeta(X), \qquad X = x - Ut,$$

where

$$\frac{1}{3} H_0^2 \left(\frac{d\zeta}{dX} \right)^2 = \zeta(\alpha - \zeta), \qquad \frac{U}{c_0} = 1 + \frac{1}{2}\alpha.$$

This leads to

$$\zeta = \alpha \, \text{sech}^2 \left[\left(\frac{3\alpha}{4H_0^2} \right)^{\frac{1}{2}} X \right],$$

where $X = x - Ut$, $U = c_0(1 + \frac{a}{2H_0})$, $\alpha = \frac{a}{H_0}$ is the maximum amplitude, and $\zeta \to 0$ as $X \to \pm\infty$. The solitary wave was observed by Scott Russell in 1834 on the Edinburgh to Glasgow canal. In the solitary wave the frequency dispersion and the nonlinear steepening are balanced to give a single-hump steady traveling wave.

13.2. Derivation of Evolution Equation in Shallow Tank

We consider the two-dimensional, irrotational motion of an inviscid, incompressible homogeneous fluid, with constant density ρ_0, subject to a constant gravitational force g. The fluid has constant undisturbed depth H_0, rests on a horizontal impermeable bed at $y = -H_0$ and has a free surface at $y = \eta(x, t)$. The fluid is contained in a tank of finite length l, closed at $x = 0$ and with an idealized wave maker oscillating with displacement d at $x = \ell - d\cos(2\pi\omega t)$. The length of the tank is much larger than the depth of the fluid and, as we are interested in resonant motion, the wavelength is the same order as the tank length. Then for a shallow water or long-wave disturbance

$$\delta = \frac{H_0}{\ell} \ll 1 \tag{13.1}$$

and for small amplitude waves the wave maker amplitude is small compared to the length of the tube, i.e.,

$$\varepsilon = \frac{d}{\ell} \ll 1. \tag{13.2}$$

The two dimensionless (independent) parameters δ and ε characterize the behavior of the fluid.

Dimensionless variables are introduced by

$$\varphi' = \frac{\varphi}{c_0\ell}, \qquad x' = \frac{x}{\ell}, \qquad \omega' = \frac{\omega\ell}{c_0},$$

$$t' = \frac{2\omega'c_0}{\ell}t, \qquad y' = \frac{y}{H_0}, \qquad \eta' = \frac{\eta}{H_0} \quad \text{and} \quad p' = \frac{p_0}{\rho_0 c_0},$$

where φ is the velocity potential, p_0 is the pressure on the free surface, $2\pi\omega$ is the frequency of the wave maker and $c_0 = (gH_0)^{\frac{1}{2}}$ is the long-wave speed. Now dropping all primes, the condition at the wave maker, which imparts a small amplitude velocity at $x = 1 - \varepsilon\cos(\pi t)$, is

$$u(x, y, t) = \frac{\partial \varphi}{\partial x}(x, y, t) = 2\pi\varepsilon\omega\sin(\pi t), \qquad -1 < y < \eta(x, t).$$

$$\tag{13.3}$$

The tank is closed at $x = 0$:

$$\frac{\partial \varphi}{\partial x}(0, y, t) = 0, \qquad -1 < y < \eta(x, t). \tag{13.4}$$

The velocity potential satisfies

$$\delta^2 \frac{\partial^2 \varphi}{\partial x^2} + \frac{\partial^2 y}{\partial y^2} = 0, \qquad 0 < x < 1 - \varepsilon \cos(\pi t), \quad -1 < y < \eta(x, t),$$
$$\tag{13.5}$$

subject to the boundary conditions on the free surface $y = \eta(x, t)$ and on the bottom and ends of the tank. The bottom is impermeable so that

$$\frac{\partial \varphi}{\partial y} = 0 \qquad \text{on } y = -1. \tag{13.6}$$

On the free surface the dynamic and kinematic conditions imply

$$2\omega \frac{\partial \varphi}{\partial t} + \frac{1}{2}\left[\left(\frac{\partial \varphi}{\partial x}\right)^2 + \delta^{-2}\left(\frac{\partial \varphi}{\partial y}\right)^2\right] + \eta = 0 \tag{13.7}$$

and

$$\frac{\partial \varphi}{\partial y} - \delta^2\left[2\omega \frac{\partial \eta}{\partial t} + \frac{\partial \eta}{\partial x}\frac{\partial \varphi}{\partial x}\right] = 0. \tag{13.8}$$

It should be noted that the presence of 2ω in (13.7) and (13.8) is a result of the scaling of t.

The experimental observations for a shallow-water tank oscillated at the first and second resonant frequencies were given by Chester and Bones [1968]. For wave maker frequencies in a band about a resonant frequency, waves produced in a tank are characterized by high peaks separated by long troughs, and the amplitude of the response, though small, is of a larger order of magnitude than the wave maker amplitude. At certain discrete frequencies abrupt changes in amplitude occur with a corresponding change in the signal shape and more than one stable oscillation is possible at some frequencies. A beating oscillation with period 18 is also observed. Outside of the resonant band the wave profile is adequately calculated from acoustic theory. Verhagen and Wijngaarden [1965] performed water wave experiments where dissipation and dispersion were neglected and so derived the shallow water equivalent of acoustic theory as in a gas. The resulting equations are formally identical with these of one-dimensional compressible flow with adiabatic index $\gamma = 2$, see Sec. 8.2.5. The evolution of some wave forms was demonstrated in Lepelletier and Raichlen [1988], and a numerical and experimental investigation is given in Boucasse, Antuono, Colagrossi and Lugni [2013].

A theory to describe the experimental observations in Chester and Bones [1968] is given in Chester [1968]. This analysis includes the effects of the boundary layer as well as dispersion and nonlinearity. Using a technique similar to that in Chester [1964] an o.d.e. was derived for the periodic resonant motions, which include the equations in Chester [1964] and Verhagen and Wijngaarden [1965]. Ockendon and Ockendon [1973] gave an asymptotic procedure that leads to Chester's equation, while in Ockendon, Ockendon and Johnson [1986] there is an asymptotic analysis of the steady state o.d.e. for periodic motions.

The theory given in Chester [1968] shows that in the periodic state there is a balance between nonlinearity and frequency dispersion, where the periodic output amplitude is $O(\varepsilon^{\frac{1}{2}})$, while the input is $O(\varepsilon)$. In the periodic state there is the similarity relation $\delta^2 = \kappa \varepsilon^{\frac{1}{2}}$, see Ockendon and Ockendon [1973]. We will derive a p.d.e., describing the evolution from rest to the final state, that includes multiple solutions for a given frequency and a beating solution.

If we assume a regular expansion in integer powers of ε, we see there is no periodic solution. Then at $O(\varepsilon)$ φ_0 satisfies

$$\frac{\partial^2 \varphi_0}{\partial t^2} - \frac{\partial^2 \varphi_0}{\partial x^2} = 0$$

with $\frac{\partial \varphi_0}{\partial x} = 0$ at $x = 0$ and $\frac{\partial \varphi_0}{\partial x} = 2\pi\omega \sin(\pi t)$ on $x = 1$. The result of solving for φ_0 gives

$$f'(t) - f'(t - 2) = \pi \sin(\pi t)$$

at the resonant frequency $\omega = \frac{1}{2}$. There is no periodic solution, but for $t > 0$

$$f'(t) = \frac{1}{2}\pi t \sin(\pi t) \qquad (13.9)$$

which shows the initial linear growth associated with acoustic resonance.

Hence, we assume an expansion for the periodic state in the form

$$\varphi(x, y, t; \varepsilon) = \varepsilon^{\frac{1}{2}}\varphi_0(x, y, t) + \varepsilon\varphi_1(x, y, t) + \cdots$$

$$\eta(x, t; \varepsilon) = \varepsilon^{\frac{1}{2}}\eta_0(x, t) + \varepsilon\eta_1(x, t) + \cdots . \tag{13.10}$$

In the neighborhood of resonance, we write

$$\omega = \frac{1}{2}(1 + \varepsilon^{\frac{1}{2}}\Delta), \tag{13.11}$$

so that $\frac{1}{2}\varepsilon^{\frac{1}{2}}\Delta$ is the detuning from resonance. The details of the perturbation procedure follow closely those of the standing wave in Sec. 8.2.3, with ε replaced by $\varepsilon^{\frac{1}{2}}$ and the forcing term comes in at $O(\varepsilon)$. The result is

$$\varphi(x, t; \varepsilon) = \varepsilon^{\frac{1}{2}}\varphi_0(x, t) + \cdots$$

$$= \varepsilon^{\frac{1}{2}}[f(t - x) + g(t + x)] + \cdots , \tag{13.12}$$

where $\frac{\partial f}{\partial t} = \frac{\partial g}{\partial t}$ and $\frac{\partial f}{\partial t}(t + 2) = \frac{\partial f}{\partial t}(t)$ and the signal $h(t) = \frac{\partial f}{\partial t}$ satisfies

$$\frac{3}{2}h\frac{dh}{dt} + \Delta\frac{dh}{dt} - \frac{\kappa}{6}\frac{d^3h}{dt^3} = \pi\omega\sin(\pi t). \tag{13.13}$$

This is the steady state KdV equation with a forcing term. Thus, the solution is a standing wave (13.12), where the signal carried by the wave is given by the forced KdV (13.13). As in the case of the unforced standing wave in Sec. 8.2.3, oppositely traveling waves do not interact since there is no integral term in (13.13).

In order to follow the evolution of the forced motion we assume an expansion of the form

$$\varphi(x, y, t; \varepsilon) = \varepsilon^{\frac{1}{2}}\varphi_0(x, y, t, \tau) + \varepsilon\varphi_1(x, y, t, \tau) + \cdots$$

$$\eta(x, t; \varepsilon) = \varepsilon^{\frac{1}{2}}\eta_0(x, t, \tau) + \varepsilon\eta_1(x, t, \tau) + \cdots , \tag{13.14}$$

where the slow time is $\tau = \varepsilon^{\frac{1}{2}}t$. Then (13.13) becomes

$$\frac{\partial h}{\partial \tau} + \frac{3}{2}h\frac{\partial h}{\partial t} + \Delta\frac{\partial h}{\partial t} - \frac{\kappa}{6}\frac{\partial^3 h}{\partial t^3} = \pi\omega\sin(\pi t), \tag{13.15}$$

where $h = \frac{\partial f}{\partial t}$. This is a periodically forced KdV equation which describes the evolution of the surface motion in the long term. It also accounts for the initial stages of the motion, since at resonance ($\omega = \frac{1}{2}$), when the nonlinearity and frequency dispersion are negligible, (13.15) reduces to

$$\frac{\partial h}{\partial \tau} = \frac{1}{2}\pi \sin(\pi t)$$

and then $h = \frac{1}{2}\pi \varepsilon^{\frac{1}{2}} t \sin(\pi t)$, which agrees with $\varepsilon^{\frac{1}{2}} f'(t)$ given by (13.9). The unforced standing wave result (7.183) is recovered from (13.15) by setting the forcing term to zero and taking the detuning parameter $\Delta = 0$.

In Cox and Mortell [1986] the evolution equation

$$\varepsilon\frac{\partial h_0}{\partial \tau} + \varepsilon\frac{3}{2}h_0\frac{\partial h_0}{\partial t} + \overline{\Delta}\frac{\partial h_0}{\partial t} - \frac{\delta^2}{6}\frac{\partial^3 h_0}{\partial t^3} = \pi\omega\sin(\pi t), \qquad (13.16)$$

with $\overline{\Delta} = \varepsilon^{\frac{1}{2}}\Delta$ and $\tau = \varepsilon t$, is derived via a functional differential equation. If we let $\tau = \varepsilon^{\frac{1}{2}} t$, $R = \varepsilon^{\frac{1}{2}} h_0$, then (13.16) becomes

$$\frac{\partial R}{\partial \tau} + \frac{3}{2}R\frac{\partial R}{\partial t} + \Delta\frac{\partial R}{\partial t} - \frac{\kappa}{6}\frac{\partial^3 R}{\partial t^3} = \pi\omega\sin(\pi t), \qquad (13.17)$$

where $\delta^2 = \kappa\varepsilon^{\frac{1}{2}}$, which is exactly (13.15). Even though (13.15) was derived with a long time expansion in power of $\varepsilon^{\frac{1}{2}}$, it contains the initial motion and so is uniformly valid. Equation (13.15) contains the nonlinear term associated with hydraulic flow, the detuning term with Δ, and the frequency dispersion term associated with κ and is the equation governing the evolution of the motion.

13.3. Nonlinearization

Equation (13.15) can be viewed as the nonlinear equation for hydraulic flow with a frequency dispersion term added. So we now consider the nonlinear differential–difference equation

$$h(t + Q(t)) - h(t) - \frac{\delta^3}{3}\frac{\partial^3 h(t+2)}{\partial t^3} = 2\pi\omega\varepsilon\sin(\pi t), \qquad (13.18)$$

where $Q(t) = 4\omega(1 + \frac{3}{2}h(t))$, $\omega = \frac{1}{2}(1 + \varepsilon^{\frac{1}{2}}\Delta)$, $h = O(\varepsilon^{\frac{1}{2}})$ and $\delta^2 = \kappa\varepsilon^{\frac{1}{2}}$. We write $h = \varepsilon^{\frac{1}{2}}h_0$, then on expanding (13.18) for $|h| \ll 1$ it becomes

$$\frac{3}{2}h_0\frac{\partial h_0}{\partial t} + \Delta\frac{\partial h_0}{\partial t} - \frac{\kappa}{6}\frac{\partial^3 h_0}{\partial t^3} = \pi\omega\sin(\pi t),$$

on requiring the standing wave condition $h_0(t + 2) = h_0(t)$. This is Eq. (13.13).

If there is an impedance condition, and thus a loss of energy, at the boundary $x = 0$, the reflection coefficient k at $x = 0$ becomes $k = 1 - 2\varepsilon^{\frac{1}{2}}\lambda$, see (8.50), with $\lambda > 0$. Then the evolution of the motion, on introducing $\tau = \varepsilon^{\frac{1}{2}}t$, is governed by

$$\frac{\partial h_0}{\partial \tau} + \frac{3}{2}h_0\frac{\partial h_0}{\partial t} + \Delta\frac{\partial h_0}{\partial t} - \frac{\kappa}{6}\frac{\partial^3 h_0}{\partial t^3} + \lambda h_0 = \pi\omega\sin(\pi t), \qquad (13.19)$$

where $\delta^2 = \kappa\varepsilon^{\frac{1}{2}}$. Equation (13.19) is a damped, forced, KdV equation, and has been shown by numerical comparison in Cox and Mortell [1986] to be a good approximation for the inclusion of boundary layer damping. Eq. (13.19) reduces to (13.15) when $\lambda = 0$.

13.4. Comparison with Experiment: Solutions of Forced KdV

The most comprehensive experimental and numerical results are, for our purpose, those of Chester and Bones [1968] for steady-state periodic sloshing in a tank. As in Cox and Mortell [1986], to present results that may be compared directly with the experiments in Chester and Bones [1968] we adjust the boundary condition (13.3) to read

$$u = 2\pi\varepsilon\omega(1 - \cos(2\pi\omega))\sin(\pi t)$$

on $x = 1 - \varepsilon\cos(\pi t)$. The free surface at $x = 0$ is given by $\eta_0|_{x=0} = -2h_0(t-1,\tau)$. With $h_0(t,\tau) = -f_0(t+1,\tau)$ and the changed boundary condition, Eq. (13.16) becomes

$$\varepsilon\frac{\partial f_0}{\partial \tau} - \varepsilon\frac{3}{2}f_0\frac{\partial f_0}{\partial t} + \overline{\Delta}\frac{\partial f_0}{\partial t} - \frac{\delta^2}{6}\frac{\partial^3 f_0}{\partial t^3} = \pi\omega(1 - \cos(2\pi\omega))\sin(\pi t),$$

$$(13.20)$$

where $h_0(t,\tau) = -f_0(t+1,\tau)$, $\tau = \varepsilon t$, $\overline{\Delta} = \varepsilon^{\frac{1}{2}}\Delta$, $\delta^2 = \kappa\varepsilon^{\frac{1}{2}}$ and $f_0(t-2,\tau) = f_0(t,\tau)$. The corresponding initial condition is $f_0(t,0) \equiv 0$. As in Cox and Mortell [1986] numerical solutions of Eq. (13.16) are used initially for comparison with experiment. By introducing the new variables $\overline{\tau} = \varepsilon^{\frac{1}{2}}t$ and $R = \varepsilon^{\frac{1}{2}}h_0$, Eq. (13.20) reduces to (13.17). The velocity of the wave motion at any point in the tank is determined by the linear relation

$$\frac{\partial \varphi_0}{\partial x} = h_0(t+x-1,\tau) - h_0(t-x-1,\tau),$$

while the signal h_0 is provided by the nonlinear equation (13.20). The steady-state equation derived from (13.20) corresponds to that derived in Chester [1968] under appropriate scalings when the integral representation for frequency dispersion is replaced by a derivative representation. To relate (13.20) to the physical problem it is necessary to evaluate $f_0(t,\tau)$ along the line $\tau = \varepsilon t$. The function $f_0(t,\tau)$ needs only be evaluated for $0 \le t \le 2$ due to the periodicity condition $f_0(t-2,\tau) = f_0(t,\tau)$. Then $f_0(t,\tau)$ is evaluated along the lines

$$\tau = \varepsilon t + 2\varepsilon(n-1), \quad n = 1,2,3\ldots, \quad 0 \le t \le 2,$$

where n is the number of cycles of the wave-maker after startup.

The numerical scheme employed to solve (13.19) and (13.20) was proposed by Fornberg [1975] and later applied by Fornberg and Whitham [1978] to the numerical study of the KdV and related equations.

The relevant experimental results are shown in Figs. 9, 15 and 16 of Chester and Bones [1968]. The evolution of the undamped signal given by Eq. (13.20) is shown here in Fig. 13.1 for the cycles 10–25 of the wavemaker operating at the resonant frequency $\omega = \frac{1}{2}$ ($\Delta = 0$) for $\varepsilon = 0.00258$ and $\delta = 0.083$. There is initial linear growth in the signal, similar to Fig. 11.1 before the shock, as a prelude to nonlinear effects becoming important. The signal exhibits amplitude

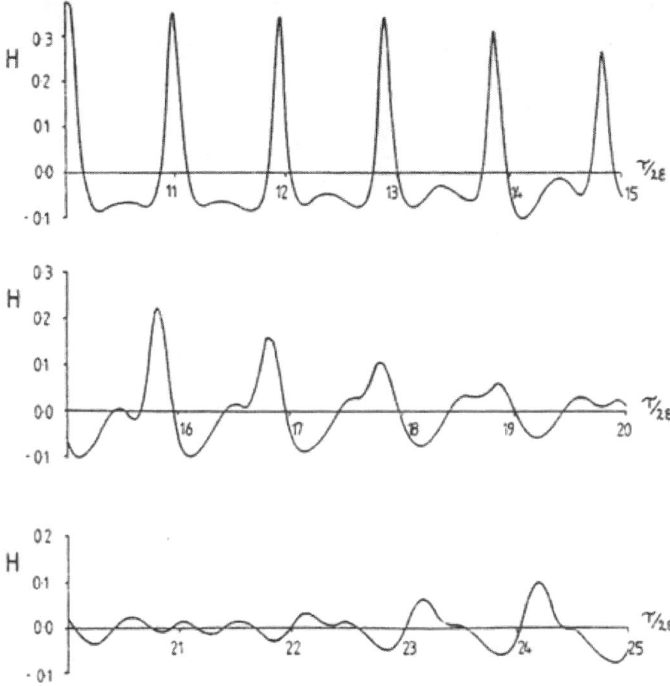

Fig. 13.1.   Evolution but no steady state, $\omega = 0.5$.

growth followed by decay until in cycle 22 the second harmonic is
dominant. This pattern of growth and decay is repeated until in cycle
66 the signal has evolved to approximately the initial behavior before
undergoing a further cycle of growth and decay. The evolution of the
energy content of the first four Fourier modes is given in Fig. 13.2
where the total energy in each cycle is used to normalize the results
displayed. There is no indication from the numerical results that the
signal settles down to a periodic steady state. Figure 13.2 is reminiscent of the Fermi, Pasta and Ulam [1955] recurrence result in the
study of oscillations of an anharmonic lattice. A similar recurrence
has been demonstrated by Yuen and Ferguson [1978] in the context
of deep-water gravity waves for a nonlinear Schrödinger equation.

It is clear that some form of damping is required for a solution to
evolve to a periodic steady state. In Chester [1968] and Ockendon and
Ockendon [1973] damping is introduced through dissipation, which

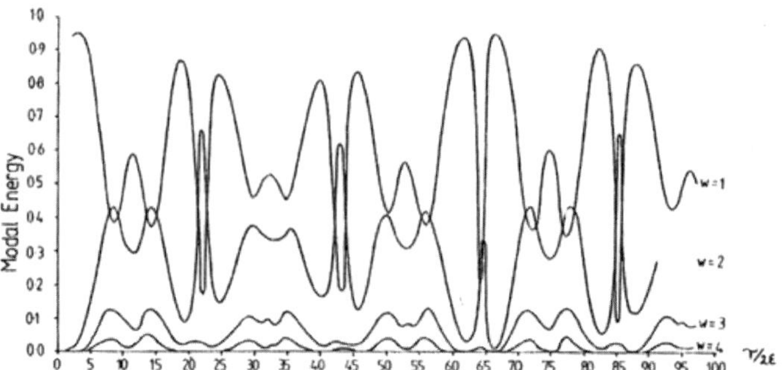

Fig. 13.2.   Evolution of modal energy for Fig. 13.1.

occurs in the boundary layer on the walls and on the base of the
tank. We introduce damping into the forced KdV equation by the
term λh_0 in Eq. (13.19). It is shown in Cox and Mortell [1986] that,
for small values of λ, Eq. (13.19) yields periodic states in close agreement with those in Chester [1968], where an integral term is used
to model boundary layer damping. Figure 13.3 shows the evolution
to a periodic steady state, when $\omega = 0.5$, $\varepsilon = 0.00258$, $\delta = 0.083$,
$\lambda = 0.025$, with $H = 3\varepsilon\omega f_0$ and $h_0(t,\tau) = -f_0(t+1,\tau)$. A single peak
is quickly established and is maintained in the periodic state. When
$\omega = 0.48$, two peaks per cycle are established and for $\omega = 0.46$ three
peaks are established. Figure 13.4 corresponds to the case $\omega = 0.43$,
which lies outside the resonant band. The signal settles down to a
periodic form well approximated by acoustic theory. When $\omega = 0.59$,
outside the resonant band, there is a beating oscillation that settles
down to a periodic form corresponding to acoustic theory. The results
given above agree with the experimental results in Fig. 13 of Chester
and Bones [1968] in that the number of peaks decreases from three
to one as ω goes from 0.46 to 0.52. Chester and Bones [1968] display
response curves which bend over near a local maximum, analogous to
a hard spring solution of Duffing's equation, so that periodic steady-
state solutions are not unique. This is displayed in their Fig. 8 for
$r = 0$ ($\omega = \frac{1}{2}$), where one solution has a single peak per period and
the other has two peaks. Figures 18 and 11 in Chester and Bones
[1968] for $\omega = 0.5$ ($r = 0$) and $r = 1.06$ respectively also show

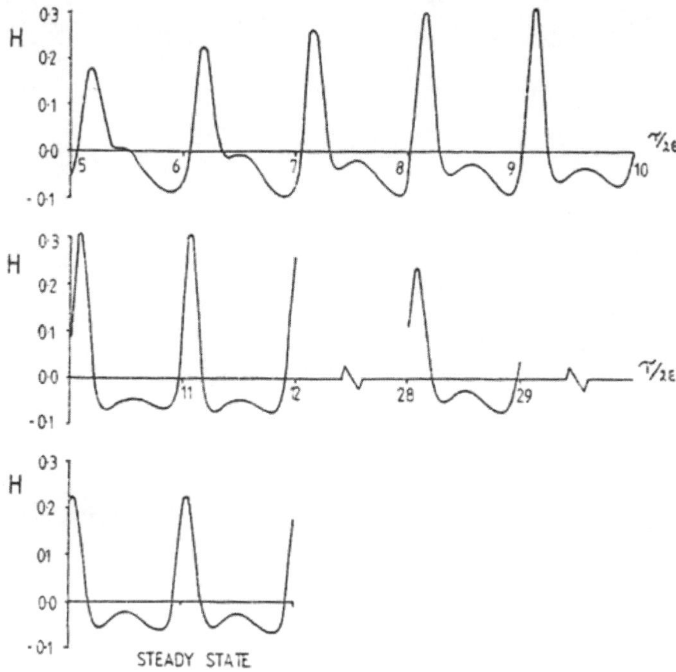

Fig. 13.3.   Evolution to periodic state with damping.

two periodic solutions for the same frequency. These correspond to solutions, from an initial state of rest, on the lower branch of the response curve. Figure 13.5 shows two possible steady-state signal profiles of $H = 3\varepsilon\omega f_0$ calculated from Eq. (13.19). (a) corresponds to the upper branch of the response curve; (b) corresponds to the lower branch when $\omega = 0.5$, $\varepsilon = 0.00258$, $\delta = 0.083$, $\lambda = 0.025$. The upper branch solution has two peaks per period, while the lower branch has one per period. We can conclude that periodic solutions of the nonlinear, dispersive Eq. (13.19) are dependent on the initial state, in contrast to the corresponding nonlinear hyperbolic equation, where shocks play the vital role of ensuring uniqueness.

As a final example of the theory, we examine the 'beat' oscillations observed experimentally and depicted in Fig. 19 of Chester and Bones [1968]. A period of 18 cycles was recorded for the forcing amplitude $\varepsilon = 0.00516$ corresponding to that in Fig. 19. Figure 13.6

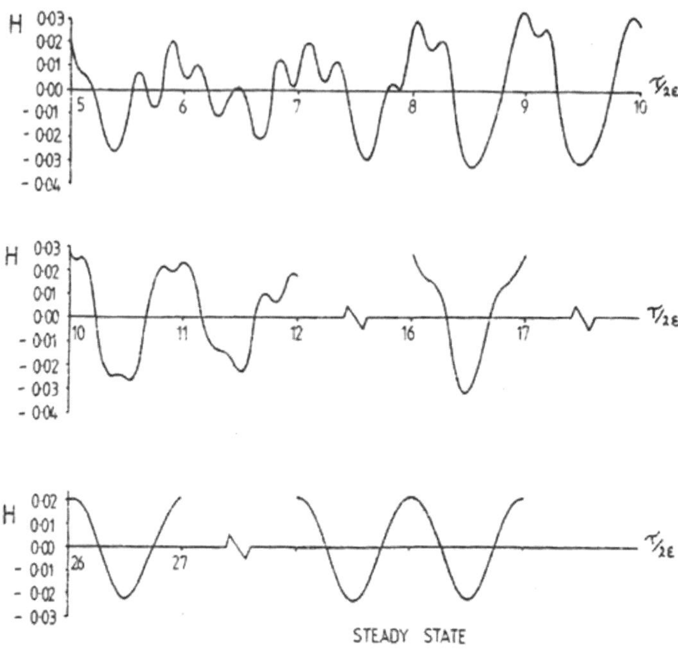

Fig. 13.4.   Evolution to periodic state outside resonant band.

is a typical beat cycle for this forcing amplitude exactly at resonance
when $\omega = 0.5$, $\delta = 0.083$ and $\lambda = 0.025$. Cycles 28–47 are shown,
but the evolution to a steady state was not calculated. In the case
of $\omega = 0.5$, $\varepsilon = 0.00258$, $\delta = 0.083$ and $\beta = 0.0287$, where β is a
boundary layer parameter, there is a 'beat' period of 21 cycles and
the signal evolves through a series of such cycles to a steady state
after 5 such cycles, see Figs. 13.3 and 13.4. Similar agreement with
the evolutionary experiments of Lepelletier and Raichlen [1988] is
shown in the theory given in Cox and Mortell [1989b].

A detailed analytic and numerical study of nonlinear sloshing
is given in Cox, Gleeson and Mortell [2005a]. There it is shown,
inter alia, that the beating phenomenon results from a "tension"
between two possible periodic states. The Burgers' parameter plays
the role of a critical parameter that determines which of the periodic
solutions (upper or lower) or a beating solution prevails.

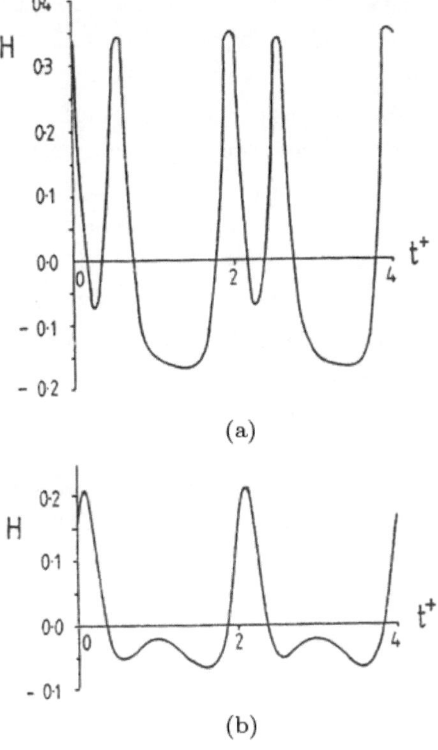

(a)

(b)

Fig. 13.5.   Two steady-state solutions.

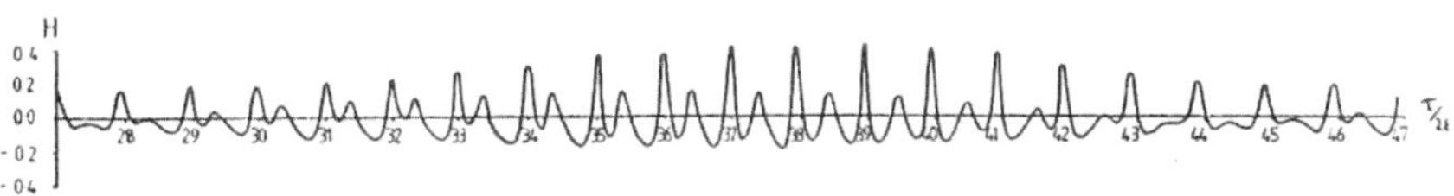

Fig. 13.6.   Beat cycle, $\omega = 0.5$, $\delta = 0.083$, $\lambda = 0.025$.

13.5. Asymptotic Solutions of the Forced Steady State KdV Equation

The evolution, under the KdV equation, of resonant forced sloshing in a tank has been analyzed in previous sections based on numerical solutions of the relevant equations. We now turn to the steady state forced KdV equation and examine the various types of waves that occur as the frequency moves through resonance. The objective is to give an analytical description of the experimental results in Fig. 9 of Chester and Bones [1968]. For wave-maker frequencies in a band about a resonant frequency, waves produced in the tank are characterized by high peaks separated by low troughs, and the amplitude of this response is of a larger order of magnitude than that of the wavemaker. Chester and Bones [1968] and Ockendon and Ockendon [1973] used numerical methods to elucidate some further features of this steady-state solution, while Cox and Mortell [1986] examined the evolution of the wave motion. A steady state was not always achieved. Some asymptotic solutions and numerical evidence, in the steady-state case, for weak dispersion were given in Ockendon, Ockendon and Johnson [1986] and Byatt-Smith [1988], while Mackey and Cox [2002] studied solutions where an additional cubic nonlinearity is present. Grimshaw and Tian [1994] and Malkov [1996] focused on the possibility of chaotic solutions, while Cox, Mortell, Pokrovskii and Rasskazov [2005] gave a rigorous proof of the existence of chaotic solutions, and the problem of passage through resonance was examined in Cox, Gleeson and Mortell [2005a].

The approach taken here follows that in Amundsen, Cox and Mortell [2007] and is based on an a matched perturbation scheme with the (dimensionless) dispersion coefficient as a small parameter. Various periodic solutions within the resonant band are constructed analytically and compared with the experimental results in Chester and Bones [1968]. See also Hattam and Clark [2015] where a similar, but distinct, approach via elliptic functions is used.

13.5.1. *Steady Solutions: Constant Forcing*

The dimensionless forced steady state KdV can be written as

$$-\gamma U_{xxx} + \Delta U_x + \frac{3}{2} U U_x = f(x), \quad -1 \leq x \leq 1, \tag{13.21}$$

where γ is the dispersion coefficient, Δ is the detuning, and $f(x)$ is the displacement of the wavemaker, see (13.13). The period is 2, i.e., $f(x + 2) = f(x)$, and $f(x)$ has zero mean over the period, i.e., $\int_{-1}^{1} f(x)dx = 0$. We seek solutions $U(x)$ such that $U(x + 2) = U(x)$. Chester and Bones [1968] have shown that in the steady, periodic

state dispersion and nonlinearity dominate the qualitative structure
of the solutions, and thus damping is neglected in (13.21).

We define $u(x)$ by

$$u = U + \frac{2}{3}\Delta \tag{13.22}$$

and integrate (13.21) to get

$$\gamma u_{xx} = \frac{3}{4}u^2 - (F(x) + c), \tag{13.23}$$

where $F = \int^x f(t)dt$ is an anti-derivative of f, c a constant of integration and, by (13.22), the zero mean condition on u is

$$\int_{-1}^{1} u(x)dx = \frac{4}{3}\Delta. \tag{13.24}$$

In the small dispersion limit, $\gamma \to 0$, the leading order in (13.23) corresponds to a dispersionless balance between nonlinearity and forcing
and gives

$$u_{\pm} = \pm\frac{2}{\sqrt{3}}\sqrt{F(x) + c}, \tag{13.25}$$

where c is chosen so that $F(x) + c > 0$ on $-1 \le x \le 1$. These
solutions are well defined and smooth, but cannot satisfy the mean
condition (13.24) for a range of Δ near $\Delta = 0$.

To begin to understand the nature of solutions in the dispersion
dominant regime, we consider the o.d.e. on $-\infty < x < \infty$

$$u'' = \frac{3}{4}u^2 - C, \tag{13.26}$$

where C is some constant. This arises from (13.23) if the forcing is
constant or the width of the dispersive layer is so small as $\gamma \to 0$
that the forcing is effectively constant in the layer.

Solutions to (13.26) may be expressed in terms of elliptic functions. However, it is instructive to construct the phase space for
(13.26), see Fig. 13.7, that shows the phase diagram for $C = 1$ and
the qualitatively distinct solutions.

There are two different regions of interest, separated by a homoclinic orbit or separatrix. The separatrix corresponds to a solution

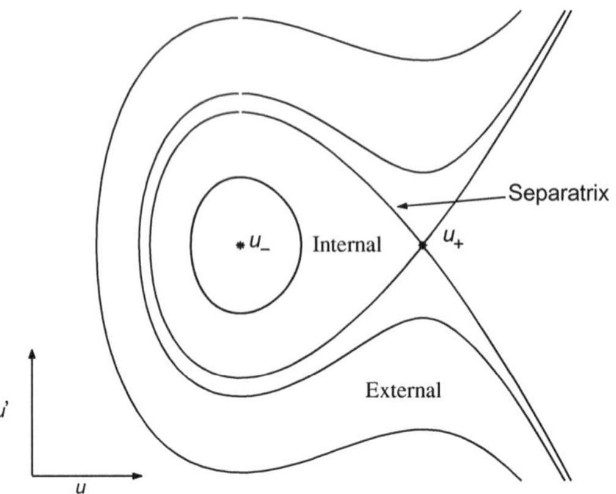

Fig. 13.7.   Phase space for (13.26).

with a single peak or trough and bounded tails. Inside the separatrix lie closed orbits which correspond to bounded oscillatory solutions, while outside the separatrix are trajectories that correspond to solutions with a single peak or trough and unbounded tails. Equation (13.26) integrates to

$$(u')^2 = \frac{u^3}{2} - 2Cu + D, \qquad (13.27)$$

where D is a constant of integration. The qualitative nature of the solution is then determined by the zeros of the the right-hand side of (13.27). Internal bounded oscillatory solutions correspond to three distinct real roots

$$|D| < \left(\frac{2}{3}C\right)^{\frac{3}{2}}.$$

External unbounded solutions correspond to one real root and $D > \left(\frac{2}{3}C\right)^{\frac{3}{2}}$. The separatrix corresponds to $D = \left(\frac{2}{3}C\right)^{\frac{2}{3}}$ and there are two real roots. In each case (13.27) may be solved in terms of elliptic functions. The internal oscillatory solutions are given by

$$u(x) = d + a\operatorname{cn}^2(k(x - \delta), m),$$

where cn is the Jacobi elliptic function and the various parameters are related to C and D, with δ an arbitrary phase shift.

For the external solutions $u(x)$ can be written implicitly [see Gradshteyn and Ryzhik, 1994, p. 277] as

$$\frac{2}{p} F\left(\frac{2\sqrt{u-\alpha}\sqrt{p}}{u-\alpha+p}, \frac{p-\frac{3}{2}\alpha}{2p}\right) = x,$$

where α is the (only) real root of $\frac{1}{2}\alpha^3 - 2C\alpha + D$, $p = \sqrt{3\alpha^2 - 4C}$ and $F(\zeta, k)$ is the incomplete elliptic integral of the first kind, see Abramowitz and Stegun [1965]. At the interface, i.e., the separatrix, the modulus m of the Jacobi elliptic function approached unity, and the limiting solution is

$$u(x) = \frac{2}{\sqrt{3}}\sqrt{C}\left(1 - 3\,\mathrm{sech}^2\left[\frac{(3C)^{\frac{1}{4}}}{2}(x-\delta)\right]\right) \tag{13.28}$$

which corresponds to a single soliton solution.

When the forcing is varied slowly, the homoclinic orbit varies and this leads to a slow variation of C in Eq. (13.28). As solutions move away from the nondispersive fixed point u_+, See Fig. 13.7, they can venture into the internal or external region. For trajectories that go into the internal region, when the forcing increases and the separatrix moves outwards, multiple periodic orbits are possible, whereas for external solutions, when the forcing decreases, only a single orbit may occur, see Figs. 13.8 and 13.9. These illustrate multiple internal trajectories and a single external trajectory. The corresponding spatial dependence of the solutions is also shown.

13.5.2. *Resonant Forcing:* $f(x) = -\pi \sin \pi x$

13.5.2.1. *External orbit*

For the resonant forcing $f(x) = -\pi \sin \pi x$, then $F(x) = \cos \pi x$. The purpose here is to show how to construct asymptotic solutions in the limit of small frequency dispersion, $\gamma \to 0$. From Eq. (13.23) we get

$$\gamma u_{xx} = \frac{3}{4}u^2 - (c + \cos(\pi x)) \tag{13.29}$$

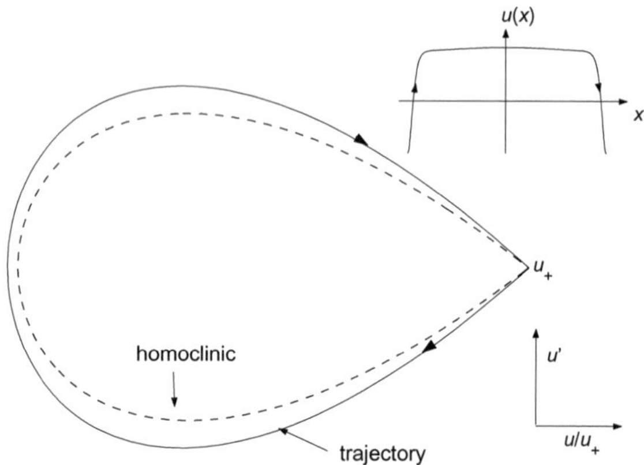

Fig. 13.8.   One external trajectory.

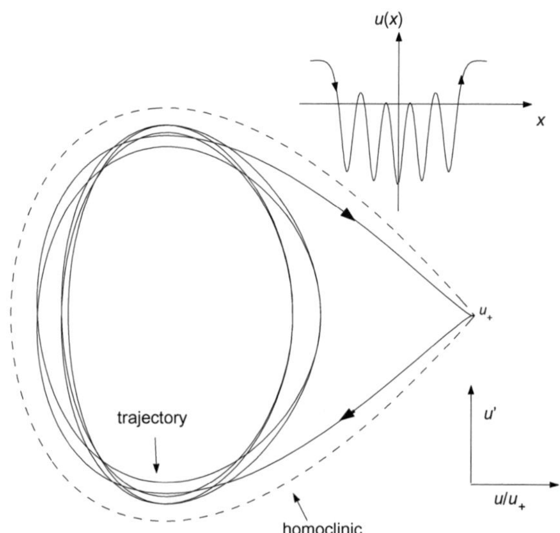

Fig. 13.9.   Multiple internal trajectories.

which, by integration and using periodicity, gives

$$c = \frac{3}{8} \int_{-1}^{1} u^2(x)\, dx. \tag{13.30}$$

This is in addition to the mean condition (13.24) and serves as a global check to ensure results are correct.

For $\gamma \ll 1$ (13.29) leads to a singular perturbation problem, where in the nondispersive region

$$u_{\pm} = \pm \frac{2}{\sqrt{3}} \sqrt{\cos(\pi x) + c}. \tag{13.31}$$

These solutions are well defined and smooth for $c > 1$, while for $c = 1$ the solution has cusps giving the edge of the resonant band; we assume $c > 1$. The local minima of u_+ are at $x = \pm 1$, and we define a stretched coordinate by $X = \frac{x \pm 1}{\sqrt{\gamma}}$ with the corresponding solution to (13.29) $V(X) = u(x)$. Then, to first order in $\sqrt{\gamma}$, the solution near $x \pm 1$ satisfies

$$V_{xx} = \frac{3}{4} V^2 - (c - 1)$$

and this corresponds to the external Eq. (13.26) with a solution given by the incomplete elliptic integral of the first kind. In the limit $\gamma \to 0$ the variation of the forcing becomes small and the variation from the separatrix approaches zero. In this limit the modulus $\sqrt{\frac{p - 3\alpha}{2p}} \to 1$ and then

$$V(X) = \frac{2}{\sqrt{3}} \sqrt{c - 1} \left(1 - 3 \operatorname{sech}^2 \left[\frac{(3(c - 1))^{\frac{1}{4}}}{2} X \right] \right).$$

Now

$$\lim_{x \to \pm 1} u_+ = \frac{2}{\sqrt{3}} \sqrt{c - 1} = \lim_{X \to \pm\infty} V(X).$$

Then the uniform external orbit solution is

$$u_{01}(x) = u(x) = \frac{2}{\sqrt{3}} \sqrt{\cos(\pi x) + c}$$

$$- \sum_{\pm} 2\sqrt{3}\sqrt{c - 1} \operatorname{sech}^2 \left[\left(\frac{3(c - 1)}{16} \right)^{\frac{1}{4}} \frac{x \pm 1}{\sqrt{\gamma}} \right]. \tag{13.32}$$

So the uniform external solution consists of solution peaks at $x = \pm 1$ superposed on the nondispersive solution (13.31) with the constant c determined by the mean condition (13.24).

13.5.2.2. *Internal orbit*

The other type of resonant solution corresponds to a departure from the homoclinic and into the internal region as in Fig. 13.9. In this case an arbitrary number of orbits may be made within the separatrix before returning to the nondispersive solution. Such solutions are centred around the local maximum of u_+ at $x = 0$. We focus first on the case when there is a single orbit before returning to the separatrix and the nondispersive solution. Now we consider solutions of

$$V_{XX} = \frac{3}{4}V^2 - (c+1).\qquad(13.33)$$

where $x = \sqrt{\gamma}X$. The solution in this case is given by a Jacobi elliptic function, see Gradshteyn and Ryzhik [1994], and as $\gamma \to 0$ the modulus $m \to 1$. Then the solution is approximated by the homoclinic, i.e.,

$$V(X) = \frac{2}{\sqrt{3}}\sqrt{c+1}\left(1 - 3\,\mathrm{sech}^2\left[\frac{(3(c+1)^{\frac{1}{4}})}{2}X\right]\right).$$

Matching with the nondispersive solution u_+

$$\lim_{x\to 0} u_+ = \frac{2}{\sqrt{3}}\sqrt{c+1} = \lim_{X\to\pm\infty} V(X)$$

and therefore the uniformly valid approximation to the leading order is

$$u_{10}(x) = u(x)$$
$$= \frac{2}{\sqrt{3}}\sqrt{\cos \pi x + c} - 2\sqrt{3}\sqrt{c+1}\,\mathrm{sech}^2\left[\left(\frac{3(c+1)}{16}\right)^{\frac{1}{4}}\frac{x}{\sqrt{\gamma}}\right].$$
$$(13.34)$$

The uniform internal solution consists of a solution peak at $x = 0$ superposed on the nondispersive solution, with c determined by the mean condition (13.24). In writing u_{01} in (13.32) and u_{10} in (13.34) we have adopted the notation u_{nm} for a solution with n internal orbits and m external orbits.

13.5.2.3. *Resonant band*

In the case of multiple orbits in the internal region, it is no longer valid to consider the forcing as constant in the limit of small dispersion. The "constants" C and D in the elliptic function Eq. (13.27) must now be considered as slowly varying parameters and a multiple scale approach is required.

For sufficiently large values of Δ, the nondispersive solutions (13.31) independently satisfy the mean condition (13.24) provided

$$\frac{4}{3}\Delta_\pm = \int_{-1}^{1} u_\pm(x)dx = \pm\frac{8\sqrt{c+1}}{\sqrt{3}\pi}E\left(\sqrt{\frac{2}{c+1}}\right) \qquad (13.35)$$

for some $c > 1$, see Gradshteyn and Ryzhik [1994], 2.576, and $E(m)$ is the complete elliptic integral of the second kind. This provides the range of detuning where solutions may independently exist. In the limiting case

$$\Delta^* = \lim_{c\to 1^+} |\Delta_\pm| = \frac{2\sqrt{6}}{\pi}. \qquad (13.36)$$

For $\Delta > \Delta^*$, $u = u_+$ is admissible while for $\Delta < -\Delta^*$, $u = u_-$ satisfies the constraint. For detuning in the range $|\Delta| < \Delta^*$ only resonant solutions affected by the dispersion will exist. Thus, we have a precise bound on the width of the resonant band.

For the external orbit u_{01} given by (13.32), the mean condition (13.24) yields

$$\Delta_{01} = \frac{2\sqrt{3}\sqrt{c+1}}{\pi}E\left(\sqrt{\frac{2}{c+1}}\right)$$

$$- 6\sqrt{\gamma}(3(c-1))^{\frac{1}{4}}\tanh\left[\left(\frac{3(c-1)}{16}\right)^{\frac{1}{4}}\frac{1}{\sqrt{\gamma}}\right], \qquad (13.37)$$

We note that as $c \to 1^+$ the amplitude of the dispersive peak vanishes and

$$\Delta_{01} = \lim_{c\to 1^+} \Delta_{01} = \frac{2\sqrt{6}}{\pi}$$

which agrees with the upper bound for the nondispersive solution u_+ given in (13.36). Thus, u_{01} is the first resonant solution that is encountered as the detuning parameter is continued from the right. As $c \to \infty$, the amplitude of the resonant peak becomes unbounded. However, in the presence of damping this growth will be bounded, see Amundsen, Cox and Mortell [2007].

For the single internal orbit the solution u_{10} is given by (13.34). The mean condition (13.24) implies

$$\Delta_{10} = \frac{2\sqrt{3}\sqrt{c+1}}{\pi} E\left(\sqrt{\frac{2}{c+1}}\right) - 6\sqrt{\gamma}(3(c+1))^{\frac{1}{4}} \tanh\left[\frac{(3(c+1))^{\frac{1}{4}}}{2\sqrt{\gamma}}\right].$$
(13.38)

The lower bound on Δ, below which solutions u_{10} do not exist, is given by

$$\Delta_{10}^* = \lim_{c \to 1^+} \Delta_{10} = \frac{2\sqrt{6}}{\pi} - 6^{\frac{5}{4}}\sqrt{\gamma}\tanh\left[\left(\frac{3}{8}\right)^{\frac{1}{4}}\frac{1}{\sqrt{\gamma}}\right].$$

An external orbit may now be added to obtain the first combined solution, on assuming that γ is sufficiently small so that there is a nondispersive region separating the two dispersive regimes. The uniform approximation is constructed from u_{10} and u_{01}

$$u_{11} = \frac{2}{\sqrt{3}}\sqrt{\cos(\pi x) + c} - 2\sqrt{3}\sqrt{c+1}\,\operatorname{sech}^2\left[\frac{(3(c+1))^{\frac{1}{4}}}{2}\frac{x}{\sqrt{\gamma}}\right]$$

$$- \sum_{\pm} 2\sqrt{3}\sqrt{c-1}\,\operatorname{sech}^2\left[\frac{(3(c-1))^{\frac{1}{4}}}{2}\frac{x \pm 1}{\sqrt{\gamma}}\right].$$
(13.39)

The associated Δ_{11} is determined from (13.39) through the mean condition (13.24) to give

$$\Delta_{11} = \frac{2\sqrt{3}\sqrt{c+1}}{\pi} E\left(\sqrt{\frac{2}{c+1}}\right)$$

$$-6\sqrt{\gamma}(3(c+1))^{\frac{1}{4}} \tanh\left[\frac{(3(c+1))^{\frac{1}{4}}}{2\sqrt{\gamma}}\right]$$

$$-6\sqrt{\gamma}(3(c-1))^{\frac{1}{4}} \tanh\left[\frac{(3(c-1))^{\frac{1}{4}}}{2\sqrt{\gamma}}\right].$$

In the limit that the external orbit vanishes, $c \to 1^+$,

$$\lim_{c \to 1^+} \Delta_{11} = \Delta_{10}^*$$

and this provides the connection to the single orbit solution u_{10} given by (13.34).

The resonant response for the undamped system (13.21) with $f(x) = -\pi\sin(\pi x)$ and $\gamma = 0.005$ is given in Fig. 13.10 that shows the full resonant response diagram determined by the approximation scheme outlined above with the norm of u defined by $\|u\|_2 = [\int_{-1}^1 u(x)^2 dx]^{\frac{1}{2}}$. Note that in Fig. 13.10 u_{50} matches smoothly to u_- and u_{01} matches to u_+.

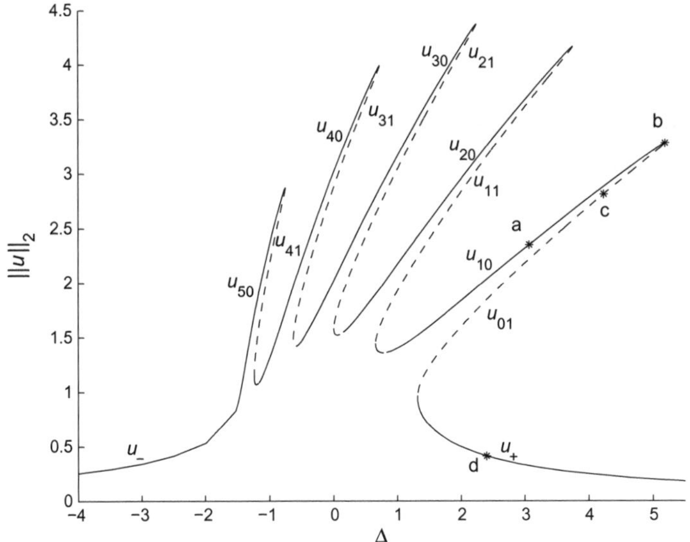

Fig. 13.10.   Response for undamped system (13.21).

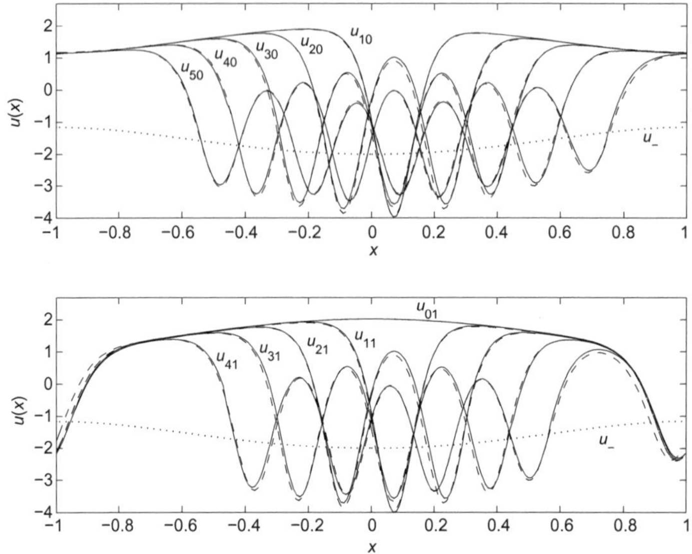

Fig. 13.11. Approximate (solid) and exact (dashed) for $c = 2$, $\gamma = 0.005$.

Figure 13.11 shows approximate solutions (solid) for $c = 2$ and $\gamma = 0.005$ as well as exact solutions (dashed) obtained by numerical integration of (13.21). It is seen that five peaks can fit in the resonant band. Finally the experimental results in Fig. 9 of Chester and Bones [1968], given here as Fig. 13.12, are compared with the asymptotic solutions, shown in Fig. 13.13, for $\gamma = 0.025$ and Burgers damping parameter $\mu = 0.015$; the details can be found in Amundsen, Cox and Mortell [2007]. It should be noted that the success of the asymptotic expansion given here requires that the various layers are separate. Depending on γ, there will be solutions at the finger tips and at the valleys between the fingers in Fig. 13.10 where the method breaks down.

13.5.3. *Subharmonic Resonant Sloshing, $\omega = 1/4$*

This extension is discussed in Amundsen, Cox, Mortell and Reck [2001]. The physical situation of the shallow water tank is described in Sec. 13.2 with the difference that the forcing velocity at

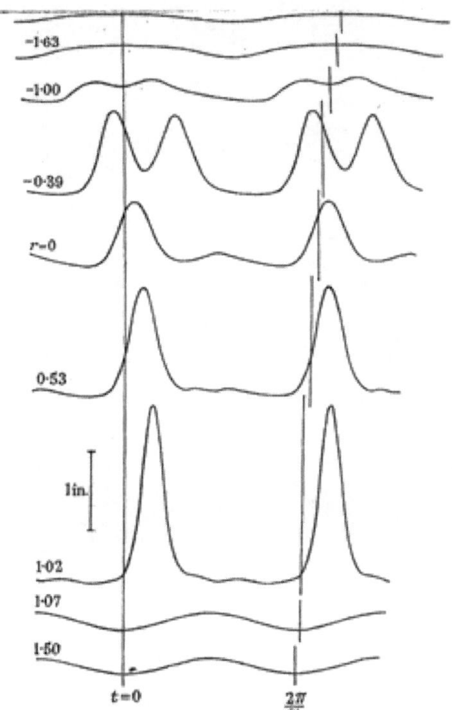

FIGURE 9. Experimental recordings of surface elevation at
$x = 0$.

$h_0 = 2$ in., $l = 0.062$ in.

Fig. 13.12.   Experimental results in Chester and Bones [1968].

$x = 1 - \varepsilon \cos(\frac{\pi}{2}t)$ is given by

$$u(x, y, t) = \frac{\partial \phi}{\partial x} = \varepsilon \left(\frac{\pi}{2} + \pi\Delta \right) \sin \left(\frac{\pi}{2}t \right). \qquad (13.40)$$

The variables have been nondimensionalized as in Sec. 13.2, the detuning is given by $\widetilde{\Delta} = 2\omega - 1/2$, where $\omega = 1/4$ is half the fundamental frequency. From the hyperbolic case of forcing at half the fundamental frequency, we know that an input at $O(\varepsilon)$ yields an output of $O(\varepsilon)$, see Sec. 9.5. Thus we set $\delta^2 = \kappa\varepsilon$ and $\widetilde{\Delta} = \varepsilon\Delta$, where $\kappa = O(1)$ and $\Delta = O(1)$. We then assume a standard multiple

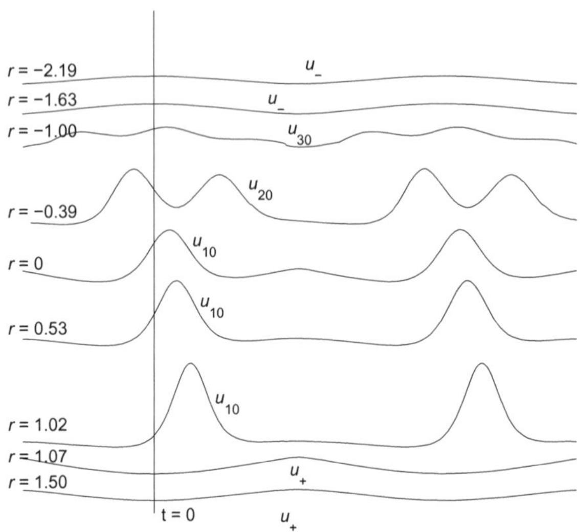

Fig. 13.13.   Asymptotic results for $\gamma = 0.025$, $\mu = 0.015$ corresponding to Fig. 13.12.

scale expansion

$$\phi = \varepsilon\phi_0(x, y, t, \tau) + \varepsilon_1^2\phi(x, y, t, \tau) + \cdots, \tag{13.41}$$

$$\eta = \varepsilon\eta_0(x, t, \tau) + \varepsilon_1^2\eta(x, t, \tau) + \cdots, \tag{13.42}$$

where the slow variable is $\tau = \varepsilon t$.

The details of the calculation are given in Amundsen, Cox, Mortell and Reck [2001]. At $O(\varepsilon)$ the linear theory leads to

$$h(t + 1, \tau) - h(t - 1, \tau) = \frac{\pi}{2} \sin \frac{\pi}{2} t, \tag{13.43}$$

where

$$\frac{\partial \phi_0}{\partial x} = -f_0'(t - x, \tau) + f_0'(t + x, \tau), \tag{13.44}$$

for an arbitrary function f_0, f_0' denotes the derivative w.r.t. the first argument, and $h(t, \tau) = f_0'(t, \tau)$. Thus, the wave response consists of two noninteracting oppositely traveling waves where f_0' is modulated on the time scale τ. The solution of (13.43) may be written

$$h(t,\tau) = \widetilde{h}(t,\tau) - \frac{\pi}{4}\,\cos\frac{\pi}{2}t, \qquad (13.45)$$

where $\widetilde{h}$ satisfies the periodicity condition

$$\widetilde{h}(t+2,\tau) = \widetilde{h}(t,\tau). \qquad (13.46)$$

The solution $h(t,\tau)$ in (13.45) is the superposition of the periodic linear forced oscillation and a nonlinear standing wave $\widetilde{h}(t,\tau)$, which will be determined at the next order. We also note from (13.43) that

$$h(t+4,\tau) - h(t,\tau) = 0,$$

i.e., $h(t,\tau)$ has the same period in t as the wave-maker and is split naturally into the sum of $\widetilde{h}$ with period 2 and $\frac{\pi}{4}\cos\frac{\pi}{2}t$ with period 4.

We determine the equation for $\widetilde{h}$ by ensuring secular terms do not arise at $O(\varepsilon^2)$, and we note that since $\widetilde{h}$ has period 2 in t, the equation for $\widetilde{h}$ must be invariant under the transformation $t \to t+2$. Then, the p.d.e for $\widetilde{h}(t,\tau)$ is

$$\widetilde{h}_\tau + 2\Delta\,\widetilde{h}_t + \frac{3}{2}\widetilde{h}\,\widetilde{h}_t - \frac{K}{6}\widetilde{h}_{ttt} = \frac{3\pi^3}{128}\,\sin\pi t. \qquad (13.47)$$

Equation (13.47) is a periodically forced KdV equation with a periodic forcing $\sin\pi t$ when the actual forcing is $\sin\frac{\pi}{2}t$. There was a similar situation for the hyperbolic case, see Eq. (8.91).

The evolution of the surface elevation, $\varepsilon\eta_0(0,t,\tau)$, at the closed end of the tank is given by,

$$\eta_0(0,t,\tau) = -2h(t,\tau) = \frac{\pi}{2}\cos\frac{\pi}{2}t - 2\widetilde{h}(t,\tau), \qquad (13.48)$$

where $\widetilde{h}(t,\tau)$ is determined by (13.47), (13.46) and zero mean initial conditions.

There are no experiments to act as a check on the theory given here. However, by setting $K = 0$ in (13.47) we get the hydraulic limit where there are experimental results available and reasonable agreement is found between theory and experiment, see Amundsen, Cox, Mortell and Reck [2001].

To include damping into (13.47), we rewrite the equation as

$$\widetilde{h}_\tau + 2\Delta\widetilde{h}_t + \frac{3}{2}\widetilde{h}\,\widetilde{h}_t + \lambda\widetilde{h} - \frac{K}{6}\widetilde{h}_{ttt}$$

$$+ \beta\sqrt{\frac{2}{\pi}}\int_{-\infty}^{\infty}(\mathrm{sgn}(r)+1)\frac{\partial\widetilde{h}}{\partial t}(t-r,\tau)|\pi r|^{-1/2}dr = \frac{3\pi^3}{128}\sin\pi t.$$

$$(13.49)$$

The term $\lambda\widetilde{h}$ arises from an impedance condition at $x = 0$, and the integral term is the effect of boundary layer damping from the sides and bottom of the tank, as formulated in Chester [1968]. In the periodic state the mean is zero. The detuning Δ is measured from the resonant frequency, $\omega = 1/4$, determined by having the reference state as the mean of the periodic state.

Figure 13.14 shows $\eta(t,\infty)$ at the closed end $x = 0$ for various values of the detuning Δ, using (13.48) and (13.49), where the initial condition is zero, $K = 0.1$, $\lambda = 0$, and $\beta = 0.05$. Figure 13.15 shows the evolution of a "beating" (nonperiodic) oscillation with zero initial conditions, $K = 0.1$, $\Delta = -0.02$, $\lambda = 0$ and $\beta = 0.05$. The corresponding evolution to a periodic state in the hyperbolic (nondispersive) case is shown in Fig. 13.16.

Figure 13.17 gives an example of two equilibrium (periodic) states of Eq. (13.49) for the same values of the parameters: $K = 0.1$,

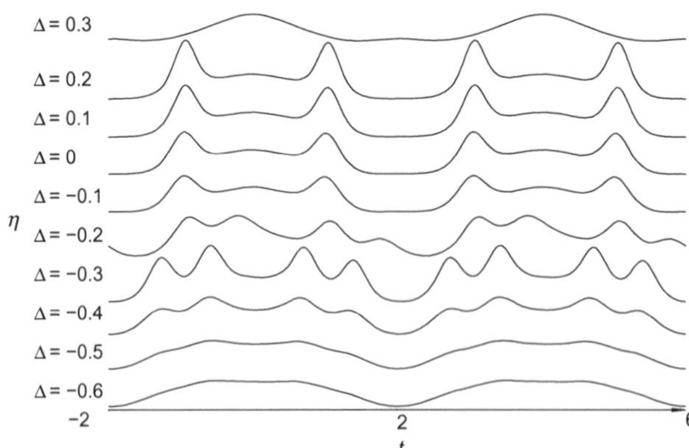

Fig. 13.14.   Steady states for $\lambda = 0$, $K = 0.01$, $\beta = 0.05$.

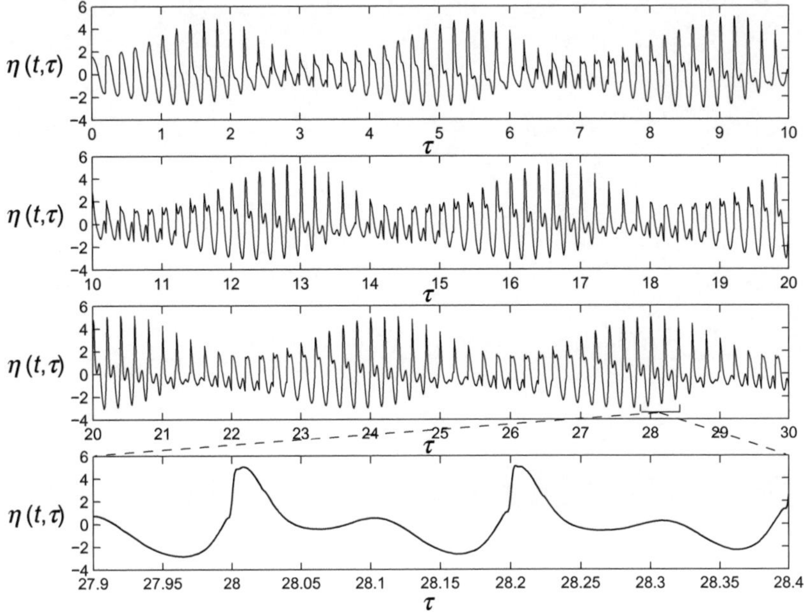

Fig. 13.15.   Nonperiodic state for $\lambda = 0$, $K = 0.1$, $\beta = 0.05$.

$\Delta = -0.02$, $\lambda = 0.1$, and $\beta = 0.05$. In fact the beating solution in Fig. 13.15 settles down to the periodic solution with one peak, shown in the upper part of Fig. 13.17, when λ is increased from $\lambda = 0$ to $\lambda = 0.1$, and the boundary layer damping coefficient $\beta = 0.05$ is held constant.

13.5.4. *Crank Drive*

For a crank drive the dimensionless form of the wave-maker location is

$$x = 1 - \varepsilon u_1 \cos \pi t/2 - \varepsilon^2 u_2 \, \cos(\pi t - \phi), \qquad (13.50)$$

where $\omega = \frac{1}{4} + \frac{1}{2}\varepsilon\Delta$, $\varepsilon u_1 = \frac{d_1}{l} \ll 1$, $\varepsilon u_2 = \frac{d_2}{l} \ll 1$, and d_1, d_1 are the dimensional displacements and ϕ is the applied phase shift. Following the previous procedure, see Amundsen, Cox, Mortell and

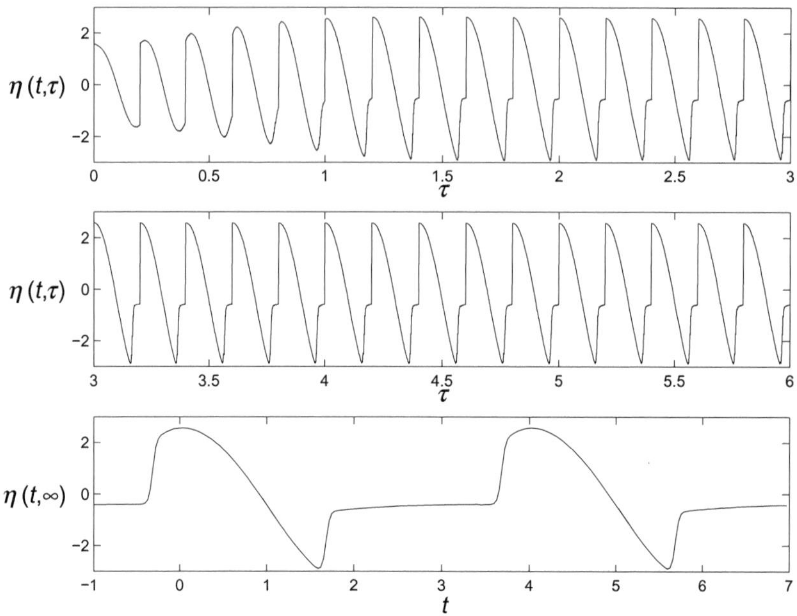

Fig. 13.16.   Evolution to periodic state.

Reck [2001], the derived p.d.e. is

$$\widetilde{h}_\tau(t,\tau) + 2\Delta\widetilde{h}_t + \frac{3}{2}\widetilde{h}\,\widetilde{h}_t + \lambda\widetilde{h} - \frac{K}{6}\widetilde{h}_{ttt}$$

$$+\beta\sqrt{\frac{2}{\pi}} \int_{-\infty}^{\infty} (\operatorname{sgn}(r) + 1)\frac{\partial\widetilde{h}}{\partial t}(t - r,\tau)|\pi r|^{-1/2}dr$$

$$= u_1^2 \frac{3\pi^3}{128} \, \sin \pi t - \frac{1}{2}\pi u_2 \, \sin(\pi t - \phi). \tag{13.51}$$

The resonating input $\varepsilon^2 u_2 \, \cos(\pi t - \phi)$ in (13.50) gives an output at $O(\varepsilon)$ and so appears on the right-hand side of (13.51). In the special case $u_2 = \frac{3\pi^2}{64}u_1^2$ and $\phi = 0$ then (13.51) reduces to the unforced,

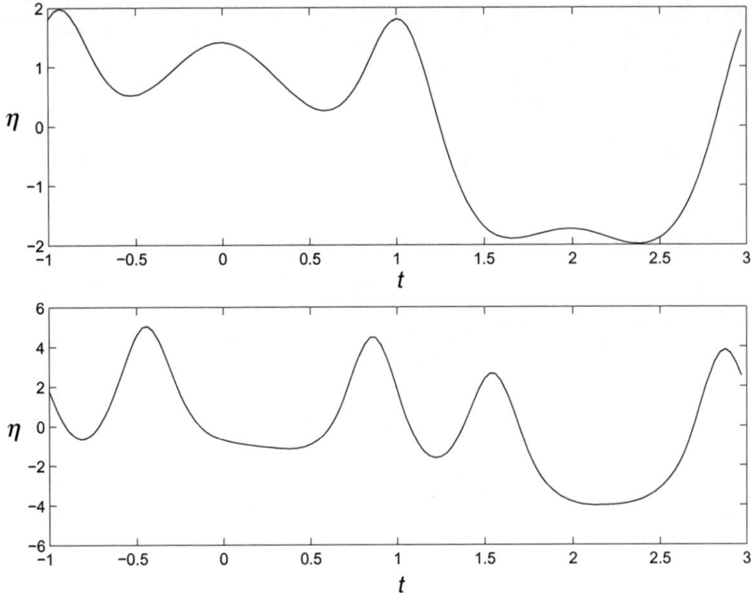

Fig. 13.17.   Two periodic states of (13.49).

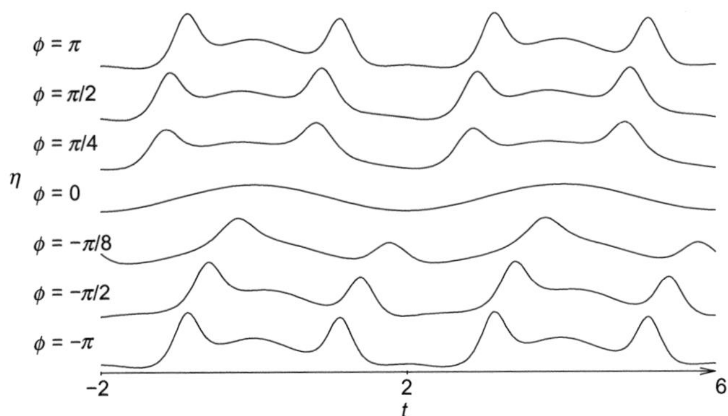

Fig. 13.18.   Steady states for various ϕ.

damped KdV

$$\widetilde{h}_\tau + 2\Delta\widetilde{h}_t + \frac{3}{2}\widetilde{h}\widetilde{h}_t + \lambda\widetilde{h} - \frac{K}{6}\widetilde{h}_{ttt}$$

$$+\beta\sqrt{\frac{2}{\pi}}\int_{-\infty}^{\infty}(\mathrm{sgn}(r)+1)\frac{\partial\widetilde{h}}{\partial t}(t-r,\tau)|\pi r|^{-1/2}dr = 0.$$

$$(13.52)$$

Figure 13.18 shows the steady state given by (13.51) for different value of the phase ϕ when $u_2 = \frac{3\pi^2}{64}u_1^2$ and $K = 0.1,\ \ \Delta = 0,\ \ \lambda = 0.1,\ \ $ and $\beta = 0.05$. This shows that for $\phi = 0$, the motion is linear in the long term, since the solution to (13.52) $\widetilde{h}(t,\tau) \to 0$ due to the damping, and the linear term in the asymptotic expansion gives, see (13.48), $\eta_0(0,t,\tau) \to \frac{\pi}{2}u_1\cos(\pi t/2)$.

Chapter 14

Nonlinear Resonance in an Open Tube

Thus far in Chapters 9 and 10 we have examined resonances in closed (or nearly closed) tubes. The primary insight to the understanding of such problems is to realize that a sufficiently accurate approximation to the travel time is essential. In the case of a closed tube the first nonlinear correction to the linear travel time, i.e., terms linear in the dependent variables, is sufficient. This nonlinearization leads to the standard mapping, defined in Eq. (9.35). However, it was noted in Sec. 8.1.1 that for a stress-free boundary, equivalent to an open end in a tube, the first nonlinear correction to the travel time cancels out over two cycles in the tube, leading to the linear result, see Eq. (8.30). Thus, to nonlinearize the travel time for an open tube the quadratic terms in the dependent variables must be included. This higher-order correction will involve the interaction of the oppositely travelling waves, and so the concept of noninteracting (modulated) simple waves no longer holds.

The problem of describing the response of a gas in a tube, open at one end, excited by an oscillating piston was studied experimentally by Lettau [1939]. He found that for the chosen piston amplitude the oscillations in the tube are small and almost harmonic at the fundamental resonance. For the second resonance there was a strong distortion, especially at the open end, but discontinuities were not observed. At the third harmonic resonance the pressure and velocity near the piston were continuous, but there were shocks at the

373

open end. The latter case is an example of a finite rate oscillation where the signal has significant distortion over one tube length. In contrast to Lettau [1939], the experiments of Sturtevant [1974] involve large amplitude oscillations of the piston and it is observed that at the fundamental frequency, though the waveform at the piston is continuous and smooth, the compressive parts of the wave steepen and break in less than one tube length. Further experiments were conducted by Zaripov and Ilgamov [1976] and Stuhlträger and Thoman [1986].

The analytical study of resonance in open tubes has focussed on the shocked regime, see Jimenez [1973], Seymour and Mortell [1973b], Chester [1981] and Cox and Kluwick [1996]. The latter three papers adopted the technique of setting up a nonlinear difference equation for a Riemann invariant where the difference involves the quadratic terms in the characteristic equations. This gives the nonlinearization of the travel time. A Taylor expansion reduces the difference equation to a nonlinear ordinary differential equation that can be integrated to give a cubic algebraic equation for the Riemann invariant. As an example, Chester [1981] gives the equation

$$g^3 + \frac{3r}{2^{2/3}}g + \sin t = 0, \tag{14.1}$$

where $g(t)$ is periodic with unit period and has zero mean over one period; r is a detuning parameter defining the resonant band. Chester claims that such a cubic equation is deducible from the work of Jimenez [1973] and Seymour and Mortell [1973b]. However, in the latter papers, as well as in Cox and Kluwick [1996], there is an interaction term involving $\int_{t-1}^{t} g^2(s)ds$ in the nonlinear travel time that is not present in Eq. (14.1).

Cox and Kluwick [1996] examine resonant gas oscillations with mixed nonlinearity (a BZT-fluid) in a closed tube. The most interesting aspect of such fluids is the possibility of shocks of expansion as well as compression. Such negative shocks were first recognized by Bethe [1942] and Zel'dovich [1946]. These account for the BZ of BZT while the T is Thompson [1971], or Thompson and Lambrakis [1973]. There is a parameter regime in Cox and Kluwick [1996] where solutions qualitatively resemble those in Chester [1981] for an ideal gas in an open tube, and we exploit this in the account given here.

The derivation of the nonlinear difference equation (2.37)–(2.39) in Cox and Kluwick [1996], or the functional equation in Chester [1981] that leads to (14.1), or the similar one in Seymour and Mortell [1973b], are all long and complicated. We observe that in the absence of the interaction term $\int_{t-1}^{t} g^2(s)ds$ the resultant travel time should arise from a simple wave. Then the cubic terms in the functional equation are derived by assuming the motion is the superposition of simple waves, and the interaction term can be added at the end of the calculation.

We follow the derivation from Secs. 8.1 and 8.1.1, which are couched in terms of stress and strain. The equations in dimensionless form are

$$u_t = \sigma_x = c^2(\lambda)\lambda_x, \text{ and } u_x = \lambda_t, \ 0 < x < 1,$$

$$\sigma(\lambda) = \Sigma(\lambda) = \lambda(1 + N\lambda + N_1\lambda^2 + O(\lambda^3)),$$

where $c^2(\lambda) = \Sigma'(\lambda) = 1 + 2N\lambda + 3N_1\lambda^2 + O(\lambda^3)$ and $|\lambda| \ll 1$. N and N_1 are $O(1)$ dimensionless material constants, and for a polytropic gas $N = -(\gamma+1)/2$ and $N_1 = (\gamma+1)(\gamma+2)/6$. The nonlinear strain measure is

$$a(\lambda) = \int_0^\lambda c(y)dy = \lambda(1 + \frac{1}{2}N\lambda + \frac{1}{2}N_1\lambda^2 - \frac{1}{6}N^2\lambda^2 + O(\lambda^3)).$$

The right (α) and left (β) traveling characteristics are defined by

$$\frac{dT}{dx}\Big|_\alpha = \frac{1}{c(\lambda)}, \ \frac{dT}{dx}\Big|_\beta = -\frac{1}{c(\lambda)},$$

where

$$u = G(\beta) + F(\alpha), \ a = G(\beta) - F(\alpha), \tag{14.2}$$

Then,

$$\frac{dT}{dx}\Big|_\alpha = 1 - N\lambda + \frac{3}{2}N_2\lambda^2 + O(\lambda^3),$$

where $N_2 = N^2 - N_1 = \frac{1}{12}(\gamma^2 - 1)$. For a simple wave traveling to the right $G \equiv 0$ and $a \smallfrown \lambda + \cdots = -F(\alpha)$, so that

$$T = \frac{\alpha}{\omega} + x + NF(\alpha)x + \frac{3}{2}N_2F^2(\alpha)x \qquad (14.3)$$

with $\alpha = \omega T$ on $x = 0$. Here ω is introduced as an arbitrary constant that will be used later for the forced problem.

For a simple wave traveling to the left, $a \smallfrown \lambda + \cdots = G(\beta)$ and $F \equiv 0$, then

$$T = \frac{\beta}{\omega} - (x - 1) + NG(\beta)(x - 1) - \frac{3}{2}N_2G^2(\beta)(x - 1) \qquad (14.4)$$

with $\beta = \omega T$ on $x = 1$.

We now consider an unforced homogeneous elastic panel with the end at $x = 0$ stress-free (or for a gas in a tube, the end is open) and the end $x = 1$ fixed (or closed for a gas). From (14.2) at $x = 0$, $G(\beta) = F(\alpha)$ and $G(\beta) = -F(\alpha)$ at $x = 1$.

At $x = 0$, from (14.4)

$$T = \frac{\alpha}{\omega} = \frac{\beta}{\omega} + 1 - NG(\beta) + \frac{3}{2}N_2G^2(\beta) \qquad (14.5)$$

with $G(\beta) = F(\alpha)$.

At $x = 1$, from (14.3)

$$T = \frac{\beta}{\omega} = \frac{\alpha}{\omega} + 1 + NF(\alpha) + \frac{3}{2}N_2F^2(\alpha) \qquad (14.6)$$

with $G(\beta) = -F(\alpha)$.

We now follow a wave leaving the fixed end, $x = 1$, at time t_0 through two cycles. It is reflected at $x = 0$ at times t_1, t_3 and at $x = 1$ at times t_2, t_4. By (14.2), (14.5) and (14.6) the times and amplitudes are related through:

$$t_1 = t_0 + 1 - NG(\eta_0) + \frac{3}{2}N_2G^2(\eta_0), \quad F(\eta_1) = G(\eta_0)$$

$$t_2 = t_1 + 1 + NF(\eta_1) + \frac{3}{2}N_2 F^2(\eta_1), \quad G(\eta_2) = -F(\eta_1),$$

$$t_3 = t_2 + 1 - NG(\eta_2) + \frac{3}{2}N_2 G^2(\eta_2), \quad G(\eta_2) = F(\eta_3),$$

$$t_4 = t_3 + 1 + NF(\eta_3) + \frac{3}{2}N_2 F^2(\eta_3), \quad G(\eta_4) = -F(\eta_3),$$

$$(14.7)$$

where $\eta_i = \omega t_i$, and $\eta = \omega t$. Now

$$G(\eta_4) = -F(\eta_3) = -G(\eta_2) = F(\eta_1) = G(\eta_0), \qquad (14.8)$$

so that the nonlinear difference equation is

$$G(\eta_4) = G(\eta_0), \qquad (14.9)$$

where the relationship between t_4 and t_0 is given by (14.7), by adding all four lines and using (14.8) to give

$$t_4 = t_0 + 4 + 6N_2 G^2(\eta_0) \qquad (14.10)$$

or

$$\eta_4 = \eta_0 + 4\omega + 6\omega N_2 G^2(\eta_0). \qquad (14.11)$$

Equations (14.9) and (14.11) constitute the nonlinear difference equation for a standing wave under the assumption that the motion is a superposition of noninteracting simple waves.

For the forced problem, the boundary condition at $x = 1$ is $G(\beta) + F(\alpha) = M\sin(2\pi\eta)$ and (14.9) is replaced by

$$G(\eta_4) - G(\eta_0) = M\sin(2\pi\eta_4) - M\sin(2\pi\omega[t_4 - 2]), \qquad (14.12)$$

where the nonlinear travel time is given by (14.10). Linear theory is described by

$$G(\eta) - G(\eta - 4\omega) = M\sin(2\pi\eta) - M\sin(2\pi\eta - 4\pi\omega), \qquad (14.13)$$

Equation (14.13) is also obtained by using (4.7) with $k = -1$ twice covering two traversals of the tube. Hence, from (4.20), (14.13) has the solution

$$G(t) = \frac{M}{2\cos(2\pi\omega)}\sin(2\pi\omega t + 2\pi\omega),$$

so that there is no periodic solution for $\omega = \omega_1 = 1/4$. The linear Eq. (14.13) is nonlinearized by using (14.11) for the difference $\eta_4 - 4\omega$.

If $\omega = \frac{1}{4}(1 + \Delta)$, where $\frac{\Delta}{4}$ is the detuning from resonance, then (14.12) becomes

$$G(\eta_4) - G(\eta_0) = 2M[\sin(2\pi\eta_4) + O(\Delta)]. \tag{14.14}$$

For the forced problem the travel time (14.10) becomes

$$t_4 = t_0 + 4 + 3N_2 G^2(\eta_0) + 3N_2(M\sin(2\pi\eta_2) - G(\eta_0))^2. \tag{14.15}$$

However, near resonance $M = o(G)$ and hence the term in M can be neglected and the standing wave result (14.11) used in the non-linearization of (14.14).

Equations (14.11) and (14.14) lead us to the equivalent of Chester's [1981] Eq. (14.1) where there is no interaction of the waves. From (14.11) with $\omega = \frac{1}{4}(1 + \Delta)$.

$$\begin{aligned}
\eta_4 &= \eta_0 + 4\omega + 6\omega N_2 G^2(\eta_0) \\
&= \eta_0 + 1 + \Delta + 6\omega N_2 G^2(\eta_0) \\
&= \eta_0 + \Delta + 6\omega N_2 G^2(\eta_0), \tag{14.16}
\end{aligned}$$

where the 1 can be discarded since G has unit period. We replace Δ in (14.16) by

$$\delta = \Delta - N_3 \int_{t-1}^{t} G^2(s)ds, \tag{14.17}$$

for a constant N_3, and this is inserted as the interaction term. The wave interaction increases $(N_3 < 0)$ or decreases $(N_3 > 0)$ the travel time of wavelets and thus alters the resonant frequency.

The nonlinear difference equation (14.14), (14.16) and (14.17) is mathematically similar to Eqs. (2.37)–(2.39) in Cox and Kluwick [1996] when $\overline{\Gamma} = 0$.

The derivation of the Eqs. (14.14) and (14.16) makes no assumption about the acceleration rate of the evolving signal, and hence this is a small amplitude, finite rate approximation for the signal. The periodic response in the small rate limit $|G'| \ll 1$ is, from (14.14), (14.16) and (14.17)

$$G'(\eta) \left[\delta + \frac{3}{2} N_2 G^2(\eta) \right] = 2M \sin(2\pi\eta) \qquad (14.18)$$

with $G(\eta + 1) = G(\eta)$ and $\int_{\eta-1}^{\eta} G(s)ds = 0$. The essential difference between (14.18) and (14.1) is the presence of δ rather than Δ in (14.18).

We can conclude from Cox and Kluwick [1996], that there is an interval for δ, $[0, \delta^+]$ in which the signal G is discontinuous and consists of two so-called sonic shocks of compression and expansion. Outside this interval the solution is continuous. At the upper end of the interval the limiting solution as δ^+ is approached from below is different from that when δ^+ is approached from above. This agrees with Chester [1981].

The effect of the inclusion of the interaction term through δ in Cox and Kluwick [1996] is to produce a continuous solution over most of the frequency range, rather than a shock solution as in Chester [1981]. Now the frequency range over which shocks occur is greatly reduced. Figure 10 in Cox and Kluwick [1996] shows some continuous and some discontinuous signal profiles. Outside the resonant band the solution is well approximated by linear theory. Discontinuities of both compression and rarefaction are shown in Fig. 2 in Chester [1981] at $r = -0.5$ arising from (14.1).

The correspondence between theory and experiment is less satisfactory for the open ended tube than for the closed end case. For the closed end the boundary condition is simple and clear and theory and experiment agree. This is not the case for the open end. We have used here the simple boundary condition that the pressure is constant at the open end. Seymour and Mortell [1973b] and Jimenez [1973] used the empirical boundary condition $u = k(p - p_0)$, where k measures the effect of a baffle. Van Wijngaarden [1968] modeled the flow at the open end as being jet-like for the outflow and sink-like for the inflow, which according to Stuhlträger and Thoman [1986] is a more realistic boundary condition. A thorough experimental and theoretical investigation of the flow near an open end is given in Disselhorst and Van Wijngaarden [1980]. Sturtevant and Keller [1978] examined the boundary conditions for a variety of mouthpiece geometries at subharmonic resonance, and use a generalization of Van Wijngaarden's model. The theory given by Chester [1981] takes into account the nonlinear effects, the boundary friction and the energy loss at the open end. Stuhlträger and Thoman [1986] find by their experiments that the theory given in Chester [1981] produces good results for the first harmonic and fair results for the second and third ones. The end result is that there is no single definitive theory agreeing with the experiments for resonance in an open tube. An extensive summary of the experiments and theories, particularly with regard to the condition at the open end, is given in Ilgamov, Zaripov, Galiullin, and Repin [1996]. Recently Amundsen, Mortell and Seymour examined continuous solutions for nonlinear resonant oscillations in open axisymmetric tubes, and compared analytical approximations with numerical simulations.

Bibliography

Abramowitz, M. and I.A. Stegun (1965) *Handbook of Mathematical Functions.* Dover.

Althaus, R. and H. Thomann (1987) Oscillations of a gas in a closed tube near half the fundamental frequency. *Journal of Fluid Mechanics*, 183: 147–161.

Amundsen, D.E., E.A. Cox, M.P. Mortell and S. Reck (2001) Evolution of nonlinear sloshing in a tank near half the fundamental resonance. *Studies in Applied Mathematics*, 107: 103–125.

Amundsen, D.E., E.A. Cox and M.P. Mortell (2007) Asymptotic analysis of the steady solutions of the KdVB equation with application to resonant sloshing. *Zeitschrift für angewandte Mathematik und Physik ZAMP*, 58: 1008–1034.

Amundsen, D.E., M.P. Mortell and B.R. Seymour (2015) Resonant radial oscillations of an inhomogeneous gas in the frustum of a cone. *Zeitschrift für angewandte Mathematik und Physik ZAMP*, 66: 2647–2663.

Asano, N. and T. Taniuti (1969) Reductive perturbation method for nonlinear wave propagation in inhomogeneous media. I. *Journal of the Physical Society of Japan*, 27(4): 1059–1062.

Bascom, W. (1964) *Waves and Beaches: The dynamics of the ocean surface.* Doubleday, New York.

Bender, C.M. and S.A. Orszag (1999) *Advanced Mathematical Methods for Scientists and Engineers — Asymptotic Methods and Perturbation Methods.* Springer, New York.

Betchov, R. (1958) Nonlinear oscillations of a column of gas. *Physics of Fluids*, 1(3): 205–212.

Bethe, H.A. (1942) The theory of shock waves for an arbitrary equation of state. *Office of Scientific Research and Development*, Report 545.

Billingham, J. and A.C. King (2000) *Wave Motion.* Cambridge Texts in Applied Mathematics, Cambridge University Press, Cambridge.

Boucasse, B., M. Antuono, A. Colagrossi and C. Lugni (2013) Numerical and experimental investigation of nonlinear shallow water sloshing. *International Journal of Nonlinear Sciences and Numerical Simulation*, 14(2): 123–238.

Boyce, W.E. and R.C. DiPrima (2012) *Elementary Differential Equations and Boundary Value Problems*, 10th Edn. John Wiley & Sons, New York.

Bremmer, H. (1951) The WKB approximation as the first term of a geometric-optical series. *Communications on Pure and Applied Mathematics*, 4(1): 105–115.

Byatt-Smith, J.G.B. (1988) Resonant oscillations in shallow water with small mean-square disturbances. *Journal of Fluid Mechanics*, 193: 369–390.

Chester, W. (1964) Resonant oscillations in closed tubes. *Journal of Fluid Mechanics*, 18: 44–64.

Chester, W. (1968) Resonant oscillations of water waves I. Theory. *Proceedings of the Royal Society of London. Series A: Mathematical and Physical Sciences*, 306: 5–22.

Chester, W. and J.A. Bones (1968) Resonant oscillations of water waves II. Experiment. *Proceedings of the Royal Society of London. Series A: Mathematical and Physical Sciences*, 306: 23–39.

Chester, W. (1981) Resonant oscillations of a gas in an open-ended tube. *Proceedings of the Royal Society of London. Series A: Mathematical and Physical Sciences*, 377: 449–467.

Chester, W. (1991) Acoustic resonance in spherically symmetric waves. *Proceedings of the Royal Society of London. Series A: Mathematical and Physical Sciences*, 434: 459–463.

Chirikov, B.V. (1979) A universal instability of many-dimensional oscillator systems. *Physics Reports*, 52(5): 263–379.

Chu, B.-T. (1963) Analysis of a self-sustained thermally driven nonlinear vibration. *Physics of Fluids*, 6(11): 1638–1644.

Chu B.-T. and S.-J. Ying (1963) Thermally driven nonlinear oscillations in a pipe with traveling shock waves. *Physics of Fluids*, 6(11): 1625–1637.

Chun, Y.-D. and Y.-H. Kim (2000) Numerical analysis for nonlinear resonant oscillations of gas in axisymmetric closed tubes. *The Journal of the Acoustical Society of America*, 108(6): 2765–2774.

Cole, J.D. (1968) *Perturbation Methods in Applied Mathematics*. Blaisdell, Waltham, MA.

Collins, W.D. (1971) Forced oscillations of systems governed by one-dimensional nonlinear wave equations. *The Quarterly Journal of Mechanics and Applied Mathematics*, 24(2): 457–466.

Courant, R. and K.O. Friedrichs (1999) *Supersonic Flow and Shock Waves*. Springer, New York.

Courant, R. and D. Hilbert (1962) *Methods of Mathematical Physics*, Vol. 2. John Wiley & Sons, New York.

Cox, E.A. and M.P. Mortell (1983) The evolution of resonant oscillations in closed tubes. *Zeitschrift für angewandte Mathematik und Physik ZAMP*, 34(6): 845–866.

Cox, E.A. and M.P. Mortell (1986) The evolution of resonant water-wave oscillations. *Journal of Fluid Mechanics*, 162: 99–116.

Cox, E.A. and M.P. Mortell (1989a) The evolution of resonant acoustic oscillations with damping. *Proceedings of the Royal Irish Academy*, 89A(2): 147–160.

Cox, E.A. and M.P. Mortell (1989b) Discussion of Nonlinear oscillations in rectangular tanks by T.G. Lepelletier and F. Raichlen. *Journal of Engineering Mechanics*, 115(7): 1585–1587.

Cox, E.A. and M.P. Mortell (1992) The evolution of nonlinear resonant oscillations in closed tubes: Solution of the standard mapping. *Chaos: An Interdisciplinary Journal of Nonlinear Science*, 2(2): 201–214.

Cox, E.A. and A. Kluwick (1996) Resonant gas oscillations exhibiting mixed nonlinearity. *Journal of Fluid Mechanics*, 318: 251–271.

Cox, E.A., M.P. Mortell, A.V. Pokrovskii and O.A. Rasskazov (2005) On chaotic wave patterns in periodically forced steady-state KdVB and extended KdVB equations, *Proceedings of the Royal Society of London. Series A: Mathematical and Physical Sciences*, 461: 2857–2885.

Cox, E.A., J.P. Gleeson and M.P. Mortell (2005) Nonlinear sloshing and passage through resonance in a shallow water tank. *Zeitschrift für angewandte Mathematik und Physik ZAMP*, 56: 645–680.

Disselhorst, J.H.M. and L. van Wijngaarden (1980) Flow in the exit of open pipes during acoustic resonance. *Journal of Fluid Mechanics*, 99: 293–319.

Drazin, P.G. and R.S. Johnson (1989) *Solitons: An Introduction.* Cambridge University Press, Cambridge.

Eisner, E. (1967) Complete solutions of the Webster horn equation. *The Journal of the Acoustical Society of America*, 41(4B): 1126–1146.

Ellermeier, W. (1993) Nonlinear acoustics in non-uniform infinite and finite layers. *Journal of Fluid Mechanics*, 257: 183–200.

Ellermeier, W. (1994a) Acoustic resonance of cylindrically symmetric waves. *Proceedings of the Royal Society of London. Series A: Mathematical and Physical Sciences*, 445: 181–191.

Ellermeier, W. (1994b) Resonant wave motion in a non-homogeneous system. *Zeitschrift für angewandte Mathematik und Physik ZAMP*, 45(2): 275–286.

Ellermeier, W. (1997) Acoustic resonance of radially symmetric waves in a thermoviscous gas. *Acta Mechanica*, 121: 97–113.

Fermi, E., J. Pasta, and S. Ulam (1955) Studies of nonlinear problems I. *Los Alamos Report LA-1940.*

Fornberg, B. (1975) On a Fourier method for the integration of hyperbolic functions. *Society for Industrial and Applied Mathematics Journal of Numerical Analysis*, 12: 504–528.

Fornberg, B. and G.B. Whitham (1978) A numerical and theoretical study of certain nonlinear wave phenomena. *Philosophical Transactions of the Royal Society of London* Series A 289: 373–404.

Friedrichs, K.O. (1948) Formation and decay of shock waves. *Communications on Pure and Applied Mathematics*, 1(3): 211–245.

Galiev, S.U., M.A. Il'Gamov and A.V. Sadykov (1970) Periodic shock waves in a gas. *Fluid Dynamics*, 5(2): 223–230.

Gardner, C.S., J.M. Greene, M.D. Kruskal and R.M. Miura (1967) Method for solving the Korteweg-deVries equation. *Physical Review Letters*, 19(19): 1095.

Gorkov, A.P. (1963) Nonlinear acoustic oscillations of a column of gas in a closed tube. *Inzhenerno-Flzicheski Jurnal*, 3: 246–250.

Gradshteyn, I.S. and I.M. Ryzhik (1994) *Table of Integrals, Series, and Products*. Academic Press.

Greene, J.M. (1979) A method for determining a stochastic transition. *Journal of Mathematical Physics*, 20(6): 1183–1201.

Grimshaw, R. and X.Tian (1994) Periodic and chaotic behaviour in a reduction of the perturbed Korteweg-de Vries equation. *Proceedings of the Royal Society of London. Series A: Mathematical and Physical Sciences*, 445: 1–21.

Guckenheimer, J. and P. Holmes (1983) *Nonlinear Oscillations, Dynamical Systems, and Bifurcations of Vector Fields*, Vol. 42. Springer Verlag, New York.

Gulyayev, A.I. and V.N. Kusnetsov (1963) Gas oscillations of large amplitude in closed tubes. *Inzh. Jurnal*, 3: 263–245.

Hamilton, M.F., Y.A. Ilinskii and E.A. Zabolotskaya (2001) Linear and nonlinear frequency shifts in acoustical resonators with varying cross sections. *The Journal of the Acoustical Society of America*, 110(1): 109–119.

Hamilton, M.F., Y.A. Ilinskii, and E.A. Zabolotskaya (2009) Nonlinear frequency shifts in acoustical resonators with varying cross sections. *The Journal of the Acoustical Society of America*, 125(3): 1309–1310.

Hattam, L.L. and S.R. Clark (2015) Modulation theory for the steady forced KdV–Burgers equation and the construction of periodic solutions. *Wave Motion*, 56: 67–84.

Hirota, R. (1972) Exact solutions of the Korteweg-deVries equation for multiple solitons, *Physical Review Letters*, 27: 1192–1194.

Hiroyuki, H. and B. Han (2015) Experimental and numerical investigation of resonance mode and wave development for an acoustic compressor. *European Journal of Scientific Research*, 133(2): 196–214.

Holmes, P.J. (1982) The dynamics of repeated impacts with a sinusoidal vibrating table. *Journal of Sound and Vibration*, 84: 173–189.

Ilgamov, M.A., R.G. Zaripov, R.G. Galiullin and V.B. Repin (1996) Nonlinear oscillations of a gas in a tube. *Applied Mechanics Reviews*, 49(3): 137–154.

Ilinskii, Y.A., B. Lipkens, T.S. Lucas, T.W. Van Doren, and E.A. Zabolotskaya (1998) Nonlinear standing waves in an acoustical resonator. *The Journal of the Acoustical Society of America*, 104(5): 2664–2674.

Jimenez, J. (1973) Nonlinear gas oscillations in pipes. Part 1. Theory. *Journal of Fluid Mechanics*, 59: 23–46.

Johnson, R.S. (2005) *Singular Perturbation Theory — Mathematical and Analytical Techniques with Applications to Engineering*. Springer, New York.

Keller, J.B. (1962) Geometrical theory of diffraction. *The Journal of the Acoustical Society of America*, 52(2): 116–130.

Keller, J.B. and L. Ting (1966) Periodic vibrations of systems governed by nonlinear partial differential equations. *Communications on Pure and Applied Mathematics*, 19(4): 371–420.

Keller, J.J. (1975) Subharmonic non-linear acoustic resonances in closed tubes. *Zeitschrift für angewandte Mathematik und Physik ZAMP*, 26(4): 395–405.

Keller, J.J. (1977) Nonlinear acoustic resonances in shock tubes with varying cross-sectional area. *Zeitschrift für angewandte Mathematik und Physik ZAMP*, 28(1): 107–122.

Kevorkian, J. (2000) *Partial Differential Equations — Analytical Solution Techniques; Texts in Applied Mathematics 35*. Springer, New York.

Kevorkian, J. and J.D. Cole (2011) *Multiple Scale and Singular Perturbation Methods*. Springer, New York.

Kluwick, A. (1981) The analytical method of characteristics. *Progress in Aerospace Sciences*, (19): 179–313, Pergamon Press.

Kruskal, M.D. and N.J. Zabusky (1964) Stroboscopic-perturbation procedure for treating a class of nonlinear wave equations. *Journal of Mathematical Physics*, 5(2): 231–244.

Kurihara, E., Y. Inoue and T. Yano (2005) Nonlinear resonant oscillations and shock waves generated between two coaxial cylinders. *Fluid Dynamics Research*, 36(1): 45–60.

Kurihara, E. and T. Yano (2006) Nonlinear analysis of periodic modulation in resonances of cylindrical and spherical acoustic standing waves. *Physics of Fluids*, 18(11): 117107.

Landau, L.D. (1945) On shock waves at large distances from the place of their origin. *Journal of Physics*, 9: 496–500.

Lamb, H. (1945) *Hydrodynamics*. Dover Publications, New York.

Lardner, R.W. (1975) The formation of shock waves in Krylov–Bogoliubov solutions of hyperbolic partial differential equations. *Journal of Sound and Vibration*, 39(4): 489–502.

Lawrenson, C.C., B. Lipkens, T.S. Lucas, D.K. Perkins and T.W. Van Doren (1998) Measurements of macrosonic standing waves in oscillating closed cavities. *The Journal of the Acoustical Society of America*, 104(2): 623–636.

Lax, P.D. (1957) Hyperbolic systems of conservation laws. *Communications on Pure and Applied Mathematics*, 10(4): 537–566.

Lebedev, N.N. (1972) *Special Functions and their Applications*. Dover Publications, New York.

Lepelletier, T.G. and F. Raichlen (1988) Nonlinear oscillations in rectangular tanks. *Journal of Engineering Mechanics*, 114(1): 1–22.

Lettau, E. (1939) Messungen un Gasschwingungen grosser amplitude in Rohrleitgunen. *Deut Kraftfahrforsch*, 39: 1–17.

Lewis, R.W. (1964) The progressing wave formalism. In *Quasi-Optics*, Vol. 1, p. 71.

Lichtenberg, A.J. and M.A. Liebermann (1992) *Regular and Chaotic Dynamics*, Springer-Verlag, New York.

Lighthill, M.J. and G.B. Whitham (1955) On kinematic waves. I. Flood movement in long rivers. *Proceedings of the Royal Society of London. Series A. Mathematical and Physical Sciences*, 229: 281–316.

Lucas, T. S. (1993) Acoustic compressor. *Proceedings of the 1993 non-fluorocarbon refrigeration and air-conditioning technology workshop,* June 23–25: pp. 43–50.

Ludford, G.S.S. (1952) On an extension of Riemann's method of integration, with applications to one-dimensional gas dynamics. *Proceedings of the Cambridge Philosophical Society,* 48: 499–510.

Mackey, D. and E.A. Cox (2002) Dynamics of a two-layer fluid sloshing problem. *Institute of Mathematics and its Applications, Journal of Applied Mathematics,* 68: 665–686.

Malkov, M.A. (1996) Spatial chaos in weakly dispersive and viscous media: A non-perturbative theory of the driven KdV–Burgers equation. *Physica D,* 95: 62–80.

Meriam, J.L. and L.G. Kraige (2012) *Engineering Mechanics — Dynamics,* 7th Edn. John Wiley & Sons, New York.

Minorsky, N. (1962) *Nonlinear Oscillations.* Van Norstrand.

Miura, R.M. (1976) The Korteweg-deVries equation: A survey of results, *Society for Industrial and Applied Mathematics Reviews,* 18: 412–459.

Mortell, M.P. (1977) The evolution of nonlinear standing waves in bounded media. *Zeitschrift für angewandte Mathematik und Physik ZAMP,* 28(1): 33–46.

Mortell, M.P., K.F. Mulchrone and B.R. Seymour (2009) The evolution of macrosonic standing waves in a resonator. *International Journal of Engineering Science,* 47(11): 1305–1314.

Mortell, M.P. and B.R. Seymour (1972) Pulse propagation in a nonlinear viscoelastic rod of finite length. *Society for Industrial and Applied Mathematics Journal of Applied Mathematics,* 22(2): 209–224.

Mortell, M.P. and B.R. Seymour (1973) The evolution of a self-sustained oscillation in a nonlinear continuous system. *Journal of Applied Mechanics,* 40(1): 53–60.

Mortell, M.P. and B.R. Seymour (1976) Wave propagation in a nonlinear laminated material: A derivation of geometrical acoustics. *The Quarterly Journal of Mechanics and Applied Mathematics,* 29(4): 457–466.

Mortell, M.P. and B.R. Seymour (1979) Nonlinear forced oscillations in a closed tube: continuous solutions of a functional equation. *Proceedings of the Royal Society of London. A. Mathematical and Physical Sciences,* 367: 253–270.

Mortell, M.P. and B.R. Seymour (1980) A simple approximate determination of stochastic transition for the standard mapping. *Journal of Mathematical Physics,* 21(8): 2121–2123.

Mortell, M.P. and B.R. Seymour (1981) A finite-rate theory of quadratic resonance in a closed tube. *Journal of Fluid Mechanics,* 112: 411–431.

Mortell, M.P. and B.R. Seymour (1988) The calculation of resonance separatrices for the near-integrable case of the standard mapping. *Zeitschrift für angewandte Mathematik und Physik ZAMP,* 39(6): 861–873.

Mortell, M.P. and B.R. Seymour (2004) Nonlinear resonant oscillations in closed tubes of variable cross-section. *Journal of Fluid Mechanics,* 519: 183–199.

Mortell, M.P. and B.R. Seymour (2005) Finite-amplitude, shockless, resonant vibrations of an inhomogeneous elastic panel. *Mathematics and Mechanics of Solids*, 10(4): 427–440.

Mortell, M.P. and B.R. Seymour (2007) Resonant oscillations of an inhomogeneous gas in a closed cylindrical tube. *Discrete and Continuous Dynamical Systems Series B*, 7(3): 619–628.

Mortell, M.P. and B.R. Seymour (2011) The propagation of small amplitude nonlinear waves in a strongly inhomogeneous medium. *Mathematics and Mechanics of Solids*, 16(6): 637–651.

Mortell, M.P. and E. Varley (1970) Finite amplitude waves in bounded media. Nonlinear free vibrations of an elastic panel. *Proceedings of the Royal Society of London. A. Mathematical and Physical Sciences*, 318: 169–196.

Murray, J.D. (1974) *Asymptotic Analysis*. Clarendon Press, Oxford.

Nayfeh, A.H. (1973) *Perturbation Methods*. John Wiley & Sons, New York.

Ockendon, J.R. and H. Ockendon (1973) Resonant surface waves. *Journal of Fluid Mechanics*, 59: 397–413.

Ockendon, H., J.R. Ockendon and A.D. Johnson (1986) Resonant sloshing in shallow water. *Journal of Fluid Mechanics*, 167: 465–479.

Ockendon, H., J.R. Ockendon, M.R. Peake and W. Chester (1993) Geometrical effects in resonant gas oscillations. *Journal of Fluid Mechanics*, 257: 201–217.

Parker, D.F. (1969) Nonlinearity, relaxation and diffusion in acoustics and ultrasonics. *Journal of Fluid Mechanics*, 39: 793–815.

Parker, D.F. (1972) Propagation of a rapid pulse through a relaxing gas. *Physics of Fluids*, 15: 256–262.

Pierce, A.D. (1989) *Acoustics: An Introduction to Its Physical Principles and Applications*. Acoustical Society of America.

Riemann, R. (1858) *Uber die Fortpflanzung ebener Luftwellen von endlicher Schwingungsweite*, Gottingun Abhandlungen, Vol. viii, p. 43.

Saenger, R.A. and G.E. Hudson (1960) Periodic shock waves in resonating gas columns. *The Journal of the Acoustical Society of America*, 32(8): 961–970.

Sansone, G. and R. Conti (1964) *Nonlinear Differential Equations*. Pergamon Press.

Seymour, B.R. and M.P. Mortell (1973a) Resonant acoustic oscillations with damping: small rate theory. *Journal of Fluid Mechanics*, 58: 353–373.

Seymour, B.R. and M.P. Mortell (1973b) Nonlinear resonant oscillations in open tubes. *Journal of Fluid Mechanics*, 60: 733–750.

Seymour, B.R. and M.P. Mortell (1975) Nonlinear geometrical acoustics. In: *Mechanics Today*. Vol. 2.(A76-14481 04-70) Pergamon Press, Inc., New York, pp. 251–312.

Seymour, B.R. and M.P. Mortell (1977) The effect of reflections on nonlinear transient pulse propagation in a laminated composite, *Journal of Applied Mechanics, ASME*, 44: 683–688.

Seymour, B.R. and M.P. Mortell (1980) A finite-rate theory of resonance in a closed tube: Discontinuous solutions of a functional equation. *Journal of Fluid Mechanics*, 99: 365–382.

Seymour, B.R. and M.P. Mortell (1981) Discontinuous solutions of a measure preserving mapping. *Society for Industrial and Applied Mathematics Journal of Applied Mathematics*, 41(1): 94–106.

Seymour, B.R. and M.P. Mortell (1985) The evolution of a finite rate periodic oscillation. *Wave Motion*, 7(5): 399–409.

Seymour, B.R., M.P. Mortell and D.E. Amundsen (2011) Resonant oscillations of an inhomogeneous gas between concentric spheres. *Proceedings of the Royal Society A: Mathematical, Physical and Engineering Science*, 467: 2149–2167.

Seymour, B.R., M.P. Mortell, and D. E. Amundsen (2012) Asymptotic solutions for shocked resonant acoustic oscillations between concentric spheres and coaxial cylinders. *Physics of Fluids*, 24(2).

Seymour, B.R. and E. Varley (1970) High frequency, periodic disturbances in dissipative systems. I. small amplitude, finite rate theory. *Proceedings of the Royal Society of London. A. Mathematical and Physical Sciences*, 314: 387–415.

Seymour, B.R. and E. Varley (1987) Exact representations for acoustical waves when the sound speed varies in space and time. *Studies in Applied Mathematics*, 76: 1–35.

Stoker, J.J. (1950) *Nonlinear Vibrations in Mechanical and Electrical Systems.* Interscience, New York.

Stoker, J.J. (1957) *Water Waves.* Interscience, New York.

Stuhlträger, E. and H. Thoman (1986) Oscillations of a gas in an open-ended tube near resonance. *Zeitschrift für angewandte Mathematik und Physik ZAMP*, 37: 155–175.

Sturtevant, B. (1974) Nonlinear gas oscillations in pipes. Part 2. Experiment. *Journal of Fluid Mechanics*, 63: 97–120.

Sturtevant, B. and J.J. Keller (1978) Subharmonic nonlinear acoustic resonances in open tubes. Part 2. Experimental investigation of the open-end boundary condition. *Zeitschrift für angewandte Mathematik und Physik ZAMP*, 29: 473–485.

Thompson, P.A. (1971) The fundamental derivative in gas dynamics. *Physics of Fluids*, 14: 1843–1849.

Thompson, P.A. and K.C. Lambrakis (1973) Negative shock waves. *Journal of Fluid Mechanics*, 60: 187–207.

Van Dyke, M.D. (1975) *Perturbation Methods in Fluid Dynamics.* Parabolic Press.

Van Wijngaarden, L. (1968) On the oscillations near resonance in open tubes. *Journal of Engineering Mathematics*, 2: 225–240.

Varley, E. and E. Cumberbatch (1966) Non-linear, high frequency sound waves. *Institute for Mathematics and its Applications Journal of Applied Mathematics*, 2(2): 133–143.

Varley, E. and E. Cumberbatch (1970) Large amplitude waves in stratified media: Acoustic pulses. *Journal of Fluid Mechanics*, 43: 513–537.

Varley, E. and T.G. Rogers (1967) The propagation of high frequency, finite acceleration pulses and shocks in viscoelastic materials. *Proceedings of the Royal Society of London. Series A. Mathematical and Physical Sciences*, 296: 498–518.

Varley, E. and B.R. Seymour (1988) A method for obtaining exact solutions to partial differential equations with variable coefficients. *Studies in Applied Mathematics*, 78(3): 183–225.

Varley, E. and B.R. Seymour, B (1998). A simple derivation of the N-soliton solution to the KdV equation, *Society for Industrial and Applied Mathematics Journal of Applied Mathematics*, 58: 904–911.

Varley, E., R. Venkataraman and E. Cumberbatch (1971) The propagation of large amplitude tsunamis across a basin of changing depth Part 1. Off-shore behaviour. *Journal of Fluid Mechanics*, 49: 775–801.

Verhagen, J.H.G. and L. van Wijngaarden (1965) Nonlinear oscillations of fluid in a container. *Journal of Fluid Mechanics*, 22: 737–752.

Wang, M. and D.R. Kassoy (1995) Nonlinear oscillations in a resonant gas column: An initial-boundary-value study. *Society for Industrial and Applied Mathematics Journal of Applied Mathematics*, 55(4): 923–995.

Webster, A.G. (1919) Acoustical impedance and the theory of horns and of the phonograph. *Proceedings of the National Academy of Sciences of the United States of America*, 5(7): 275–282.

Whitham, G.B. (1950) The behaviour of supersonic flow past a body of revolution, far from the axis. *Proceedings of the Royal Society of London. Series A. Mathematical and Physical Sciences*, 201: 89–109.

Whitham, G.B. (1952) The flow pattern of a supersonic projectile. *Communications on Pure and Applied Mathematics*, 5(3): 301–348.

Whitham, G.B. (1974) *Linear and Nonlinear Waves*. John Wiley & Sons, New York.

Wood, L.A. and K.P. Byrne (1981) Analysis of a random repeated impact process. *Journal of Sound and Vibration*, 82: 329–345.

Yuen, C. and W.E. Ferguson (1978) Fermi-Pasta-Ulam recurrence in the two-space dimensional nonlinear Schrödinger equation. *Physics of Fluids*, 21: 2116–2118.

Zabusky, N.J. (1967) A synergetic approach to problems of nonlinear dispersive wave propagation and interaction. *Nonlinear Partial Differential Equations: A Symposium on Methods of Solution*. Academic Press.

Zabusky, N.J. and M.D. Kruskal (1965) Interaction of solitons in a collisionless plasma and the recurrence of initial states. *Physical Review Letters*, 15(6): 240–243.

Zarembo, L.K. (1967) Forced vibrations of finite amplitude in a tube. *Soviet Physics Acoustic-Ussr*, 13(2): 257.

Zaripov, R.G. and M.A. Ilgamov (1976) Non-linear gas oscillations in a pipe. *Journal of Sound and Vibration*, 46(2): 245–257.

Zel'dovich, Y (1946) On the possibility of rarefaction shock waves. *Zhurnal Eksperimental'noi i Teoreticheskoi Fiziki*, 4: 363–364.

Index